B. BOLLOBÁS

177

A Higher-Dimensional Sieve Method

HAROLD G. DIAMOND

H. HALBERSTAM

With Procedures for Computing Sieve Functions

WILLIAM F. GALWAY

CAMBRIDGE UNIVERSITY PRESS
Cambridge, New York, Melbourne, Madrid, Cape Town, Singapore, São Paulo, Delhi

Cambridge University Press
The Edinburgh Building, Cambridge CB2 8RU, UK

Published in the United States of America by Cambridge University Press, New York

www.cambridge.org
Information on this title: www.cambridge.org/9780521894876

© H. Diamond, H. Halberstam and W. Galway 2008

This publication is in copyright. Subject to statutory exception
and to the provisions of relevant collective licensing agreements,
no reproduction of any part may take place without
the written permission of Cambridge University Press.

First published 2008

Printed in the United Kingdom at the University Press, Cambridge

A catalog record for this publication is available from the British Library

ISBN 978-0-521-89487-6 hardback

Cambridge University Press has no responsibility for the persistence or
accuracy of URLs for external or third-party internet websites referred to
in this publication, and does not guarantee that any content on such
websites is, or will remain, accurate or appropriate.

In memory of Hans-Egon (Ted) Richert

Contents

List of Illustrations		*page* xi
List of Tables		xiii
Preface		xv
Notation		xvii

	Part I Sieves	1
1	**Introduction**	3
	1.1 The sieve problem	3
	1.2 Some basic hypotheses	4
	1.3 Prime g-tuples	6
	1.4 The $\Omega(\kappa)$ condition	8
	1.5 Notes on Chapter 1	11
2	**Selberg's sieve method**	13
	2.1 Improving the Eratosthenes–Legendre sieve	13
	2.2 A new parameter	14
	2.3 Notes on Chapter 2	17
3	**Combinatorial foundations**	19
	3.1 The fundamental sieve identity	19
	3.2 Efficacy of the Selberg sieve	22
	3.3 Multiplicative structure of modifying functions	25
	3.4 Notation: $\mathcal{P}$, $S(\mathcal{A}, \mathcal{P}, z)$, and V	26
	3.5 Notes on Chapter 3	27
4	**The Fundamental Lemma**	29
	4.1 A start: an asymptotic formula for $S(\mathcal{A}_q, \mathcal{P}, z)$	29
	4.2 A lower bound for $S(\mathcal{A}_q, \mathcal{P}, z)$	33
	4.3 Notes on Chapter 4	42

Contents

5	**Selberg's sieve method (continued)**	**43**
5.1	A lower bound for $G(\xi, z)$	43
5.2	Asymptotics for $G^*(\xi, z)$	52
5.3	The j and σ functions	56
5.4	Prime values of polynomials	64
5.5	Notes on Chapter 5	66
6	**Combinatorial foundations (continued)**	**67**
6.1	Statement of the main analytic theorem	67
6.2	The $S(\chi)$ functions	70
6.3	The "linear" case $\kappa = 1$	71
6.4	The cases $\kappa > 1$	73
6.5	Notes on Chapter 6	79
7	**The case $\kappa = 1$: the linear sieve**	**81**
7.1	The theorem and first steps	81
7.2	Bounds for $V\Sigma^{\pm}$: the set-up	85
7.3	Bounds for $V\Sigma^{\pm}$: conclusion	88
7.4	Completion of the proof of Theorem 7.1	92
7.5	Notes on Chapter 7	95
8	**An application of the linear sieve**	**97**
8.1	Toward the twin prime conjecture	97
8.2	Notes on Chapter 8	102
9	**A sieve method for $\kappa > 1$**	**103**
9.1	The main theorem and start of the proof	103
9.2	The S_{21} and S_{22} sums	107
9.3	Bounds on $\Sigma^{\pm}$	110
9.4	Completion of the proof of Theorem 9.1	120
9.5	Notes on Chapter 9	122
10	**Some applications of Theorem 9.1**	**125**
10.1	A Mertens-type approximation	125
10.2	The sieve setup and examples	129
11	**A weighted sieve method**	**135**
11.1	Introduction and additional conditions	135
11.2	A set of weights	137
11.3	Arithmetic interpretation	140
11.4	A simple estimate	144
11.5	Products of irreducible polynomials	147
11.6	Polynomials at prime arguments	149
11.7	Other weights	150
11.8	Notes on Chapter 11	151

Part II Proof of the Main Analytic Theorem — 153

12 Dramatis personae and preliminaries — 155
- 12.1 P and Q and their adjoints — 155
- 12.2 Rapidly vanishing functions — 158
- 12.3 The Π and Ξ functions — 160
- 12.4 Notes on Chapter 12 — 161

13 Strategy and a necessary condition — 163
- 13.1 Two different sieve situations — 163
- 13.2 A necessary condition — 164
- 13.3 A program for determining F and f — 166

14 Estimates of $\sigma_\kappa(u) = j_\kappa(u/2)$ — 169
- 14.1 Lower bounds on σ — 169
- 14.2 Differential relations — 173
- 14.3 The adjoint function of j — 177
- 14.4 Inequalities for $1 - j$ — 178
- 14.5 Relations between σ' and $1 - \sigma$ — 183
- 14.6 The ξ function — 184
- 14.7 An improved upper bound for $1 - j$ — 190
- 14.8 Notes on Chapter 14 — 190

15 The p_κ and q_κ functions — 193
- 15.1 The p functions — 193
- 15.2 The q functions — 195
- 15.3 Zeros of the q functions — 196
- 15.4 Monotonicity and convexity relations — 197
- 15.5 Some lower bounds for ρ_κ — 199
- 15.6 An upper bound for ρ_κ — 201
- 15.7 The integrands of $\widetilde{\Pi}$ and $\widetilde{\Xi}$ — 202

16 The zeros of $\widetilde{\Pi} - 2$ and $\widetilde{\Xi}$ — 207
- 16.1 Properties of the Π and Ξ functions — 207
- 16.2 Solution of some Π and Ξ equations — 209
- 16.3 Estimation of $\widetilde{\Pi}(2.7\kappa)$ — 214

17 The parameters α_κ and β_κ — 217
- 17.1 The cases $\kappa = 1, 1.5$ — 217
- 17.2 The cases $\kappa = 2, 2.5, 3, \ldots$ — 220
- 17.3 Proof of Proposition 17.3 — 222
- 17.4 Notes on Chapter 17 — 227

18 Properties of F_κ and f_κ — 229
- 18.1 F_κ and $f_\kappa \to 1$ at ∞ — 229
- 18.2 $Q_\kappa(u) > 0$ for $u > 0$ — 229

Appendix 1	**Procedures for computing sieve functions**	233
A1.1	DDEs and the Iwaniec inner product	234
A1.2	The upper and lower bound sieve functions	235
A1.3	Using the Iwaniec inner product	236
A1.4	Some features of *Mathematica*	239
A1.5	Computing $F_\kappa(u)$ and $f_\kappa(u)$	240
A1.6	The function $\mathrm{Ein}(z)$	241
A1.7	Computing the adjoint functions	242
A1.8	Computing $j_\kappa(u)$	250
A1.9	Computing α_κ and β_κ	254
A1.10	Weighted-sieve computations	255
Bibliography		259
Index		265

List of Illustrations

6.1	$F_2(u)$, f_2, and $1/\sigma_2(u)$	79
14.1	The function $\xi(t)$	184
14.2	The differences $\ell_2(t) - \xi(t)$ and $\ell_1(t) - \xi(t)$	186
A1.1	$F_\kappa(u;\alpha,\beta)$ and $f_\kappa(u;\alpha,\beta)$ for two choices of α and β	236
A1.2	α_κ and β_κ	236
A1.3	$P_\kappa(u;\alpha,\beta)$ and $Q_\kappa(u;\alpha,\beta)$ for two choices of α and β	237
A1.4	$p_\kappa(u)$ for two values of κ	238
A1.5	$q_\kappa(u)$ for two values of κ	238
A1.6	$\langle Q, q \rangle_\kappa$ for two choices of α and β	239
A1.7	Two views of the integrand for computing $q_\kappa(u)$	245
A1.8	Integrand for computing $q_\kappa(u)$ for three values of u/κ	248
A1.9	$j_\kappa(u)$	250
A1.10	Integrand for computing $j_\kappa(u)$, "left-hand" path	251
A1.11	Integrand for computing $j_\kappa(u)$, "right-hand" path	253
A1.12	Region determined by the constraints of Theorem 11.1	256
A1.13	Lower bound function for the weighted sieve	257

List of Tables

11.1	r values for small g	148
11.2	r values for small g and h	148
11.3	r values for small G	149
11.4	r values for small g and k and prime arguments	150
11.5	r values for small G and prime arguments	150
11.6	Values of τ_r for which $r \approx N_{\min}(g, gk/\tau_r, \tau_r)$	152
14.1	Values of ξ, $2\log t$, ℓ_1, and ℓ_2	185
15.1	The first few q functions	196
15.2	Values of ρ_κ	197
15.3	$\rho_\kappa + 1$ compared with 2.7κ	201
16.1	Bounds on $\widetilde{\Pi}_\kappa(v_\kappa)$	213
17.1	Values of α_κ, β_κ, and $\rho_\kappa + 1$	227
A1.1	Accuracy of approximations to $q_\kappa(u)$	249

Preface

Nearly a hundred years have passed since Viggo Brun invented his famous sieve, and yet the use of sieve methods is still evolving. At one time it seemed that, as analytic tools improved, the use of sieves would decline, and only their role as an auxiliary device would survive. However, as probability and combinatorics have penetrated the fabric of mathematical activity, so have sieve methods become more versatile and sophisticated, especially in conjunction with other theories and methods, until, in recent years, they have played a part in some spectacular achievements that herald new directions in mathematical discovery.

An account of all the exciting and diverse applications of sieve ideas, past and present, has yet to be written. In this monograph our aim is modest and narrowly focused: we construct (in Chapter 9) a hybrid of the Selberg [Sel47] and Rosser–Iwaniec [Iwa80] sieve methods to deal with problems of sieve dimension (or density) that are integers or half integers. This theory achieves somewhat sharper estimates than either of its ancestors, the former as given by Ankeny and Onishi [AO65]. The sort of application we have in mind is to show that a given polynomial with integer coefficients (some obvious cases excluded) assumes at integers or at primes infinitely many almost-prime values, that is, values that have few prime factors relative to the degree of the polynomial. To describe our procedure a little more precisely, we extend the pioneering method of Jurkat and Richert [JR65] for dimension 1 (that combined the Selberg sieve method with infinitely many iterations of the Buchstab identity) to higher dimensions by means of the Rosser–Iwaniec approach; in the process we give an alternative account of that approach.

The restriction we make to integer and half integer dimensions simplifies the analytic component of our method; an account avoiding this constraint exists [DHR88]–[DHR96], but is much more complicated. A justification for our restriction is that most sieve applications of the above kind occur in this context. We include an account of the case of dimension 1 because it serves as a model for what is to come and involves little extra work. While our treatment of that special case is not quite as sharp as in the classical exposition of Iwaniec [Iwa80] or that given more recently by Greaves [Grv01], it is somewhat simpler.

It should be said that our results for higher dimensions, unlike the case of dimension 1, are almost certainly not best possible, not even in a single instance; and that our approach might not be the right one there. Nevertheless, our method does have good applications, is simple to use, and, despite some complications of detail, rests solely on elementary combinatorial inequalities and relatively simple analysis. The combinatorics we have developed may in due course find other applications.

The first comprehensive account of sieve methods, by the second author and H.-E. Richert [HR74], appeared in 1974 and has been long out of print. Although it is also out of date in some important respects, we have tended to follow its overall design, and we have drawn on it for examples and applications.

We are happy to express our thanks to the many who have contributed to this work: the aforementioned authors, on whose ideas we have built; H.-E. Richert, who shared in our discoveries; our former students Ferrell S. Wheeler and David M. Bradley for their extensive computational assistance; our patrons, the University of Illinois and the National Science Foundation, who supported our research; our colleague A. J. Hildebrand for LaTeX and mathematical advice; Sidney Graham and Craig Franze for help in rooting out errors; and Cherri Davison, who skillfully and cheerfully converted our manuscript into LaTeX. Also, we thank our wives for their support during the preparation of this book.

The *Mathematica*®[†] package of sieve-related functions described in Appendix 1, as well as a list of comments and corrigenda, will be maintained at http://www.math.uiuc.edu/SieveTheoryBook. Finally, we request that readers advise us of any errors or obscurities they find. Our e-mail address is sievetheorybook@math.uiuc.edu.

---

† *Mathematica* is a registered trademark of Wolfram Research, Inc.

Notation

Standard terminology

$[x]$ denotes the largest integer not exceeding x.

$a \mid b$ means a divides b evenly, i.e., $b \equiv 0 \mod a$.

(a, b) denotes the greatest common divisor of the integers a and b (when no confusion with notation for an open interval is possible) and $\{a, b\}$ their least common multiple (see p. 14).

The symbols for the classical arithmetic functions have their usual meaning: $\mu(\cdot)$ is the Moebius function, $\tau(\cdot)$ the divisor function, $\phi(\cdot)$ Euler's totient function, $\pi(x)$ the number of primes not exceeding x, and $\pi(x, k, \ell)$ the number of primes not exceeding x and congruent to ℓ modulo k.

We use $\nu(\cdot)$ for the number of distinct prime divisors and $\Omega(\cdot)$ for the number of prime divisors counted according to multiplicity. Throughout Part I of this manuscript, $p(\cdot)$ and $p^+(\cdot)$ are the least and largest prime factors respectively of an integer (see p. 19).

The constants π and e have their usual meanings, and γ is always Euler's constant.

$O(\cdot)$ and $o(\cdot)$ have their usual meanings relating to the size of a function, and $O_z(\cdot)$ indicates dependence of the implied constant upon z.

$\mathcal{A}, \mathcal{B}, \mathcal{C}, \ldots$ denote integer sequences or sets, and $|\mathcal{A}|, |\mathcal{B}|, |\mathcal{C}|, \ldots$ their cardinalities; $\mathcal{A}_d$ denotes the sequence of multiples of d in $\mathcal{A}$. That is, $\mathcal{A}_d := \{a \in \mathcal{A} : d \mid a\} = \{a \in \mathcal{A} : a \equiv 0 \mod d\}$.

$\mathcal{P}$ is always a set of primes, the variable p denotes a prime throughout Part I of this book, and $\mathcal{P}^c$ is the set of primes not in $\mathcal{P}$.

$\mathbb{N}$ is the sequence of natural numbers, $\mathbb{Q}$ the set of rationals.

Sieve notation

The following lists indicate where sieve functions and sieve terminology are introduced and defined:

The notions of a function being *divisor closed* and/or *combinatorial* are defined on p. 27.

P_r denotes an integer having at most r prime factors, counted according to multiplicity; thus n is a P_r if $\Omega(n) \leq r$ (see p. 141).

Multiplicative functions

ω	p. 5
ω^*	p. 49
g	p. 15
g^*	p. 44
ρ	pp. 37, 39, 64, 125

Remainder terms

$r_\mathcal{A}(d)$	p. 5
R	p. 14 (2.5)
$R_\mathcal{A}(Y, z)$	p. 109 (9.27)

Summatory functions

$G, G_\xi(P)$	p. 15 (2.10)
$G(\xi, z), G(\xi)$	pp. 30 (4.2), §5.1, 61 (5.44)
$G^*(\xi, z)$	p. 44 (5.7), §5.2
$G^*(\xi)$	pp. 45, 52 (5.22), 54 (5.25)
$D(w_1, w)$	p. 139
$\mathcal{E}(x, d)$	p. 97

Notation

Integrals (and associated expressions)

$T(\xi)$	p. 53
$T(\xi, z)$	p. 56 (5.28)
$U(\xi, z)$	p. 58
$\langle G, G \rangle_\kappa$	pp. 158, 178, 234 (A1.3)
$\Pi_\kappa(u,v), \Xi_\kappa(u,v)$	p. 160 (12.24), (12.25)
$\Pi(u), \Xi(u)$	p. 161
$\widetilde{\Pi}(u), \widetilde{\Xi}(u)$	p. 161

Products

P	p. 3
$V(\mathcal{P})$	p. 5
$P(z)$	p. 26
$V(z)$	p. 26
$V^*(z)$	p. 56

Sifting functions

$S(\mathcal{A}, \mathcal{P})$	p. 4
$S(\mathcal{A}, \mathcal{P}, z)$	pp. 7, 26
$S_i(\chi), S_{2i}(\chi), i = 1, 2$	p. 70, (6.9) (6.10) (6.11)
$E_r^{(-)^\nu}$	pp. 89, 110 (9.38), 116 (9.50)
$E_\infty^{(-)^\nu}$	p. 89 (7.14)
$\Sigma_1^{(-)^\nu}$	p. 106 (9.16)
$\Sigma_2^{(-)^\nu}$	p. 109 (9.26)
$D_\kappa^{(-)^r}(x, w, z_0)$	p. 110
$W(\mathcal{A}, \mathcal{P}, z, y)$	p. 135 (11.1)
$W_0(\mathcal{A}, \mathcal{P}, z, y)$	p. 135
$W(\mathcal{A}, \mathcal{P}, z, y, \lambda)$	p. 137 (11.6)

Transcendental functions

$\ell(w)$	p. 50 (5.19)
$\sigma_\kappa(\cdot)$	pp. 56, 68 (6.7) (6.8)
$j_\kappa(\cdot)$	pp. 56 (5.29) (5.30), 250
$j_\kappa^{(-1)}(u)$	p. 57 (5.32)
$F_\kappa(u), f_\kappa(u)$	pp. 67 Theorem 6.1, 235
$\phi^{(-)^\tau}(\cdot)$	p. 84
$\delta_1(y), \delta_2(y)$	pp. 106 (9.19), 108 (9.23)
$\Phi(t)$	p. 139
$P_\kappa(u), Q_\kappa(u)$	p. 155 (12.1)
$p_\kappa(u)$	pp. 157 (12.11), 193 §15.1, 242 §A1.7
$q_\kappa(u)$	pp. 157 (12.16), 195 §15.2, 242 §A1.7
$\text{Ein}(t)$	pp. 157 (12.12) (12.13), 241
$r_\kappa(u)$	p. 177 (14.21)
$\psi(u)$	p. 183
$\xi(t)$	p. 184

Weight functions

$w(a), w_0(a)$	p. 135

Constants/parameters

Δ	p. 7
κ, A	p. 8 Definition 1.3
$\alpha_\kappa, \beta_\kappa$	p. 67 Theorem 6.1
u_κ	p. 173 Lemma 14.4
μ_0	p. 136 (11.4)
τ	p. 136 (11.3)
Λ_r	p. 142 (11.16) Corollary 11.2
$N(u, v; \kappa, \mu_0, \tau)$	pp. 142, 255
ρ_κ	p. 196 §15.3

Basic conditions

$\Omega(\kappa)$	p. 8, §1.4
$\Omega^*(\kappa)$	p. 44
Q_0, R_0, M_0	p. 136 (11.2) (11.3) (11.4)

"Modifying" functions

$\chi(\cdot)$	p. 13
$\overline{\chi}(\cdot)$	p. 19 (3.1)
$\chi^{\pm}(\cdot), \eta^{\pm}(\cdot)$ ($\kappa = 1$)	p. 71 §6.3 (6.15)–(6.18)
$\chi^{\pm}(\cdot), \eta^{\pm}(\cdot)$ ($\kappa > 1$)	p. 74 §6.4 (6.20)–(6.22)

For cases $\kappa > 1$, see also

$\chi^{\pm}(\cdot), \eta^{\pm}(\cdot)$	pp. 103–104 (9.5)–(9.7)
$\overline{\chi^{\pm}}(d)$	p. 104 (9.8)

Part I
Sieves

1
Introduction

1.1 The sieve problem

Let $\mathcal{P}$ be a finite set of primes $\{p\}$ (the symbol p denotes a prime throughout Part I of this book) and let

$$P := \prod_{p \in \mathcal{P}} p.$$

(Later, starting in Chapter 3, we shall let $\mathcal{P}$ denote an infinite set of primes and use $\mathcal{P}_z$ to denote the finite set $\mathcal{P} \cap [2, z)$, i.e., $\mathcal{P}$ truncated at z.) The indicator function of the set of all integers n coprime with P, that is, having no divisors in $\mathcal{P}$, is expressed in terms of the Moebius μ function by

$$(1.1) \qquad \sum_{d \mid (n,P)} \mu(d) = \begin{cases} 1, & (n, P) = 1, \\ 0, & \text{otherwise.} \end{cases}$$

We call $\mathcal{P}$ a *sieve* and say that $\mathcal{P}$ *sifts out* an integer n if $(n, P) > 1$.

Let $\mathcal{A}$ be a finite integer sequence, taking account of possible repetitions. An example of such a sequence is

$$\mathcal{A} = \{n^2 + 1 : -9 \le n \le 11\}.$$

When we apply the sieve $\mathcal{P}$ to $\mathcal{A}$—we might say alternatively, when we put, or filter, $\mathcal{A}$ through $\mathcal{P}$, or *sift* $\mathcal{A}$ by $\mathcal{P}$—the elements of $\mathcal{A}$ that remain *unsifted* are those that are coprime with P, and their number

$S(\mathcal{A}, \mathcal{P})$ is given by

$$S(\mathcal{A}, \mathcal{P}) := |\{a \in \mathcal{A} \colon (a, P) = 1\}| = \sum_{a \in \mathcal{A}} \sum_{d \mid (a, P)} \mu(d)$$

$$= \sum_{\substack{a \in \mathcal{A} \\ d \mid P}} \sum_{d \mid a} \mu(d) = \sum_{d \mid P} \mu(d) \sum_{\substack{a \in \mathcal{A} \\ d \mid a}} 1.$$

Writing $\mathcal{A}_d := \{a \in \mathcal{A} : d \mid a\}$, we have the *Eratosthenes–Legendre formula*

(1.2) $$S(\mathcal{A}, \mathcal{P}) = \sum_{d \mid P} \mu(d) |\mathcal{A}_d|.$$

Example 1.1. Take $\mathcal{A} = \{n \in \mathbb{N} \colon n \leq x\}$ and take $\mathcal{P}$ to be the set of all primes p not exceeding $x^{1/2}$. Then $|\mathcal{A}_d| = [x/d]$ and, by the famous observation of Eratosthenes, the identity for $S(\mathcal{A}, \mathcal{P})$ yields the prime counting formula

$$\pi(x) - \pi(x^{1/2}) + 1 = \sum_{d \mid P} \mu(d) \left[\frac{x}{d}\right], \qquad P = \prod_{p \leq x^{1/2}} p.$$

We can think of two natural ways to write the sum: either as

$$x \prod_{p \leq x^{1/2}} \left(1 - \frac{1}{p}\right) + \sum_{d \mid P} \mu(d) \left(\left[\frac{x}{d}\right] - \frac{x}{d}\right),$$

or as

$$x \sum_{\substack{d \mid P \\ d \leq x}} \frac{\mu(d)}{d} + \sum_{\substack{d \mid P \\ d \leq x}} \mu(d) \left(\left[\frac{x}{d}\right] - \frac{x}{d}\right).$$

In the first way, the leading term does suggest the correct order of magnitude of $\pi(x)$, but it turns out that the sum of the "remainders" has the same order of magnitude. The second way appears to be more promising, but it turns out that here we do not know how to handle either sum!

1.2 Some basic hypotheses

In the above example we know, of course, how the sequence $\mathcal{A}$ is distributed in the residue classes $0 \mod d$, $d \mid P$; in fact, the corresponding information is available for many integer sequences $\mathcal{A}$ occurring in the

1.2 Some basic hypotheses

literature and takes the form, which henceforward we assume, that *there exists a convenient approximation X to $|\mathcal{A}|$ and a non-negative multiplicative arithmetic function $\omega(\cdot)$ such that*

(1.3) $\qquad 0 \leq \omega(p) < p \quad (p \in \mathcal{P}), \qquad \omega(p) = 0 \quad (p \notin \mathcal{P}),$

and such that the remainder terms

(1.4) $\qquad r_{\mathcal{A}}(d) := |\mathcal{A}_d| - \dfrac{\omega(d)}{d} X \quad (d \mid P)$

are suitably small, at least on average, over some restricted range of values of d. (In a naive sense, the number $\omega(p)/p$ is the probability that the prime p of $\mathcal{P}$ is a divisor of elements in $\mathcal{A}$.) With this assumption, we obtain

$$S(\mathcal{A}, \mathcal{P}) = X \sum_{d \mid P} \mu(d) \frac{\omega(p)}{p} + \sum_{d \mid P} \mu(d) r_{\mathcal{A}}(d)$$

$$= X \prod_{p \in \mathcal{P}} \left(1 - \frac{\omega(p)}{p}\right) + \sum_{d \mid P} \mu(d) r_{\mathcal{A}}(d).$$

Again, unless $\mathcal{P}$ is *very* sparse, we expect the remainder sum to be too large to derive asymptotics for $S(\mathcal{A}, \mathcal{P})$, but we have the impression nevertheless that $S(\mathcal{A}, \mathcal{P})$ should be measured in terms of the magnitude of the "leading" term

$$X \prod_{p \in \mathcal{P}} \left(1 - \frac{\omega(p)}{p}\right) =: XV(\mathcal{P}),$$

say. *The aim of a sieve method is to modify the Moebius function in the indicator function* (1.1) *in a way that allows us to approximate $S(\mathcal{A}, \mathcal{P})$ from above, and sometimes from below, with some accuracy, and to obtain asymptotics for $S(\mathcal{A}, \mathcal{P})$ when $\mathcal{P}$ is sparse.*

It is instructive to see why we assume that $\omega(p) < p$ holds for all $p \in \mathcal{P}$. Otherwise—that is, if there existed a prime $p^* \in \mathcal{P}$ for which $\omega(p^*)/p^*$ equals (or is very near to) 1—we would have

$$|\mathcal{A}_{p^*}| - X \approx |\mathcal{A}_{p^*}| - X\omega(p^*)/p^*,$$

and the last quantity is small by hypothesis, as is $X - |\mathcal{A}|$ as well. It follows that $|\mathcal{A}| - |\mathcal{A}_{p^*}|$ is small, i.e., most members of $\mathcal{A}$ are multiples of p^*. After these elements are sifted out, little would be left in $\mathcal{A}$—or for us to say.

Appeal to probabilistic thinking is often helpful in arithmetic investigations but tends to fall short when it comes to supplying proofs. The usual reason is that such thinking is based upon a probabilistic model involving a sequence of *independent* events, whereas the actual arithmetical "events" being modeled—in our case, "divisibility of elements of $\mathcal{A}$ by primes p from $\mathcal{P}$"—have a poor independence relation for sets of primes whose products have a size comparable to X. If these events were independent, then indeed we should expect $XV(\mathcal{P})$ to be a true measure of $S(\mathcal{A}, \mathcal{P})$; instead, we have seen in the classical case of sifting the interval $[1, x]$ by the primes not exceeding $x^{1/2}$, that there

$$XV(\mathcal{P}) = x \prod_{p \leq x^{1/2}} \left(1 - \frac{1}{p}\right) \sim x \frac{e^{-\gamma}}{\log x^{1/2}} = 2e^{-\gamma} \frac{x}{\log x} \quad \text{as} \quad x \to \infty,$$

by the well-known Mertens' product formula ([HW79], Theorem 429) and $2e^{-\gamma} = 1.122918\ldots$, whereas by the Prime Number Theorem

$$\pi(x) \sim \frac{x}{\log x} \quad \text{as} \quad x \to \infty.$$

In contrast, suppose we sift $[1, x]$ by a "thin" infinite sequence of primes $\mathcal{P}: p_1 < p_2 < \ldots$ such that

$$\sum_{j=1}^{\infty} \frac{1}{p_j} < \infty.$$

In this case the density of integers divisible by none of the primes of $\mathcal{P}$ is indeed

$$\prod_{j=1}^{\infty} \left(1 - \frac{1}{p_j}\right).$$

1.3 Prime g-tuples

Before we begin our account in earnest, we consider another example more relevant to our main objective. The inspiration for this example is the famous *twin prime conjecture*, which asserts that there are infinitely many pairs of positive integers $(n, n+2)$, which are both prime numbers. The sieve method of Brun broke new ground by producing an upper bound for the number of pairs of twin primes in any interval $[1, x]$, but the original conjecture remains unproved.

Except for the example (2, 3), there is no other pair of primes of the form $(n, n+a)$ for a an odd number, since one member of the pair is then even. Similar reasoning shows that (3, 5, 7) is the only triple of primes of the form $(n, n+2, n+4)$. There are analogues of the twin prime conjecture for pairs or triples of primes that are not ruled out by congruential reasoning, such as $(n, n+4)$ or $(n, n+2, n+6)$. More generally, the *prime g-tuples conjecture* asserts that, absent any congruential obstruction, there exist infinitely many prime g-tuples $(n, n+a_1, \ldots, n+a_{g-1})$ (with fixed integers $a_1, \ldots, a_{g-1}$).

As a first attempt at detecting twin primes, take

$$\mathcal{A} = \{n(n+2) \colon 1 \leq n \leq X\}$$

and $\mathcal{P}$ as the set of all primes. The number of twin primes $(p, p+2)$ with $\sqrt{X+2} < p \leq X$ is provided by

$$S(\mathcal{A}, \mathcal{P}, \sqrt{X+2}),$$

where $S(\mathcal{A}, \mathcal{P}, z)$ denotes the number of elements in $\mathcal{A}$ coprime with the primes of $\mathcal{P}$ that are less than z. Here, as in Example 1.1, we are not able to approximate the S expression effectively. However, it provides a framework for our investigations.

Example 1.2. Let

$$L(n) := \prod_{i=1}^{g} (a_i n + b_i),$$

where the coefficients are integers satisfying $(a_i, b_i) = 1$ $(i = 1, \ldots, g)$ and the discriminant

$$\Delta = \prod_{i=1}^{g} a_i \prod_{1 \leq r < s \leq g} (a_r b_s - a_s b_r)$$

is non-zero. The non-vanishing of Δ ensures that the linear factors of L are not constant and that none is a linear multiple of another. Now let $\mathcal{P}$ be the set of all primes less than z and

$$\mathcal{A} = \{L(n) \colon x - y < n \leq x\}, \qquad 1 < y \leq x.$$

Here $X = y$, $\omega(d)$ is the number of incongruent solutions modulo d of the congruence $L(n) \equiv 0 \mod d$, and $|r_\mathcal{A}(d)| \leq \omega(d)$. From elementary number theory, $\omega(p) \leq g$ for all primes p, with equality when $p \nmid \Delta$.

When $p \mid \Delta$, $\omega(p)$ may take on any integer value in $[0, g)$. Let $\nu(d)$ denote the number of distinct prime divisors of d. Then, for squarefree d,

$$\omega(d) \leq g^{\nu(d)}$$

with equality when $(d, \Delta) = 1$. We shall come back to this example, basic to the "prime g-tuples" conjecture, and estimate $S(\mathcal{A}, \mathcal{P})$ in several applications later. It would be a great triumph for sieve theory to show that $L(n) = P_{g+\ell}$ infinitely often for some positive integer $\ell < g$; for that would imply that one of the factors $a_i n + b_i$ is a prime!

1.4 The $\Omega(\kappa)$ condition

We introduce at this point a weak average condition on $\omega(\cdot)$ that is to hold throughout.

Definition 1.3. We say that a sieve problem satisfies *the $\Omega(\kappa)$ condition* provided there exist constants $\kappa \geq 1$, $A > 1$ such that

$$(1.5) \qquad \prod_{w_1 \leq p < w} \left(1 - \frac{\omega(p)}{p}\right)^{-1} \leq \left(\frac{\log w}{\log w_1}\right)^{\kappa} \left(1 + \frac{A}{\log w_1}\right), \quad 2 \leq w_1 < w.$$

The parameter κ is clearly not unique—if $\Omega(\kappa)$ holds for some number κ, then it holds for any $\kappa' > \kappa$. Nevertheless, in most sieve problems the minimal κ is known and we refer to it as the *dimension*, or *sifting density*, of the problem. Problems of dimension 1 are especially important and we refer to them as *linear*. Note that $\Omega(\kappa)$ implies that

$$\prod_{w_1 \leq p < w} \left(1 - \frac{\omega(p)}{p}\right)^{-1} \leq \prod_{2 \leq p < w} \left(1 - \frac{\omega(p)}{p}\right)^{-1} \ll (\log w)^{\kappa}.$$

We pause here to check that $\Omega(\kappa)$ holds in Example 1.2 with $\kappa = g$. By adjusting the bound A if necessary, we may assume that $w_1 \geq g + 1$. Then, since $\omega(p) \leq g$, we have

$$\prod_{w_1 \leq p < w} \left(1 - \frac{\omega(p)}{p}\right)^{-1} \leq \exp\left\{\sum_{w_1 \leq p < w} -\log\left(1 - \frac{g}{p}\right)\right\}$$

$$= \exp\left\{g \sum_{w_1 \leq p < w} \frac{1}{p} + \sum_{w_1 \leq p < w} \sum_{r=2}^{\infty} \frac{1}{r}\left(\frac{g}{p}\right)^r\right\}.$$

Thus
$$\prod_{w_1 \le p < w} \left(1 - \frac{\omega(p)}{p}\right)^{-1} \le \exp\left\{g \sum_{w_1 \le p < w} \frac{1}{p} + \frac{1}{2}g^2 \sum_{p \ge w_1} \frac{1}{p(p-g)}\right\}$$
$$\le \exp\left\{g \sum_{w_1 \le p < w} \frac{1}{p} + \frac{1}{2}g^2(g+1) \sum_{p \ge w_1} \frac{1}{p^2}\right\}$$
$$= \exp\left\{g \log \frac{\log w}{\log w_1} + O\left(\frac{1}{\log w_1}\right) + O\left(\frac{1}{w_1}\right)\right\}$$
$$= \left(\frac{\log w}{\log w_1}\right)^g \exp\left\{O\left(\frac{1}{\log w_1}\right)\right\},$$

which implies $\Omega(g)$; at the next to last stage we used Mertens' sum formula ([HW79], Theorems 427, 428) that

$$(1.6) \quad \sum_{w_1 \le p < w} \frac{1}{p} = \log \frac{\log w}{\log w_1} + O\left(\frac{1}{\log w_1}\right), \quad 2 \le w_1 < w.$$

Note that the preceding argument shows incidentally that $\Omega(\kappa)$ holds with $\kappa = A_0$ whenever $\omega(p) \le A_0$ holds for all primes $p \in \mathcal{P}$.

As an immediate consequence of $\Omega(\kappa)$, on taking logarithms, we have

$$(1.7) \quad \sum_{w_1 \le p < w} \frac{\omega(p)}{p} \le \kappa \log\left(\frac{\log w}{\log w_1}\right) + \frac{A}{\log w_1}, \quad 2 \le w_1 < w.$$

Several useful variants of the last inequality follow by partial summation, and we note them here for later use.

Lemma 1.4. *Assume $\Omega(\kappa)$, and let f be a continuous nonnegative monotone function on an interval $[w_1, w]$, $w_1 \ge 2$. If f is increasing on $[w_1, w]$, then*

$$(1.8) \quad \sum_{w_1 \le p < w} \frac{\omega(p)}{p} f(p) \le \frac{Af(w)}{\log w} + \int_{w_1}^w f(t)\left(\frac{\kappa}{\log t} + \frac{A}{\log^2 t}\right) \frac{dt}{t}.$$

If f is decreasing on $[w_1, w]$, then

$$(1.9) \quad \sum_{w_1 \le p < w} \frac{\omega(p)}{p} f(p) \le \frac{Af(w_1)}{\log w_1} + \kappa \int_{w_1}^w \frac{f(t)\, dt}{t \log t}.$$

Proof. We have

$$L(s,t) := \sum_{s \le p < t} \frac{\omega(p)}{p} \le \kappa \log\left(\frac{\log t}{\log s}\right) + \frac{A}{\log s}, \quad 2 \le s \le t.$$

For f increasing,

$$\sum_{w_1 \leq p < w} \frac{\omega(p)}{p} f(p) - \int_{w_1}^{w} f(t)\left(\frac{\kappa\, dt}{t \log t} + \frac{A\, dt}{t \log^2 t}\right)$$

$$= -\int_{w_1}^{w} f(t)\, d\left\{L(t,w) - \kappa \log\left(\frac{\log w}{\log t}\right) - \frac{A}{\log t}\right\}$$

$$= -f(t)\left\{L(t,w) - \kappa \log\left(\frac{\log w}{\log t}\right) - \frac{A}{\log t}\right\}\Big|_{w_1}^{w}$$

$$+ \int_{w_1}^{w} \left\{L(t,w) - \kappa \log\left(\frac{\log w}{\log t}\right) - \frac{A}{\log t}\right\} df(t)$$

$$\leq A\, f(w)/\log w.$$

For f decreasing,

$$\sum_{w_1 \leq p < w} \frac{\omega(p)}{p} f(p) - \int_{w_1}^{w} \frac{f(t)\, dt}{t \log t}$$

$$= \int_{w_1}^{w} f(t)\, d\left\{L(w_1,t) - \kappa \log\left(\frac{\log t}{\log w_1}\right) - \frac{A}{\log w_1}\right\}$$

$$= f(t)\left\{L(w_1,t) - \kappa \log\left(\frac{\log t}{\log w_1}\right) - \frac{A}{\log w_1}\right\}\Big|_{w_1}^{w}$$

$$- \int_{w_1}^{w} \left\{L(w_1,t) - \kappa \log\left(\frac{\log t}{\log w_1}\right) - \frac{A}{\log w_1}\right\} df(t)$$

$$\leq A f(w_1)/\log w_1. \qquad \square$$

Corollary 1.5. *Assume* $\Omega(\kappa)$ *and* $2 \leq w_1 < w$. *Then*

(1.10) $$\sum_{w_1 \leq p < w} \frac{\omega(p)}{p} \log p \leq \kappa \log \frac{w}{w_1} + A\left(1 + \log \frac{\log w}{\log w_1}\right),$$

(1.11) $$\sum_{p < w} \frac{\omega(p)}{p} (p^\epsilon - 1) \leq \frac{\kappa(w^\epsilon - 1)}{\epsilon \log w} + \frac{Aw^\epsilon}{\log w} + O\left(\frac{w^\epsilon}{1 + \epsilon^2 \log^2 w}\right), \quad 0 < \epsilon \leq 1,$$

(1.12) $$\sum_{w_1 \leq p < w} \frac{\omega(p)}{p \log p} \leq \frac{A}{\log^2 w_1} + \frac{\kappa}{\log w_1} - \frac{\kappa}{\log w}.$$

Proof. The first and third inequalities follow at once from the lemma. We show that the second inequality holds uniformly for $\epsilon > 0$. The first term on the right side of (1.8) is bounded above by $Aw^\epsilon/\log w$; it remains to estimate the integral, which is in this case

$$\mathcal{I} := \int_2^w (t^\epsilon - 1)\left\{\frac{\kappa}{\log t} + \frac{A}{\log^2 t}\right\}\frac{dt}{t} = \int_{\epsilon \log 2}^{\epsilon \log w} (e^v - 1)\left(\frac{\kappa}{v} + \frac{A\epsilon}{v^2}\right)dv$$

$$\leq \kappa \int_0^{\epsilon \log w} \frac{e^v - 1}{v} dv + A\epsilon \int_{\epsilon \log 2}^{\epsilon \log w} \frac{e^v - 1}{v^2} dv =: \mathcal{I}_1 + \mathcal{I}_2, \text{ say.}$$

We estimate the integrals explicitly, for possible numerical applications.

$$\mathcal{I}_1 = \kappa \frac{(e^v - v - 1)}{v}\Big|_0^{\epsilon \log w} + \kappa \int_0^{\epsilon \log w} \frac{e^v - v - 1}{v^2} dv$$

$$= \frac{\kappa(w^\epsilon - 1 - \epsilon \log w)}{\epsilon \log w} + \int_0^{\epsilon \log w} \kappa \sum_{r=1}^\infty \frac{v^{r-1}}{(r+1)!} dv.$$

The last integral equals

$$\kappa \sum_{r=1}^\infty \frac{(\epsilon \log w)^r}{(r+1)! \, r} = \frac{\kappa}{(\epsilon \log w)^2} \sum_{r=1}^\infty \frac{(\epsilon \log w)^{r+2}}{(r+2)!} \frac{r+2}{r}$$

$$\leq \frac{3\kappa(w^\epsilon - 1 - \epsilon \log w)}{(\epsilon \log w)^2}.$$

Thus

$$\mathcal{I}_1 \leq \frac{\kappa(w^\epsilon - 1)}{\epsilon \log w} + \frac{3\kappa(w^\epsilon - 1 - \epsilon \log w)}{(\epsilon \log w)^2}.$$

In the same manner,

$$\mathcal{I}_2 \leq A\epsilon \frac{(e^v - 1 - v)}{v^2}\Big|_0^{\epsilon \log w} + 2A\epsilon \int_{\epsilon \log 2}^{\epsilon \log w} \frac{e^v - 1 - v}{v^3} dv$$

$$\leq A\epsilon \frac{(w^\epsilon - 1 - \epsilon \log w)}{(\epsilon \log w)^2} + \frac{2A}{\log 2} \int_0^{\epsilon \log w} \frac{e^v - 1 - v}{v^2} dv$$

$$\leq \frac{A(w^\epsilon - 1 - \epsilon \log w)}{\epsilon \log^2 w} + \frac{6A(w^\epsilon - 1 - \epsilon \log w)}{(\log 2)\, \epsilon^2 \log^2 w},$$

(the last by using the integral estimate from $\mathcal{I}_1$). The error term of (1.11) covers the cases of both small and large values of $\epsilon \log w$. □

1.5 Notes on Chapter 1

With minor exceptions, we use the notation introduced in [HR74].

Overviews of sieve methods, useful examples, and many problems are given in the books [HR74], [BaD04], and [MV06].

We shall treat the case $\kappa = 1$ in Chapter 7. However, our main thrust is to deal with integer or half integer dimensions that exceed 1, and we analyze that case in Chapter 9.

Bateman and Horn [BH62] conjectured that

$$|\{n \leq x \colon \Omega(L(n)) = g\}| \sim Cx(\log x)^{-g}, \quad x \to \infty,$$

with an explicit constant C depending on the coefficients a_i and b_j. This conjecture has not been confirmed for any $g \geq 2$. Approximations take the form

$$|\{n \leq x \colon \Omega(L(n)) \leq r_g\}| \gg x(\log x)^{-g},$$

where $r_g \sim g \log g$ ([HR74], Theorem 10.5). Better values for r_g for small g are given in Table 11.1 below.

In connection with the remarks following Example 1.2 on prime g-tuples, there are the recent spectacular results of Goldston *et al.* ([GPY, GPY06, GGPY]) about gaps between primes and many related results, some conditional. These results will be the subject of a forthcoming book by those authors.

The condition $\Omega(\kappa)$ could be weakened slightly by replacing the factor $1 + A/\log w_1$ with $\exp(A/\log w_1)$, as some authors have done. We retain the original formulation of Iwaniec.

2
Selberg's sieve method

2.1 Improving the Eratosthenes–Legendre sieve

To circumvent shortcomings of the Eratosthenes–Legendre formula (1.2), one searches for approximations

$$(2.1) \qquad \sum_{d|(n,P)} \mu(d)\chi(d)$$

to the indicator function (1.1), where the arithmetic functions $\chi(\cdot)$ are real, satisfy $\chi(1) = 1$, and otherwise are constructed to modify the behavior of the Moebius function in ways that lead to good bounds for $S(\mathcal{A}, \mathcal{P})$ when (2.1) is substituted for (1.1). This approach was pioneered by V. Brun almost a century ago, and his earliest idea will be described in the next chapter.

In this chapter we set out instead the enormously successful and versatile *upper bound* method of A. Selberg, which is based on the observation that for *any* such function χ,

$$(2.2) \qquad \sum_{d|(n,P)} \mu(d) \leq \Big(\sum_{d|(n,P)} \mu(d)\chi(d) \Big)^2.$$

Indeed, the left side of the formula is 0 in all cases except when n is relatively prime to P, in which case each side equals 1, and the right side is always nonnegative. It follows at once from (2.2) and (1.4) that

$$(2.3) \quad S(\mathcal{A},\mathcal{P}) \leq \sum_{d_1|P}\sum_{d_2|P} \mu(d_1)\chi(d_1)\mu(d_2)\chi(d_2)|\mathcal{A}_{\{d_1,d_2\}}| \leq X\Sigma + R,$$

where $\{d_1, d_2\}$ denotes the least common multiple of d_1 and d_2,

$$(2.4) \qquad \Sigma := \sum_{d_1|P}\sum_{d_2|P} \mu(d_1)\chi(d_1)\mu(d_2)\chi(d_2)\frac{\omega(\{d_1,d_2\})}{\{d_1,d_2\}},$$

and

$$(2.5) \qquad R := \sum_{d_1|P}\sum_{d_2|P} |\chi(d_1)\chi(d_2)r_{\mathcal{A}}(\{d_1,d_2\})|$$
$$= \sum_{d|P} |r_{\mathcal{A}}(d)| \sum_{\{d_1,d_2\}=d} |\chi(d_1)\chi(d_2)|.$$

We point out that the squared expression on the right of (2.2) actually has the same form as (2.1), for it is equal to

$$\sum_{d|(n,P)} \mu(d)\chi^*(d),$$

where, since $\mu(d_1)\mu(d_2) = \mu(\{d_1,d_2\})\mu((d_1,d_2))$,

$$(2.6) \qquad \chi^*(d) := \sum_{\{d_1,d_2\}=d} \mu((d_1,d_2))\chi_1(d_1)\chi_2(d_2).$$

2.2 A new parameter

Ideally, one would like to find a function χ to minimize the right side of (2.3), but no one knows how to do that. Instead, one limits the size of R by introducing a new parameter $\xi \geq 2$ and stipulating that

$$\chi(d) = 0 \quad \text{for} \quad d \geq \xi,$$

so that we can further restrict the sum in R extending over $d \mid P$ with the new condition $d < \xi^2$; and then finding χ to minimize Σ. For the moment we leave R aside and focus on Σ. Finally, we restrict the sum in Σ to numbers $d_1, d_2 \mid P$ for which $\omega(\{d_1,d_2\}) \neq 0$, since the other terms make no contribution. Here, since ω is multiplicative and d_1, d_2 are squarefree,

$$\frac{\omega(\{d_1,d_2\})}{\{d_1,d_2\}} = \frac{\omega(d_1)}{d_1}\frac{\omega(d_2)}{d_2}\frac{(d_1,d_2)}{\omega((d_1,d_2))} = \frac{\omega(d_1)}{d_1}\frac{\omega(d_2)}{d_2}\prod_{p|(d_1,d_2)}\frac{p}{\omega(p)}$$
$$= \frac{\omega(d_1)}{d_1}\frac{\omega(d_2)}{d_2}\prod_{\substack{p|d_1\\p|d_2}}\left(1+\frac{p-\omega(p)}{\omega(p)}\right) = \frac{\omega(d_1)}{d_1}\frac{\omega(d_2)}{d_2}\sum_{\substack{d|d_1\\d|d_2}}\frac{1}{g(d)},$$

2.2 A new parameter

where g is the multiplicative function given at prime numbers by

(2.7) $$g(p) := w(p)/(p - w(p)).$$

Note by (1.3) that $p - w(p) > 0$ for all p and $g(p) = 0$ when p is not in $\mathcal{P}$. On substituting in Σ we see at once that

$$\Sigma = {\sum_{d|P}}^{*} \frac{1}{g(d)} x_d^2, \quad \text{where} \quad x_d := \sum_{d|m|P} \mu(m)\chi(m) \frac{w(m)}{m}$$

and the star indicates that the sum is restricted to numbers d with $w(d) \neq 0$. At the same time we observe that

(2.8) $$\sum_{d|P} \mu(d) x_d = \sum_{m|P} \mu(m)\chi(m) \frac{w(m)}{m} \sum_{d|m} \mu(d) = 1$$

and that $x_d = 0$ when $d \geq \xi$. Thus Σ is a positive quadratic form in the x_d, and in light of (2.8) we rewrite Σ in the form

(2.9) $$\Sigma = {\sum_{\substack{d<\xi \\ d|P}}}^{*} \frac{1}{g(d)} \{x_d - \mu(d)g(d)C\}^2 + 2C \sum_{d|P} \mu(d) x_d - C^2 \sum_{\substack{d<\xi \\ d|P}} g(d)$$

$$= {\sum_{\substack{d<\xi \\ d|P}}}^{*} \frac{1}{g(d)} \{x_d - \mu(d)g(d)C\}^2 + 2C - C^2 \sum_{\substack{d<\xi \\ d|P}} g(d).$$

On choosing $C = 1/G$, where

(2.10) $$G = G_\xi(P) := \sum_{d<\xi,\, d|P} g(d),$$

we conclude that $\Sigma = 1/G$ provided that the values to be taken by χ can be chosen so that

$$x_d = \mu(d)g(d)/G \quad \text{when} \quad d < \xi \quad \text{and} \quad d \mid P.$$

But such a choice can be achieved with the Moebius inversion formulas

$$u(d) = \sum_{\substack{m|P \\ d|m}} v(m) \quad \text{if and only if} \quad v(d) = \sum_{t|(P/d)} \mu(t) u(dt).$$

We take

$$u(d) := \sum_{\substack{m|P \\ d|m}} \mu(m)\chi(m) \frac{w(m)}{m} = x_d = \mu(d) \frac{g(d)}{G}, \quad d < \xi,$$

and then, by inversion, since $u(d) = x_d = 0$ when $d \geq \xi$,

$$v(d) := \mu(d)\chi(d)\frac{\omega(d)}{d} = \sum_{t\mid(P/d)} \mu(t)u(dt)$$

$$= \sum_{\substack{t\mid(P/d) \\ dt<\xi}} \mu(t)\mu(dt)\frac{g(dt)}{G} = \mu(d)\frac{g(d)}{G}\sum_{\substack{t\mid P,\,(t,d)=1 \\ t<\xi/d}} g(t).$$

Thus we obtain the Selberg choice

(2.11) $$\chi(d) = \chi_S(d) := \frac{d}{\omega(d)}\frac{g(d)}{G}\sum_{\substack{t\mid P,\,(t,d)=1 \\ t<\xi/d}} g(t), \quad d\mid P.$$

We see that $\chi_S(1) = 1$, and also $\chi_S(d) = 0$ when $d \geq \xi$, since then the sum on the right is empty.

We may now conclude that

(2.12) $$S(\mathcal{A},\mathcal{P}) \leq \frac{X}{G} + R,$$

where G is given by (2.10) and R by (2.5).

One final comment about the modifying factors χ_S: we show that

(2.13) $$0 \leq \chi_S(d) \leq 1, \quad d\mid P.$$

Indeed, the left-hand inequality is obvious from (2.11), and the right-hand estimate is true trivially if $g(d) = 0$, also by (2.11); otherwise it holds since, for any positive integer d,

$$G = \sum_{\delta\mid d}\sum_{\substack{n<\xi,\,n\mid P \\ (n,d)=\delta}} g(n) = \sum_{\delta\mid d} g(\delta) \sum_{\substack{t<\xi/\delta,\,t\mid P \\ (t,d/\delta)=(t,\delta)=1}} g(t)$$

$$= \sum_{\delta\mid d} g(\delta) \sum_{\substack{t<\xi/\delta \\ t\mid P,\,(t,d)=1}} g(t) \geq \Big(\sum_{\delta\mid d} g(\delta)\Big) \sum_{\substack{t<\xi/d \\ t\mid P,\,(t,d)=1}} g(t),$$

and when $d\mid P$

$$\sum_{\delta\mid d} g(\delta) = \prod_{p\mid d}(1+g(p)) = g(d)\prod_{p\mid d}\Big(\frac{1}{g(p)}+1\Big)$$

$$= g(d)\prod_{p\mid d}\Big(\frac{p-\omega(p)}{\omega(p)}+1\Big) = \frac{dg(d)}{\omega(d)}.$$

By (2.11) this completes the proof of (2.13).

It remains to estimate R as given by (2.5). We simply apply (2.13) in (2.5) and use the estimate

$$\sum_{\{d_1,d_2\}=d} 1 \leq 3^{\nu(d)}, \quad d \mid P.$$

To see this, note that there are just three possibilities for a prime p to divide $\{d_1, d_2\}$: p divides exactly one of d_1, d_2 or p divides both.

This leads us, by (2.12), to

Theorem 2.1. (SELBERG) *Let ξ be an arbitrary positive parameter. Suppose $\mathcal{A}$ and $\mathcal{P}$ are as described in Section 1.1 (see (1.3) and (1.4)), and that G is given by (2.10) and (2.7). Then*

$$S(\mathcal{A}, \mathcal{P}) \leq \frac{X}{G} + \sum_{\substack{d < \xi^2 \\ d \mid P}} 3^{\nu(d)} |r_\mathcal{A}(d)|,$$

where X and $r_\mathcal{A}(d)$ are as in (1.4).

In applications, ξ is to be chosen so that the remainder sum on the right is of smaller order, or no larger than, the other term. In particular, we will nearly always take $\xi < X^{1/2}$.

2.3 Notes on Chapter 2

See [Sel47] for the original account of A. Selberg's sieve method, also [Sel91]. Here we have followed the presentation in [HR74], including introduction of the parameter ξ. There are, of course, other accounts of this famous method in the literature; for one such approach see the proof of Theorem 12.9 in [BaD04].

Selberg used the notation $\lambda_d = \mu(d)\chi(d)$ in (2.2), and his method is often referred to as the λ^2-method. Simple and elegant as his approach is, the estimation of the sum G presents, as we shall see, some technical problems when studied on the basis of $\Omega(\kappa)$.

Selberg developed also a weighted form of the λ^2 method (described in [Sel91]), and this has since been generalized, refined, and extended in [H-B97, HoTs06], and notably, in [GPY].

3
Combinatorial foundations

3.1 The fundamental sieve identity

We return to (2.1) and derive a very useful expression for the difference
$$\sum_{d|(n,P)} \mu(d) - \sum_{d|(n,P)} \mu(d)\chi(d)$$
that helps to suggest good choices for χ. Given an integer $d > 1$, we write $p(d)$ for the smallest prime factor of d and $p^+(d)$ for the largest prime factor. Also, we set $p(1) := \infty$ and $p^+(1) = 1$. We associate with χ the *complementary* function $\overline{\chi}$ given by

(3.1) $\quad \overline{\chi}(1) := 0, \qquad \overline{\chi}(d) := \chi(d/p(d)) - \chi(d) \quad (d > 1).$

Then χ and $\overline{\chi}$ are connected by the following relation.

Lemma 3.1. (THE FUNDAMENTAL SIEVE IDENTITY) *Suppose $\chi(1) = 1$ and that $\chi(d)$ is arbitrary for $d > 1$; also let $\overline{\chi}(\cdot)$ be as in (3.1). Then, for any arithmetic function h and any squarefree natural number n,*

(3.2) $\quad \displaystyle\sum_{d|n} \mu(d)h(d) = \sum_{d|n} \mu(d)\chi(d)h(d) + \sum_{d|n} \mu(d)\overline{\chi}(d) \sum_{\substack{t|n \\ p^+(t) < p(d)}} \mu(t)h(dt).$

In particular, when h is multiplicative we have

(3.3) $\quad \displaystyle\sum_{d|n} \mu(d)h(d) = \sum_{d|n} \mu(d)\chi(d)h(d) + \sum_{d|n} \mu(d)\overline{\chi}(d)h(d) \prod_{\substack{p|n \\ p<p(d)}} (1-h(p)),$

and hence, when $h = 1$ identically,

(3.4) $\quad \displaystyle\sum_{d|n} \mu(d) = \sum_{d|n} \mu(d)\chi(d) + \sum_{\substack{d|n \\ p(d)=p(n)}} \mu(d)\overline{\chi}(d).$

Before proving the identity we illustrate its uses with two important applications. First, let

$$\chi^{(k)}(d) = \begin{cases} 1, & \nu(d) \le k, \\ 0 & \text{otherwise,} \end{cases}$$

where k is a non-negative integer. Then $\overline{\chi^{(k)}}(d) = 1$ if and only if $\nu(d) = k+1$ and is otherwise 0; and therefore

$$\sum_{d|n} \mu(d) = \sum_{\substack{d|n \\ \nu(d) \le k}} \mu(d) + (-1)^{k+1} \sum_{\substack{d|n,\, \nu(d)=k+1 \\ p(d)=p(n)}} 1.$$

In particular, take $n = (a, P)$ and let k be any even integer, and ℓ any odd integer. Then

$$\sum_{\substack{d|(a,P) \\ \nu(d) \le \ell}} \mu(d) \le \sum_{d|(a,P)} \mu(d) \le \sum_{\substack{d|(a,P) \\ \nu(d) \le k}} \mu(d),$$

inequalities that are the foundations of Brun's first and simplest sieve method.

The second application is really the case $k = 0$ of the first: we have

$$\sum_{d|(a,P)} \mu(d) = 1 - \sum_{\substack{p|(a,P) \\ p((a,P))=p}} 1$$

and hence

(3.5) $$S(\mathcal{A}, \mathcal{P}) = |\mathcal{A}| - \sum_{p \in \mathcal{P}} S(\mathcal{A}_p, \mathcal{P}_p),$$

where $\mathcal{P}_p$ denotes the set of primes in $\mathcal{P}$ that are less than p. This is a form of Buchstab's famous identity [Buc37]. Clearly, applying an upper bound estimate to each term in the sum on the right leads to a *lower bound* for $S(\mathcal{A}, \mathcal{P})$. This remark is the basis of the Ankeny–Onishi lower bound sieve estimate [AO65].

Proof of the Fundamental Sieve Identity. Write a typical divisor d of n in the form

$$d = p_1 p_2 \cdots p_r, \quad p_1 > p_2 > \cdots > p_r,$$

so that $p^+(d) = p_1$ and $p(d) = p_r$. Then

$$1 - \chi(d) = \sum_{j=1}^{r} (\chi(p_1 \cdots p_{j-1}) - \chi(p_1 \cdots p_j)) = \sum_{j=1}^{r} \overline{\chi}(p_1 \cdots p_j),$$

3.1 The fundamental sieve identity

where $\chi(p_1 \cdots p_{j-1})$ at $j = 1$ is interpreted as $\chi(1) = 1$. We restate the preceding identity as

$$1 - \chi(d) = \sum_{\substack{t \mid d \\ p^+(d/t) < p(t)}} \overline{\chi}(t),$$

so that

$$\sum_{d \mid n} \mu(d) h(d)(1 - \chi(d)) = \sum_{d \mid n} \mu(d) h(d) \sum_{\substack{t \mid d \\ p^+(d/t) < p(t)}} \overline{\chi}(t).$$

Since d is squarefree, upon writing $d = t\delta$ we have $\mu(d) = \mu(t)\mu(\delta)$ and, after inversion of order,

$$\sum_{t \mid n} \mu(t) \overline{\chi}(t) \sum_{\substack{\delta \mid (n/t) \\ p^+(\delta) < p(t)}} \mu(\delta) h(t\delta) = \sum_{t \mid n} \mu(t) \overline{\chi}(t) \sum_{\substack{\delta \mid n \\ p^+(\delta) < p(t)}} \mu(\delta) h(t\delta),$$

since $\delta \mid n$ and $p^+(\delta) < p(t)$ imply $\delta \mid (n/t)$. This proves the identity, apart from a change of notation, and the two particular cases are then immediate consequences. □

As one more illustration of the Fundamental Sieve Identity, we apply it to the Selberg sieve method and show that

$$(3.6) \quad \sum_{d \mid (n,P)} \mu(d) = \left(\sum_{d \mid (n,P)} \mu(d) \chi(d) \right)^2 - \left(\sum_{\substack{d \mid (n,P) \\ p(d) = p((n,P))}} \mu(d) \overline{\chi}(d) \right)^2$$

$$=: I^2 - II^2,$$

say; here, we recall, $\chi(1) = 1$ and for $d > 1$, the real numbers $\chi(d)$ are initially arbitrary. Of course, (2.2) follows at once from (3.6). In Section 3.2 we derive an expression for II in terms of G and g. Together, these formulas provide a measure of the efficacy of the Selberg sieve.

By (2.6) and by the Fundamental Sieve Identity,

$$E := E_n := \sum_{d \mid (n,P)} \mu(d) - \sum_{d \mid (n,P)} \mu(d) \chi^*(d) = \sum_{\substack{d \mid (n,P) \\ p(d) = p((n,P))}} \mu(d) \overline{\chi^*}(d),$$

where

$$\overline{\chi^*}(d) = \chi^*(d/p(d)) - \chi^*(d) \quad (d > 1), \quad \overline{\chi^*}(1) = 0.$$

Hence
$$E = - \sum_{\substack{d|(n,P) \\ p(d)=p((n,P))}} \{\mu(d/p(d))\chi^*(d/p(d)) + \mu(d)\chi^*(d)\},$$

and, by (2.6),
$$\mu(d)\chi^*(d) = \sum_{\{d_1,d_2\}=d} \mu(d_1)\mu(d_2)\chi(d_1)\chi(d_2)$$

also, writing $q = p((n, P))$,
$$\mu(d/q)\chi^*(d/q) = \sum_{\{t_1,t_2\}=d/q} \mu(t_1)\mu(t_2)\chi(t_1)\chi(t_2).$$

Now when $d \mid (n, P)$, $p(d) = q$ and $\{d_1, d_2\} = d$, the pairs d_1, d_2 may be written

$$d_1 = qt_1,\ d_2 = t_2;\quad d_1 = t_1,\ d_2 = qt_2;\quad \text{or}\quad d_1 = qt_1,\ d_2 = qt_2$$

with $\{t_1, t_2\} = d/q$ in each case. It follows that

$$E = - \sum_{\substack{d|(n,P) \\ p(d)=q}} \sum_{\{t_1,t_2\}=d/q} \{\mu(t_1)\mu(t_2)\chi(t_1)\chi(t_2) + \mu(qt_1)\mu(t_2)\chi(qt_1)\chi(t_2)$$
$$+ \mu(t_1)\mu(qt_2)\chi(t_1)\chi(qt_2) + \mu(qt_1)\mu(qt_2)\chi(qt_1)\chi(qt_2)\}$$

$$= - \sum_{\substack{d|(n,P) \\ p(d)=q}} \sum_{\{t_1,t_2\}=d/q} \{\mu(t_1)\chi(t_1)+\mu(qt_1)\chi(qt_1)\}\{\mu(t_2)\chi(t_2)+\mu(qt_2)\chi(qt_2)\}$$

$$= -\left\{\sum_{\substack{d|(n,P) \\ p(d)=q}} \mu(d)(\chi(d/p(d)) - \chi(d))\right\}^2 = -\left\{\sum_{\substack{d|(n,P) \\ p(d)=q}} \mu(d)\overline{\chi}(d)\right\}^2,$$

and this proves (3.6).

3.2 Efficacy of the Selberg sieve

We can provide a measure of the amount by which the Selberg estimate overshoots the truth. To do this, we show first that

$$(3.7) \qquad \overline{\chi}(d) = \overline{\chi}_S(d) = \frac{1}{G} \prod_{p|(d/q)} \left(1 - \frac{\omega(p)}{p}\right)^{-1} \sum_{\substack{t|P,\ (t,d)=1 \\ \xi/d < t < (\xi/d)q}} g(t),$$

3.2 Efficacy of the Selberg sieve

where $\overline{\chi_S}(d)$ is the complementary Selberg modifying function, g and G are defined in (2.7) and (2.10) respectively, and $q = p(d)$.

Proof of (3.7). The Selberg function is defined in (2.11) and its complement in (3.1). We have

$$\overline{\chi_S}(d) = \chi_S(d/q) - \chi_S(d)$$
$$= \frac{d/q}{w(d/q)} \frac{g(d/q)}{G} \sum_{\substack{t \mid P, (t,d/q)=1 \\ t < \xi/(d/q)}} g(t) - \frac{d}{w(d)} \frac{g(d)}{G} \sum_{\substack{t \mid P, (t,d)=1 \\ t < \xi/d}} g(t).$$

Now $g(p) = w(p)/(p - w(p))$, whence

$$\frac{dg(d)}{w(d)} = \frac{qg(q)}{w(q)} \cdot \frac{d/q\, g(d/q)}{w(d/q)}, \quad \frac{qg(q)}{w(q)} = \left(1 - \frac{w(q)}{q}\right)^{-1},$$

and so

$$(3.8)\quad \overline{\chi_S}(d) = \frac{dg(d)}{w(d)} \cdot \frac{1}{G} \left\{ \left(1 - \frac{w(q)}{q}\right) \sum_{\substack{t \mid P, (t,d/q)=1 \\ t < (\xi/d)q}} g(t) - \sum_{\substack{t \mid P, (t,d)=1 \\ t < \xi/d}} g(t) \right\}.$$

Consider the first sum. Setting $d/q = s$, $q < p(s)$, it becomes

$$\sum_{\substack{t \mid P, (t,s)=1 \\ t < \xi/s}} g(t) = \left({\sum_{\substack{t \mid (P/q), (t,s)=1 \\ t < \xi/s}}}' + {\sum_{\substack{t \mid P, (t,s)=1 \\ t < \xi/s,\, q \mid t}}}'' \right) g(t)$$

$$= \sum_{\substack{t \mid P, (t,d)=1 \\ t < (\xi/d)q}} g(t) + g(q) \sum_{\substack{m \mid P, (m,d)=1 \\ m < \xi/d}} g(m)$$

on writing $t = qm$ in Σ'', so that $(m, q) = 1$ and $(m, s) = 1$. Since

$$\left(1 - \frac{w(q)}{q}\right) g(q) = \frac{w(q)}{q},$$

the bracketed expression on the right of (3.8) is

$$\left(1 - \frac{w(q)}{q}\right) \sum_{\substack{t \mid P, (t,d)=1 \\ t < (\xi/d)q}} g(t) + \left(1 - \frac{w(q)}{q}\right) g(q) \sum_{\substack{m \mid P, (m,d)=1 \\ m < \xi/d}} g(m) - \sum_{\substack{t \mid P, (t,d)=1 \\ t < \xi/d}} g(t)$$

$$= \left(1 - \frac{w(q)}{q}\right) \left\{ \sum_{\substack{t \mid P, (t,d)=1 \\ t < (\xi/d)q}} g(t) - \sum_{\substack{t \mid P, (t,d)=1 \\ t < \xi/d}} g(t) \right\} = \left(1 - \frac{w(q)}{q}\right) \sum_{\substack{t \mid P, (t,d)=1 \\ \xi/d < t < (\xi/d)q}} g(t).$$

Inserting the last expression in (3.8), we obtain (3.7). □

Now, to measure the "overshoot" in our Selberg estimate, consider the second sum, II, in (3.6). Taking $q = p((n, P))$, and then applying (3.7), we get

$$II := \sum_{\substack{d|(n,P) \\ p(d)=q}} \mu(d)\overline{\chi}(d)$$

$$= \sum_{\substack{d|n,\, d|P \\ p(d)=q}} \mu(d) \prod_{p|(d/q)} \left(1 - \frac{\omega(p)}{p}\right)^{-1} \frac{1}{G} \sum_{\substack{t|P,\, (t,d)=1 \\ \xi/d < t < (\xi/d)q}} g(t).$$

Let $s = d/q$, so $p(s) > q$. Then

$$II = -G^{-1} \sum_{\substack{s|(n/q) \\ s|(P/q)}} \mu(s) \prod_{p|s} \left(1 - \frac{\omega(p)}{p}\right)^{-1} \sum_{\substack{t|(P/q),\, (t,s)=1 \\ \xi/(sq) < t < \xi/s}} g(t),$$

or, on writing $m = st$, so that both of s, t divide P/q, $(t, s) = 1$, and $s \mid (n/q)$,

(3.9) $$II = -G^{-1} \sum_{\substack{\xi/q < m < \xi \\ m|(P/q)}} g(m) \sum_{\substack{s|m \\ s|(n,P)/q}} \frac{\mu(s)}{g(s)} \prod_{p|s} \left(1 - \frac{\omega(p)}{p}\right)^{-1}.$$

Now, by the definition of g,

$$\frac{1}{g(s)} \prod_{p|s} \left(1 - \frac{\omega(p)}{p}\right)^{-1} = \prod_{p|s} \frac{p}{\omega(p)},$$

and therefore, summing over s, remembering that $m \mid (P/q)$, we obtain

(3.10) $$\sum_{s|(m,n/q)} \mu(s) \prod_{p|s} \frac{p}{\omega(p)} = \prod_{p|(m,n/q)} \left(1 - \frac{p}{\omega(p)}\right) = \frac{\mu((m,n/q))}{g((m,n/q))}.$$

Finally, incorporating (3.10) into (3.9) and then taking absolute values, we obtain

$$|II| = \left| G^{-1} \sum_{\substack{\xi/q < m < \xi \\ m|(P/q)}} \mu((m,n/q))\, g\!\left(\frac{m}{(m,n/q)}\right) \right| \leq 1,$$

where the upper bound of 1 follows from the definition of G (equation (2.10)). Using this bound in (3.6) provides our measure of the amount by which the Selberg estimate overshoots the truth.

3.3 Multiplicative structure of modifying functions

Sums of type (2.1) that are upper bounds for (1.1) have an interesting algebraic structure: if $\mathcal{U}$ is the set of all such sums, then $\mathcal{U}$ is closed with respect to multiplication. Indeed, if

$$\sum_{d|(n,P)} \mu(d) \le \sum_{d|(n,P)} \mu(d)\chi_i(d), \quad i=1,2,$$

then, since the left side is 1 when $(n,P) = 1$ and is 0 otherwise, the product

(3.11)
$$\sum_{d|(n,P)} \mu(d)\chi_1(d) \cdot \sum_{d|(n,P)} \mu(d)\chi_2(d)$$

$$= \sum_{d|(n,P)} \sum_{\{d_1,d_2\}=d} \mu(d_1)\chi_1(d_1)\mu(d_2)\chi_2(d_2)$$

$$= \sum_{d|(n,P)} \mu(d)\chi^*(d),$$

where

$$\chi^*(d) := \sum_{\{d_1,d_2\}=d} \mu((d_1,d_2))\chi_1(d_1)\chi_2(d_2),$$

also lies in $\mathcal{U}$. Writing

$$\chi^{(0)}(d) = \begin{cases} 1, & d=1, \\ 0, & \text{otherwise,} \end{cases}$$

the sum

$$\sum_{d|(n,P)} \mu(d)\chi^{(0)}(d) \ (=1)$$

is the multiplicative identity of $\mathcal{U}$.

On the other hand, if $\chi_1(\cdot) = \chi^-(\cdot)$ determines a sum of type (2.1) that is a *lower* bound for (1.1) and $\chi_2(\cdot) = \chi^+(\cdot)$ gives an upper bound for (1.1), then clearly the product on the left side of (3.11) is less than or equal to 0 whenever $(n,P) \ge 2$, and hence $\chi^*(\cdot)$ is a new modifying function of χ^- type, that is,

$$\sum_{d|(n,P)} \mu(d)\chi^*(d) \le \sum_{d|(n,P)} \mu(d).$$

This relation is interesting, because it provides, in principle, a way of generating new lower bounds for (1.1). For example, if $\chi^-(\cdot)$ is some lower modifying function and $\chi(\cdot)$ is *any* real arithmetic function satisfying $\chi(1) = 1$, then

$$\left(\sum_{d|(n,P)} \mu(d)\chi^-(d) \right)\left(\sum_{d|(n,P)} \mu(d)\chi(d) \right)^2$$

determines another lower modifying function.

3.4 Notation: $\mathcal{P}$, $S(\mathcal{A}, \mathcal{P}, z)$, and V

To continue, we make a change of notation that we shall adhere to for the remainder of this monograph: from now on, $\mathcal{P}$ is an *infinite set of primes* and as a (finite) sieve we take $\mathcal{P}_z$, the set $\mathcal{P}$ truncated at z; to be precise, $\mathcal{P}_z = \{p \in \mathcal{P}: p < z\}$, and we write

$$P(z) = \prod_{\substack{p \in \mathcal{P} \\ p < z}} p, \quad P(z_0, z) = P(z)/P(z_0) = \prod_{\substack{p \in \mathcal{P} \\ z_0 \leq p < z}} p \quad (2 \leq z_0 < z),$$

$$S(\mathcal{A}, \mathcal{P}, z) := S(\mathcal{A}, \mathcal{P}_z) = |\{a \in \mathcal{A}: (a, P(z)) = 1\}|,$$

$$V(z) := V(\mathcal{P}_z) = \prod_{\substack{p \in \mathcal{P} \\ p < z}} \left(1 - \frac{\omega(p)}{p}\right).$$

For any choice of z_0 satisfying $2 \leq z_0 < z$, we have, on writing $P(z) = P(z_0)P(z_0, z)$,

$$\begin{aligned}
(3.12) \quad S(\mathcal{A}, \mathcal{P}, z) &= \sum_{n|P(z)} \mu(n)|\mathcal{A}_n| = \sum_{d|P(z_0,z)} \sum_{t|P(z_0)} \mu(dt)|\mathcal{A}_{dt}| \\
&= \sum_{d|P(z_0,z)} \mu(d) \sum_{t|P(z_0)} \mu(t)|(\mathcal{A}_d)_t| \\
&= \sum_{d|P(z_0,z)} \mu(d) S(\mathcal{A}_d, \mathcal{P}, z_0)
\end{aligned}$$

by (1.2); when now we apply the Fundamental Identity (Lemma 3.1) with $n = P(z)$ and

$$h(d) = \begin{cases} S(\mathcal{A}_d, \mathcal{P}, z_0), & (d, P(z_0)) = 1, \\ 0, & (d, P(z_0)) > 1 \end{cases}$$

to the sum on the right, we obtain

$$\text{(3.13)} \quad S(\mathcal{A},\mathcal{P},z) = \sum_{d|P(z_0,z)} \mu(d)\chi(d)S(\mathcal{A}_d,\mathcal{P},z_0)$$
$$+ \sum_{d|P(z_0,z)} \mu(d)\overline{\chi}(d)S(\mathcal{A}_d,\mathcal{P},p(d)).$$

Only the second sum on the right needs a word of explanation; by the Identity it is

$$\sum_{d|P(z_0,z)} \mu(d)\overline{\chi}(d) \sum_{t|P(z_0,p(d))} \mu(t)S(\mathcal{A}_{dt},\mathcal{P},z_0)$$

and the inner sum is equal to (cf. (3.12)) $S(\mathcal{A}_d,\mathcal{P},p(d))$.

However χ is chosen later, we require from now on that it is *divisor closed* in the sense that whenever $\chi(d) = 1$ and $t \mid d$, then $\chi(t) = 1$ too. Next, we call χ *combinatorial* if it takes only the values 0 or 1. If χ is divisor closed and combinatorial, then it is easy to see that $\overline{\chi}$ is combinatorial too and that $\overline{\chi}(d) = 0$ whenever $\chi(d) = 1$. Thus one may view χ and $\overline{\chi}$ as the indicator functions of two disjoint subsets of divisors of $P(z_0,z)$.

A word of explanation is in order for the parameter z_0; its role is two-fold. First, we shall choose z_0 small enough so that, apart from remainder sums that occur, $S(\mathcal{A},\mathcal{P},z)$ is asymptotic to

$$XV(z) := X \prod_{\substack{p \in \mathcal{P} \\ p < z}} \left(1 - \frac{\omega(p)}{p}\right),$$

for $2 \leq z \leq z_0$, as naive probability suggests. Also, as we shall see, there is some technical advantage in a preliminary sifting by the primes of $\mathcal{P}$ that are smaller than z_0. We shall return to this development in Chapter 6.

3.5 Notes on Chapter 3

A form of Lemma 3.1 first appeared in [HR74], Chapter 2 (1.6), but Iwaniec was the first to make effective use of it. The result was given the name "Fundamental Sieve Identity" in [DHR88], Lemma 2.1. Brun's "pure" sieve, the simplest sieve method devised by Brun, is our first application of Lemma 3.1; otherwise we do not discuss Brun's method, but instead refer the reader to the Brun–Hooley method [Hoo94, FH00], a simple but effective extension of it.

Balog [Bal85] refers to Lemma 3.1 as Richert's Fundamental Identity and derives from it, as does Harman [Hrm96], diophantine inequalities for primes that improve on results given by the better-known Vaughan identity. (Details are given in [Bal85].) Thus Lemma 3.1 joins a group of simple identities that have played an important part in modern prime number theory. The Buchstab identity, the second application of the lemma, is another such relation.

The linear ($\kappa = 1$) sieve method of Jurkat and Richert [JR65], may be said to rest on infinitely many iterations of Buchstab's identity (3.5) and Selberg's sieve method.

4
The Fundamental Lemma

4.1 A start: an asymptotic formula for $S(\mathcal{A}_q, \mathcal{P}, z)$

The aim of this chapter is to prove Theorem 4.1 below, which gives, under quite general conditions, an asymptotic formula for $S(\mathcal{A}_q, \mathcal{P}, z)$ provided that z is small enough relative to X. This theorem serves as a springboard for the principal results of this monograph; it has also many applications in the literature. Separate arguments are given for the upper and the lower estimates. Along the way we establish Lemma 4.4, a simple and useful O-bound for $S(\mathcal{A}, \mathcal{P}, z)$.

Theorem 4.1. (FUNDAMENTAL LEMMA) *Assume (1.3) and condition $\Omega(\kappa)$ hold and recall that $r_\mathcal{A}$ is defined by (1.4). Further, suppose that $z \geq 2$, $v \geq 1$, and q is any natural number whose prime factors lie in $\mathcal{P} \smallsetminus \mathcal{P}_z$. Then we have*

$$S(\mathcal{A}_q, \mathcal{P}, z) = \frac{\omega(q)}{q} XV(z)\{1 + O(\exp(-v\log v - 3v/2))\}$$
$$+ \theta \sum_{\substack{n < z^{2v} \\ n \mid P(z)}} 3^{\nu(n)} |r_\mathcal{A}(qn)|,$$

where $V(z)$ was defined in Section 3.4, $|\theta| \leq 1$, and the O-constant may depend on κ and A.

Proof. We start by establishing the upper bound, using Selberg's estimate from Chapter 2 with $\mathcal{P}_z$ in place of $\mathcal{P}$ and quote from there: for ξ a positive parameter,

$$(4.1) \qquad S(\mathcal{A}, \mathcal{P}, z) \leq \frac{X}{G(\xi, z)} + \sum_{\substack{d \mid P(z) \\ d < \xi^2}} 3^{\nu(d)} |r_\mathcal{A}(d)|,$$

where

(4.2) $$G(\xi, z) := \sum_{\substack{d < \xi \\ d \mid P(z)}} g(d).$$

Note that if $\xi \leq z$, the summation condition $d \mid P(z)$ on the right is superfluous and can be omitted. In that case

$$G(\xi, z) = \sum_{d < \xi} |\mu(d)| g(d) =: G(\xi).$$

Writing

$$V(z) := \prod_{p < z} \left(1 - \frac{\omega(p)}{p}\right) = \prod_{p < z} (1 + g(p))^{-1},$$

we have

$$G(\xi, z) = \sum_{\substack{d < \xi \\ d \mid P(z)}} g(d) \leq \sum_{d \mid P(z)} g(d) = \prod_{p < z} (1 + g(p)) = V(z)^{-1},$$

but for our purpose (see (4.1)) we need a lower estimate for $G(\xi, z) V(z)$. We obtain a first such estimate on the basis of condition $\Omega(\kappa)$ when $z \leq \xi$, actually when z is *very* much smaller than ξ.

We begin our argument by deploying a device known in the literature as Rankin's method. Since $g(\cdot) \geq 0$, we have

$$1/V(z) - G(\xi, z) = \sum_{d \mid P(z)} g(d) - G(\xi, z) = \sum_{\substack{d \geq \xi \\ d \mid P(z)}} g(d) \leq \sum_{\substack{d \geq \xi \\ d \mid P(z)}} g(d) \left(\frac{d}{\xi}\right)^{1-s}$$

for any s satisfying $s \leq 1$, whence

$$1/V(z) - G(\xi, z) \leq \xi^{s-1} \sum_{d \mid P(z)} g(d) d^{1-s} = \xi^{s-1} \prod_{p < z} (1 + g(p) p^{1-s})$$

$$= \frac{\xi^{s-1}}{V(z)} \prod_{p < z} \left(1 - \frac{\omega(p)}{p} + \frac{\omega(p)}{p^s}\right)$$

$$= \frac{\xi^{s-1}}{V(z)} \prod_{p < z} \left(1 + \frac{\omega(p)}{p} \left(p^{1-s} - 1\right)\right)$$

and therefore

$$1 - V(z) G(\xi, z) \leq \exp\left\{-(1-s)\log \xi + \sum_{p < z} \frac{\omega(p)}{p} \left(p^{1-s} - 1\right)\right\}$$

$$\leq \exp\left\{-(1-s)\log \xi + \frac{e^{(1-s)\log z}}{(1-s)\log z} \left(\kappa + O(1-s) + O\left(\frac{1}{(1-s)\log z}\right)\right)\right\}$$

4.1 A start: an asymptotic formula for $S(\mathcal{A}_q, \mathcal{P}, z)$

by (1.11). Now set

$$v := \log \xi / \log z, \quad \lambda := (1-s)\log z.$$

The last inequality can be restated as

(4.3) $\quad 1 - V(z)G(\xi, z) \leq \exp\left\{ -\lambda v + \left(e^\lambda / \lambda\right)\left(\kappa + O\left(\frac{\lambda}{\log z} + \frac{1}{\lambda}\right)\right)\right\}.$

If $v > z$, we have

$$\log \xi > z \log z > \sum_{p<z} \log p \geq \log P(z),$$

so that $\xi > P(z)$ and here

$$G(\xi, z) = \sum_{d \mid P(z)} g(d) = V(z)^{-1},$$

i.e., $1/G(\xi, z) = V(z)$.

Now suppose that $v \leq z$ but that v is large. Choose

$$\lambda = \log v + \log \log v.$$

Then, in (4.3),

$$\frac{\lambda}{\log z} + \frac{1}{\lambda} \ll \frac{\log v}{\log z} + \frac{1}{\log v} \ll 1 \quad \text{and} \quad \frac{e^\lambda}{\lambda} < v,$$

and we obtain

$$1 - V(z)G(\xi, z) \leq \exp\left\{-v(\log v + \log \log v) + O(v)\right\},$$

or

$$\frac{1}{G(\xi, z)} \leq \frac{V(z)}{1 - \exp\left\{-v \log v - v \log \log v + O(v)\right\}}$$
$$= V(z)(1 + \exp\left\{-v \log v - v \log \log v + O(v)\right\}).$$

(The higher-order power series terms of the series

$$1/(1-\epsilon) = 1 + \epsilon + \epsilon^2 + \cdots$$

have been absorbed in the $\exp O(v)$ error term.)

It follows that for all sufficiently large values of v,

(4.4) $\quad \dfrac{1}{G(\xi, z)} \leq V(z)\left\{1 + O\left(\exp\left\{-v \log v - 2v\right\}\right)\right\}.$

For small values of $v := \log \xi / \log z$, $v \geq 1$, we can assert that

(4.5) $$G(\xi, z)^{-1} \ll V(z).$$

(For large values of v, this relation follows from (4.4).) To show (4.5), choose a number v' sufficiently large that, by (4.4),

$$V(z')G(\xi', z') \geq 1/2$$

for $z' = z^{1/v'}$ and $\xi' = z$. Then

$$G(\xi, z) \geq G(z, z) \geq G(z, z').$$

Also, by condition $\Omega(\kappa)$,

$$V(z)/V(z') = \prod_{z' \leq p < z} \left(1 - \frac{\omega(p)}{p}\right) \gg 1.$$

It follows that

$$G(\xi, z)V(z) \geq G(z, z')V(z') \cdot V(z)/V(z') \gg G(z, z')V(z') \gg 1.$$

Thus we have shown that if $z \leq \xi$ and condition $\Omega(\kappa)$ holds, then (4.1) implies that

(4.6) $$S(\mathcal{A}, \mathcal{P}, z) \leq XV(z)\left\{1 + O(v^{-v}e^{-2v})\right\} + \sum_{\substack{d | P(z) \\ d < \xi^2}} 3^{\nu(d)} |r_\mathcal{A}(d)|.$$

We restate (4.6) in a more general form that we shall need at the next stage.

Lemma 4.2. *Suppose that condition $\Omega(\kappa)$ (Definition 1.3) holds, that the parameter ξ of (4.6) satisfies $\xi \geq z$, and that q is a natural number such that $(q, P(z)) = 1$. Then, writing $v = (\log \xi)/\log z$ (so that $v \geq 1$),*

$$S(\mathcal{A}_q, \mathcal{P}, z) \leq \frac{\omega(q)}{q} XV(z)\left\{1 + O(v^{-v}e^{-2v})\right\} + \sum_{\substack{d | P(z) \\ d < z^{2v}}} 3^{\nu(d)} |r_\mathcal{A}(qd)|.$$

The preceding argument is unchanged; we have only to observe that

$$|\mathcal{A}_{qd}| = \frac{\omega(qd)}{qd} X + r_\mathcal{A}(qd) = \frac{\omega(d)}{d}\left(\frac{\omega(q)}{q} X\right) + r_\mathcal{A}(qd)$$

for $d | P(z)$ and $(q, P(z)) = 1$, so that $\omega(q)X/q$ replaces X and $r_\mathcal{A}(qd)$ appears in place of $r_\mathcal{A}(d)$. Lemma 4.2 yields an upper bound estimate implicit in Theorem 4.1; because of its frequent application, we have stated it explicitly.

4.2 A lower bound for $S(\mathcal{A}_q, \mathcal{P}, z)$

Returning to the proof of Theorem 4.1, it remains to establish the lower bound for $S(\mathcal{A}_q, \mathcal{P}, z)$. We start with Buchstab's identity (3.5) with $\mathcal{P}$ replaced by $\mathcal{P}_z$ (that is, $\mathcal{P}$ truncated at z) and $\mathcal{A}_q$ in place of $\mathcal{A}$ for $(q, P(z)) = 1$. The altered formula reads

$$S(\mathcal{A}_q, \mathcal{P}, z) = |\mathcal{A}_q| - \sum_{\substack{p<z \\ p \in \mathcal{P}}} S(\mathcal{A}_{qp}, \mathcal{P}, p).$$

We combine this relation with the identity

(4.7) $$V(z) = 1 - \sum_{p<z} \frac{\omega(p)}{p} V(p).$$

We establish the last formula as we did (3.5): take $h(d) = \omega(d)/d$ in (3.3) so that the sum on the left side of (3.3) is $V(z)$, and on the right take $\chi(d) = 1$ when $d = 1$ and 0 otherwise, so that $\overline{\chi}(d) = 1$ when d is prime, and is 0 otherwise. Alternatively, (4.7) is equivalent to the simple identity

$$\prod_{j=1}^{m}(1 - x_j) = 1 - \sum_{j=1}^{m} x_j \prod_{i<j}(1 - x_i).$$

Together, the Buchstab formula and (4.7) yield

$$S(\mathcal{A}_q, \mathcal{P}, z) - \frac{\omega(q)}{q} XV(z)$$
$$= |\mathcal{A}_q| - \frac{\omega(q)}{q} X - \sum_{\substack{p<z \\ p \in \mathcal{P}}} \left\{ S(\mathcal{A}_{qp}, \mathcal{P}, p) - \frac{\omega(qp)}{qp} XV(p) \right\}.$$

We shall be led to a lower bound for $S(\mathcal{A}_q, \mathcal{P}, z)$ by applying Lemma 4.2 to each term in the sum on the right with $\xi p^{-1/2}$ in place of ξ (in order to take care of the accumulation of error terms in this sum), p in place of z, and v replaced by

$$v_p := \frac{\log(\xi p^{-1/2})}{\log p} = \frac{\log \xi}{\log p} - \frac{1}{2} > \frac{\log \xi}{\log z} - \frac{1}{2} = v - \frac{1}{2}.$$

The condition "$\xi \geq z$" in the lemma translates into $\xi p^{-1/2} \geq p$ or $v_p \geq 1$ and holds when $v - 1/2 \geq 1$, that is, when

$$v \geq 3/2 \text{ or, equivalently, when } z \leq \xi^{2/3}.$$

Subject to this assumption, we obtain

(4.8) $\quad S(\mathcal{A}_q,\mathcal{P},z) - \dfrac{w(q)}{q} XV(z) \geq$

$$- \sum_{\substack{p<z \\ p \in \mathcal{P}}} \left\{ \dfrac{w(qp)}{qp} XV(p)\, O\left(\exp(-v_p \log v_p - 2v_p)\right) \right.$$

$$\left. + \sum_{\substack{d < \xi^2/p \\ d \mid P(p)}} 3^{\nu(d)} |r_{\mathcal{A}}(qpd)| \right\} - |r_{\mathcal{A}}(q)|.$$

On the right, the remainder terms contribute

$$|r_{\mathcal{A}}(q)| + \sum_{\substack{p<z \\ p \in \mathcal{P}}} \sum_{\substack{d<\xi^2/p \\ d \mid P(p)}} 3^{\nu(d)} |r_{\mathcal{A}}(qpd)| \leq \sum_{\substack{n<\xi^2 \\ n \mid P(z)}} 3^{\nu(n)} |r_{\mathcal{A}}(qn)|.$$

Let Σ_0 denote the remaining sum on the right side of (4.8). Then

$$\Sigma_0 \ll \dfrac{w(q)}{q} XV(z) \sum_{p<z} \dfrac{w(p)}{p} \dfrac{V(p)}{V(z)} \exp(-v_p \log v_p - 2v_p)$$

$$\ll \dfrac{w(q)}{q} XV(z) \sum_{p<z} \dfrac{w(p)}{p} \left(\dfrac{\log z}{\log p}\right)^{\kappa} \exp(-v_p \log v_p - 2v_p)$$

$$= \dfrac{w(q)}{q} XV(z) \dfrac{1}{\log z} \sum_{p<z} \dfrac{w(p)}{p} \log p$$

$$\times \exp\left\{ -v_p \log v_p - 2v_p + (\kappa+1) \log \dfrac{\log z}{\log p} \right\}$$

$$\ll \dfrac{w(q)}{q} XV(z) \exp\left\{ -v_p \log v_p - 2v_p + (\kappa+1) \log \dfrac{\log z}{\log p} \right\},$$

the last by (1.10).

The expression in the exponent, when regarded as a function of p, has partial derivative with respect to p equal to

$$\dfrac{1}{p \log p} \left\{ \left(v_p + \dfrac{1}{2}\right)(3 + \log v_p) - \kappa - 1 \right\} > 0$$

if (remember that $v_p > v - \tfrac{1}{2}$)

$$v \geq (1/3)(\kappa+4) \ (\geq 5/3 > 3/2).$$

Hence this expression can be estimated from above by replacing p with z and replacing v_p by $v - \tfrac{1}{2}$. It follows that

$$\Sigma_0 \ll \dfrac{w(q)}{q} XV(z) \exp\left\{ -\left(v - \dfrac{1}{2}\right) \log\left(v - \dfrac{1}{2}\right) - 2v \right\}.$$

4.2 A lower bound for $S(\mathcal{A}_q, \mathcal{P}, z)$

By the mean value theorem,

$$v \log v - \left(v - \frac{1}{2}\right) \log\left(v - \frac{1}{2}\right) \leq \frac{1}{2}(1 + \log v) \leq v/2, \quad v > 1/2.$$

At the end we used the inequality $1 + t \leq e^t$. We conclude that

$$\Sigma_0 \ll \frac{\omega(q)}{q} X V(z) \exp\left(-v \log v - \frac{3}{2}v\right) \quad \text{if } v \geq \frac{1}{3}(\kappa + 4).$$

In summary, by (4.8) and provided that $v \geq (\kappa + 4)/3$,

$$S(\mathcal{A}_q, \mathcal{P}, z) \geq \frac{\omega(q)}{q} X V(z)\{1 + O(\exp(-v \log v - 3v/2))\}$$
$$- \sum_{\substack{n < z^{2v} \\ n \mid P(z)}} 3^{\nu(n)} |r_\mathcal{A}(qn)|.$$

On the other hand, if $1 \leq v < (\kappa + 4)/3$, the inequality holds trivially in the form $S(\mathcal{A}_q, \mathcal{P}, z) \geq 0$ provided that the constant implied by the O-notation is large enough, depending on κ and A from condition $\Omega(\kappa)$. While the lower bound in the Fundamental Lemma is formally valid even for small values of v, it is of interest only when v is large enough for the expression on the right to be positive.

When combined with the upper bound of Lemma 4.2, the last inequality concludes the proof of the Fundamental Lemma. □

A sum involving the remainder term $r_\mathcal{A}$ occurs in both Theorem 2.1 and the Fundamental Lemma. In many applications $|r_\mathcal{A}(\cdot)|$ is of size at most $\omega(\cdot)$, which yields a simple estimate for the remainder sum.

Lemma 4.3. *Suppose that $\Omega(\kappa)$ holds and*

(4.9) $$|r_\mathcal{A}(d)| \leq \omega(d) \text{ whenever } d \mid P(z).$$

Let K be a fixed positive integer. Then for any $z \geq 2$,

$$E := \sum_{\substack{d \mid P(z) \\ d < z^2}} K^{\nu(d)} |r_\mathcal{A}(d)| \leq z^2 \, V(z)^{-K}.$$

Proof. We estimate E using another application of Rankin's method:

recalling the hypothesis on $|r_A(d)|$, we have

$$E \le \sum_{\substack{d|P(z) \\ d<z^2}} \frac{z^2}{d} K^{\nu(d)} |r_A(d)| \le \sum_{d|P(z)} \frac{z^2}{d} K^{\nu(d)} \omega(d) = z^2 \prod_{p<z} \left(1 + \frac{K\omega(p)}{p}\right)$$

$$\le z^2 \prod_{p<z} \left(1 + \frac{\omega(p)}{p}\right)^K \le z^2 \prod_{p<z} \left(1 - \frac{\omega(p)}{p}\right)^{-K} = z^2 V(z)^{-K}. \qquad \square$$

Applying the preceding result, we have a useful special case of the upper bound estimate of the Fundamental Lemma.

Lemma 4.4. *Suppose that $\Omega(\kappa)$ holds and also that $|r_A(d)| \le \omega(d)$ whenever $d \mid P(z)$. Then for any positive constant $A_1 < 1/2$,*

$$S(\mathcal{A}, \mathcal{P}, z) \ll XV(z) \text{ if } z \le X^{A_1}$$

and

$$S(\mathcal{A}, \mathcal{P}, z) \ll XV(X) \text{ if } z \ge X^{A_1},$$

where the constants implied by the $\ll$-notation may depend on κ and A_1.

Proof. Take $q = 1$ in Lemma 4.2 and also $v = 1$, so that $\xi = z$. Then the first term in the formula of that lemma is $\ll XV(z)$. We estimate the second term by the last lemma with $K = 3$.

If $z \le X^{A_1}$, then

$$E \ll X^{2A_1} V(z)^{-3} \ll XV(z),$$

as claimed, since $V(z)^{-1}$ is at most a logarithmic power of X.

Now suppose that $z > X^{A_1}$. Trivially, $S(\mathcal{A}, \mathcal{P}, z) \le S(\mathcal{A}, \mathcal{P}, X^{A_1})$ and therefore (with X^{A_1} in place of z in the preceding estimate)

$$S(\mathcal{A}, \mathcal{P}, X^{A_1}) \ll XV(X^{A_1}) + X^{2A_1} V(X^{A_1})^{-3}$$
$$\ll XV(X^{A_1}) \ll XV(X).$$

The last relation is a consequence of $\Omega(\kappa)$:

$$V(X^{A_1})/V(X) = \prod_{X^{A_1} \le p < X} \left(1 - \frac{\omega(p)}{p}\right)^{-1} \ll 1.$$

This establishes the second bound for $S(\mathcal{A}, \mathcal{P}, z)$. $\qquad \square$

Here is an application of Lemma 4.4.

4.2 A lower bound for $S(\mathcal{A}_q, \mathcal{P}, z)$

Corollary 4.5. *Let $\mathcal{A}$ be the integer sequence described in Example 1.2, and let $\rho(p)$ denote the number of incongruent solutions of $L(n) \equiv 0 \mod p$. Let $\mathcal{P}$ be a (thin) set of primes such that*

(4.10) $$\sum_{\substack{p < y \\ p \in \mathcal{P}}} \frac{1}{p} \geq \delta \log \log y - A_0$$

for some constants $\delta \in (0,1)$ and $A_0 > 0$. Then

$$|\{n: x - y < n \leq x, \, (L(n), \mathcal{P}) = 1\}| \ll \prod_{\substack{p \mid \Delta \\ p \in \mathcal{P}}} \left(1 - \frac{1}{p}\right)^{\rho(p) - g} \frac{y}{(\log y)^{\delta g}},$$

where the constant implied by the $\ll$-notation depends only on g and A_0 (but not on the coefficients of $L(n)$).

Proof. In Example 1.2,

$$\omega(p) = \begin{cases} \rho(p) & \text{if } p \in \mathcal{P}, \\ 0 & \text{if } p \notin \mathcal{P}, \end{cases}$$

and we may assume that

$$\rho(p) < p \quad \text{for all} \quad p \in \mathcal{P},$$

since otherwise the expression on the left side of our result is zero and the inequality holds trivially. Also, $\Omega(g)$ holds (with $A = A(g)$) because, from above, $\omega(p) \leq g$. Hence the quantity being estimated is at most

$$|\{n: x - y < n \leq x, \, (L(n), P(y)) = 1\}| =: S(\mathcal{A}, \mathcal{P}, y) \ll y \prod_{\substack{p < y \\ p \in \mathcal{P}}} \left(1 - \frac{\rho(p)}{p}\right),$$

where the $\ll$-constant depends only on g.

A small calculation shows the product in the last formula to be at most

$$\prod_{\substack{p < y \\ p \in \mathcal{P}}} \left(1 - \frac{1}{p}\right)^{\rho(p)} = \prod_{\substack{p < y \\ p \in \mathcal{P}}} \left(1 - \frac{1}{p}\right)^g \prod_{\substack{p < y \\ p \in \mathcal{P}}} \left(1 - \frac{1}{p}\right)^{\rho(p) - g}.$$

The first product on the right does not exceed

$$\exp\left(-g \sum_{\substack{p < y \\ p \in \mathcal{P}}} \frac{1}{p}\right) \leq \frac{e^{A_0 g}}{(\log y)^{\delta g}}$$

by hypothesis, whereas the second product is, by the discussion in Example 1.2, equal to

$$\prod_{\substack{p<y \\ p\mid\Delta,\, p\in\mathcal{P}}} \left(1-\frac{1}{p}\right)^{\rho(p)-g} \leq \prod_{p\mid\Delta,\, p\in\mathcal{P}} \left(1-\frac{1}{p}\right)^{\rho(p)-g}.$$

We remark that the last product is finite since $\Delta \neq 0$. □

We state, without proofs, several special cases of Corollary 4.5.

Example 4.6. Suppose that $1 < y \leq x$ and $\mathcal{P}$ is a set of primes satisfying (4.10) for some positive constants δ and A_0. Then

$$|\{n\colon x-y < n \leq x, (n,\mathcal{P})=1\}| \ll \frac{y}{\log^\delta y},$$

where the $\ll$-constant depends on A_0 only.

A special case of this example arises when $\mathcal{P}$ consists of all primes lying in one or more distinct residue classes with respect to some modulus k. If r is the number of these residue classes, $\ell_i \bmod k$ ($i=1,\ldots,r$), and $(\ell_i, k) = 1$ for each i, then $\delta = r/\varphi(k)$ and $A_0 = O_k(1)$.

An instance of this last case occurs in representations of numbers as sums of two squares. It is known that a number n is representable as a sum of two squares if and only if all prime factors p of n with $p \equiv 3 \bmod 4$ occur with even multiplicity. So, for example, 45 is a sum of two squares, while 47 is not. The main contribution to an estimate of

$$|\{n \in (x-y, x]\colon n = a^2+b^2,\ \text{some}\ a, b \in \mathbb{N}\}|$$

arises by sieving $(x-y, x]$ by the collection $\mathcal{P}$ of primes in the progression $3 \bmod 4$. Here $k=4$, $\varphi(k)=2$ and $r=1$, so

$$|\{n\colon x-y < n \leq x, (n,\mathcal{P})=1\}| \ll \frac{y}{\sqrt{\log y}},$$

which is the correct order of magnitude for the preceding counting function. There are many other specific examples of Corollary 4.5 in which $L(n)$ is a particular product of linear polynomials and $\mathcal{P}$ is the set of primes lying in the union of distinct arithmetic progressions.

A Fundamental Lemma is especially useful for determining precise information about the cardinality of a sequence whose elements have no *very* small prime factors. As a first illustration, let u and x be real numbers such that $u \geq 1$ and $x^{1/u} \geq 2$, and let $q = q(x, u) > 1$ denote

4.2 A lower bound for $S(\mathcal{A}_q, \mathcal{P}, z)$

a number having no prime factors less than $x^{1/u}$ and satisfying $\log q \ll \log x$; such a number q is sometimes referred to as a *quasi-prime* (relative to x and u). If we take $u = u(x)$ to be a function of x tending arbitrarily slowly to infinity with x and suppose that $1 \ll (\log X)/\log x \ll 1$, then, by Theorem 4.1 (with $q = 1$ in the statement of the theorem) and $\mathcal{A} = \{n: 1 \le n \le X\}$, we have

$$|\{q: q \le X\}| \sim \left(ue^{-\gamma}\frac{\log X}{\log x}\right)\pi(X) \quad \text{as} \quad X \to \infty.$$

To see this, take $z = x^{1/u}$ and $v = u/4$, say, in Theorem 4.1; we obtain in fact a good quality "quasi-prime number theorem," which shows that the quasi-primes up to X are hardly more dense than the primes.

The following is a more elaborate application of Theorem 4.1 (again with $q = 1$) to quasi-primes.

Example 4.7. Let $h_1(n), \ldots, h_g(n)$ be distinct irreducible polynomials with integer coefficients and write $H(n) = h_1(n) \cdots h_g(n)$. Let $\rho(d)$ denote the number of solutions of the congruence $H(n) \equiv 0 \bmod d$ that are incongruent mod d, and assume that $\rho(p) < p$ for all primes p. With u and x as described above, we have

$$|\{n: 1 \le n \le x, \ h_i(n) = q_i \ (i = 1, \ldots, g)\}|$$
$$= x \prod_{p < x^{1/u}} \left(1 - \frac{\rho(p)}{p}\right)\left\{1 + O\left(\exp\left(-\frac{1}{4}u\log u\right)\right)\right\} + O(x^{1/2}\log^{3g} x),$$

where $q_1, \ldots, q_g$ are all quasi-primes relative to x and u. Moreover, the expression on the right is

$$(ue^{-\gamma})^g \prod_p \left(1 - \frac{\rho(p) - 1}{p - 1}\right)\left(1 - \frac{1}{p}\right)^{-g+1} \frac{x}{\log^g x}\left\{1 + O_H\left(\exp\left(-\frac{u}{4}\log u\right)\right)\right\}$$
$$+ O_H(x^{1/2}\log^{3g} x),$$

where all the O_H-constants depend at most on the coefficients and degrees of $h_1, \ldots, h_g$.

Proof. Let $\mathcal{A} = \{H(n): 1 \le n \le x\}$ and take $\mathcal{P}$ to be the set of all primes. We may assume that all the polynomials h_i have positive degree; for if any one had zero degree, then the condition that $\rho(p) < p$ for all p would imply that the polynomial was identically 1, and the expression on the left in the statement of the example would then be 0 and the result trivial.

We have

$$|\mathcal{A}_d| := |\{n : 1 \le n \le x, \; H(n) \equiv 0 \bmod d\}| = \rho(d)(x/d + \theta), \quad |\theta| < 1,$$

so that $X = x$, $\omega(d) = \rho(d)$, and

$$r_{\mathcal{A}}(d) = |\mathcal{A}_d| - (\rho(d)/d)\,x, \quad |r_{\mathcal{A}}(d)| \le \rho(d).$$

By the Chinese Remainder Theorem ([HW79], Theorem 121), ρ is a multiplicative function. Further, by Lagrange's theorem, $\rho(p)$ is at most the degree of H for every prime p; of course, by hypothesis, $\rho(p) < p$ always. Also, by the formula (10.1) of Landau, we have the Mertens-type relation

$$\sum_{p < y} \frac{\rho(p)}{p} \log p = g \log y + O_H(1).$$

This result is proved in Chapter 10.

From the last formula we can show that $\Omega(g)$ holds and therefore we can take here $\kappa = g$. Indeed (cf. Definition 1.3 and the subsequent proof that $\Omega(\kappa)$ holds for the sequence generated by L),

$$\prod_{w_1 \le p < w} \left(1 - \frac{\rho(p)}{p}\right)^{-1} = \exp\left(\sum_{w_1 \le p < w} \frac{\rho(p)}{p} + \sum_{w_1 \le p < w}\left\{\log\left(1-\frac{\rho(p)}{p}\right)^{-1} - \frac{\rho(p)}{p}\right\}\right),$$

and we have to evaluate the two sums on the right. By Landau's formula and summation by parts,

$$(4.11) \qquad \sum_{w_1 \le p < w} \frac{\rho(p)}{p} = g \log \frac{\log w}{\log w_1} + O_H\left(\frac{1}{\log w_1}\right).$$

Then, taking $w_1 = p$ and letting $w \to p + 0$, we obtain

$$\rho(p)/p \ll_H 1/\log p.$$

Next, for $w > w_1 \ge 2$, we have

$$\sum_{w_1 \le p < w} \left\{\log\left(1-\frac{\rho(p)}{p}\right)^{-1} - \frac{\rho(p)}{p}\right\} = \sum_{w_1 \le p < w} \sum_{m=2}^{\infty} \frac{1}{m}\left(\frac{\rho(p)}{p}\right)^m$$

$$\le \sum_{w_1 \le p < w} \frac{1}{2}\left(\frac{\rho(p)}{p}\right)^2 \frac{1}{1 - \rho(p)/p}$$

$$\ll_H \sum_{w_1 \le p < w} \left(\frac{\rho(p)}{p}\right)^2 \ll_H \sum_{w_1 \le p < w} \frac{\rho(p)}{p \log p} \ll \frac{1}{\log w_1},$$

4.2 A lower bound for $S(\mathcal{A}_q, \mathcal{P}, z)$

the last from (4.11), by another summation by parts. Hence

$$(4.12) \quad \prod_{w_1 \leq p < w} \left(1 - \frac{\rho(p)}{p}\right)^{-1} = \left(\frac{\log w}{\log w_1}\right)^g \exp\left\{O_H\left(\frac{1}{\log w_1}\right)\right\},$$

which implies $\Omega(g)$.

Now taking $z = x^{1/u}$ and $v = u/4$ (as before) and using the fact that $|r_\mathcal{A}(n)| \leq w(n) = \rho(n)$, Theorem 4.1 and Lemma 4.4 give

$$(4.13) \quad S(\mathcal{A}, \mathcal{P}, x^{1/u}) = x \prod_{p < x^{1/u}} \left(1 - \frac{\rho(p)}{p}\right) \cdot \left\{1 + O\left(\exp\left(-\frac{1}{4}u \log u\right)\right)\right\}$$

$$+ O(x^{1/2} \log^{3g} x).$$

Equation (4.12) suggests a connection between the last product and Mertens' product formula with a remainder,

$$(4.14) \quad \prod_{p < x} \left(1 - \frac{1}{p}\right) = e^{-\gamma} (\log x)^{-1} \{1 + O(1/\log x)\}.$$

By Mertens' formula,

$$\prod_{w_1 \leq p < w} \left(1 - \frac{1}{p}\right)^{-g} = \left(\frac{\log w}{\log w_1}\right)^g \left\{1 + O\left(\frac{1}{\log w_1}\right)\right\},$$

and this and (4.14) together imply that the product

$$\prod_{p \geq t} \left(1 - \frac{\rho(p)}{p}\right)\left(1 - \frac{1}{p}\right)^{-g}$$

converges and equals $1 + O_H(1/\log t)$. It follows that the product on the right side of (4.13) is

$$\prod_{p < x^{1/u}} \left(1 - \frac{1}{p}\right)^g \cdot \prod_{p < x^{1/u}} \left(1 - \frac{\rho(p)}{p}\right)\left(1 - \frac{1}{p}\right)^{-g}$$

$$= \frac{(e^{-\gamma}u)^g}{(\log x)^g} \cdot \prod_p \left(1 - \frac{\rho(p)}{p}\right)\left(1 - \frac{1}{p}\right)^{-g} \cdot \left\{1 + O_H\left(\frac{u}{\log x}\right)\right\}.$$

For large x, the error term in the last formula is $\ll \exp(-(u/4)\log u)$, so we drop it when inserting the formula into (4.13). Finally, we change the form of the product slightly by writing

$$\left(1 - \frac{\rho(p)}{p}\right)\left(1 - \frac{1}{p}\right)^{-1} = 1 - \frac{\rho(p) - 1}{p - 1},$$

and the proof of Example 4.7 is complete. □

4.3 Notes on Chapter 4

Theorem 4.1 was given the name "Fundamental Lemma" by Kubilius in [Kub64]. There are other versions of this result, arising from different sieve approaches, of which that of [FI78] is particularly sharp and attractive.

There are forms of the Fundamental Lemma in which all constants are explicit. For one version and an interesting application, see [Hal00].

The bound on E given in Lemma 4.3 can be regarded as a quantitative "quasi-independence" result for $\mathcal{A}$.

Examples 4.6 and 4.7 come from [HR74]. There is a discussion of quasi-primes in Chapter 1.5 of [Lin63].

5
Selberg's sieve method (continued)

5.1 A lower bound for $G(\xi, z)$

This is a long chapter. Its goal is to establish Theorem 5.6, in Section 5.3, the version of Selberg's sieve that is the starting point for our work in the pivotal Chapter 9. We extend Theorem 2.1 by giving a (sharp) lower bound for

$$G(\xi, z) := \sum_{\substack{d < \xi \\ d | P(z)}} g(d),$$

where g is the multiplicative function whose value on the primes is

(5.1) $$g(p) := \frac{\omega(p)}{p - \omega(p)} = \frac{\omega(p)}{p} + \frac{\omega(p)}{p} g(p).$$

(g and $G(\xi, z)$ occurred earlier in (2.10) and (4.2).) Since the Fundamental Lemma, Theorem 4.1, is effective only when $v \to \infty$, that is, when $\log z$ is small compared with $\log \xi$, a more precise lower bound for $G(\xi, z)$ is needed for smaller values of v.

The main results encountered along the way are the inequalities (5.8) and (5.44). The biggest task is to establish the first of these, which connects $G(\xi, z)$ with a related sum. The combinatorial device used here, Lemma 5.1, is of some independent interest. The second inequality is the desired lower bound for $G(\xi, z)$, from which the theorem follows readily. The important functions j (or their equivalent σ expressions) make their first appearance in this chapter.

Since

$$1 + g(p) = (1 - \omega(p)/p)^{-1},$$

condition $\Omega(\kappa)$ can be restated in the form

(5.2) $$\prod_{w_1 \leq p < w} (1 + g(p)) \leq \left(\frac{\log w}{\log w_1}\right)^\kappa \left(1 + \frac{A}{\log w_1}\right), \quad 2 \leq w_1 < w.$$

Letting $w_1 = p$ and $w \to p + 0$, we obtain

(5.3) $$g(p) \leq \frac{A}{\log p} \quad \text{and} \quad \frac{\omega(p)}{p} \leq \frac{A}{A + \log p}.$$

Hence

$$0 \leq g(p) - \frac{\omega(p)}{p} \leq A \frac{\omega(p)}{p \log p}$$

and therefore, by (1.7),

(5.4) $$0 \leq \sum_{w_1 \leq p < w} \left(g(p) - \frac{\omega(p)}{p}\right) \log p \leq A \sum_{w_1 \leq p < w} \frac{\omega(p)}{p}$$
$$\leq A\kappa \log \frac{\log w}{\log w_1} + \frac{A^2}{\log w_1}, \quad 2 \leq w_1 < w.$$

By (5.2) and (4.14) (Mertens' product approximation with a remainder), we obtain the estimate

(5.5) $$\prod_{w_1 \leq p < w} (1 + g(p))\left(1 - \frac{1}{p}\right)^\kappa \leq \exp\left(\frac{A_1}{\log w_1}\right), \quad 2 \leq w_1 < w,$$

where A_1 can depend on A and κ.

We derive a refined lower bound for $G(\xi, z)$ by means of

Lemma 5.1. (THE TOPPING-UP LEMMA) *With $g(\cdot)$ as defined in (5.1) above, there exists a function $g^*(\cdot)$ defined on the primes such that*

$$g^*(p) \geq g(p)$$

for all primes p and

(5.6) $$\exp\left(-\frac{2A_1}{\log u}\right) \leq \prod_{u \leq p < v} (1 + g^*(p))\left(1 - \frac{1}{p}\right)^\kappa \leq \exp\left(\frac{3A_1}{\log u}\right)$$

holds for all pairs of numbers u, v satisfying $2 \leq u < v$.

We call (5.6) the $\Omega^*(\kappa)$ condition.

Corollary 5.2. *Let $G^*(\xi, z)$ be the summatory function given by*

(5.7) $$G^*(\xi, z) := \sum_{\substack{d < \xi \\ d | P(z)}} g^*(d).$$

5.1 A lower bound for $G(\xi, z)$

where $g^*(d) := \prod_{p|d} g^*(p)$ when $d \mid P(z)$. Then, for all positive ξ and z,

(5.8) $$G(\xi, z) \prod_{p<z}(1+g(p))^{-1} \geq G^*(\xi, z) \prod_{p<z}(1+g^*(p))^{-1}.$$

In particular, for $\xi \leq z$, we have $G^*(\xi, z) =: G^*(\xi)$, $G(\xi, z) =: G(\xi)$ and

$$G(\xi)V(\xi) \geq G^*(\xi) \prod_{p<\xi}(1+g^*(p))^{-1}.$$

It follows from the corollary that if we can find asymptotics for the function on the right side of (5.8), we shall have determined a lower bound for the function on the left.

Proof of the Corollary. The argument is very simple. We first prove the inequality for the case $g^*(p) = g(p)$ for all primes $p < z$ except one, p_0 say, where $g^*(p_0) > g(p_0)$. We have

$$G(\xi, z) = \sum_{\substack{d<\xi \\ d|P(z)/p_0}} g(d) + g(p_0) \sum_{\substack{d<\xi/p_0 \\ d|P(z)/p_0}} g(d) =: S_1 + g(p_0)S_2,$$

say, where obviously $S_1 \geq S_2$, and similarly $G^*(\xi, z) = S_1 + g^*(p_0)S_2$. Hence

$$\frac{G(\xi, z)}{1+g(p_0)} - \frac{G^*(\xi, z)}{1+g^*(p_0)} = \frac{S_1 + g(p_0)S_2}{1+g(p_0)} - \frac{S_1 + g^*(p_0)S_2}{1+g^*(p_0)}$$

$$= \frac{(S_1 - S_2)(g^*(p_0) - g(p_0))}{(1+g(p))(1+g^*(p))} \geq 0.$$

By iterating this procedure as often as is necessary, we complete the proof. $\square$

Proof of Lemma 5.1. Let

$$b_p := \log(1+g(p)) + \kappa \log\left(1-p^{-1}\right)$$

for all primes p. Then, by (5.5)

(5.9) $$\sum_{w_1 \leq p < w} b_p \leq \frac{A_1}{\log w_1}, \quad 2 \leq w_1 < w.$$

Our aim is to construct a sequence $\{b_p^*\}$ such that $b_p^* \geq b_p$ for all primes p and

(5.10) $$-\frac{2A_1}{\log u} \leq \sum_{u \leq p < v} b_p^* \leq \frac{3A_1}{\log u}, \quad 2 \leq u < v.$$

Then we define $g^*(p)$ by

(5.11) $\qquad b_p^* =: \log\left(1 + g^*(p)\right) + \kappa \log\left(1 - p^{-1}\right).$

Since this construction is not simple, here is an outline of our argument: we identify on the real line a sequence of non-overlapping intervals $I = [C, D]$ and associate with them *blocks* $B_I = \{b_p^* : p \in I\}$ of terms from the sequence $\{b_p^*\}$ of two kinds according to whether $D < C^2$ or $D \geq C^2$. In the first case we construct a "short block" B_I such that (among other things) $\Sigma_{p \in I} b_p^* = 0$. In the remaining case we construct a "long block" B_I in the form $\{b_p^* : q \leq p < q^2\}$ with $b_p^* = b_p$. In each case we give upper and lower estimates of sums of b_p^*.

Now for the details. As is implicit in these remarks, the numbers b_p are not necessarily all positive. Indeed, from the definition of b_p, (5.1), and (5.3) we have

$$b_p \leq g(p) - \frac{\kappa}{p} \leq \frac{\omega(p) - \kappa}{p} + \left(\frac{A}{\log p}\right)^2.$$

Condition $\Omega(\kappa)$ implies that $\omega(p) \leq \kappa$ on average, and $b_p < 0$ can occur. If $b_p \leq 0$ for *all* p, take $b_p^* = 0$ for all p. In this case, by (5.11),

$$\prod_{u \leq p \leq v} (1 + g^*(p))\left(1 - p^{-1}\right)^\kappa = 1$$

for $2 \leq u \leq v < \infty$. Also, $g^*(p) \geq g(p)$ for all p, and the lemma holds trivially.

Next, let q be the least prime such that $b_q > 0$. In case $q > 2$, define $b_p^* = 0$, $p < q$. Let Q be the minimal prime exceeding q for which

(5.12) $\qquad \sum_{q \leq p \leq Q} b_p \leq 0,$

or let $Q = \infty$ in case (5.12) holds for no finite number Q. We have

(5.13) $\qquad \sum_{q \leq p \leq r} b_p > 0, \quad q \leq r < Q,$

and in case Q is finite, $b_Q < 0$. We break the argument into two cases.

Case I. $Q < q^2$. Here define

$$b_p^* = b_p, \quad q \leq p < Q, \quad \text{and} \quad b_Q^* = -\sum_{q \leq p < Q} b_p.$$

5.1 A lower bound for $G(\xi, z)$

By (5.12), $b_Q^* \geq b_Q$. Suppose $[u, v] \subset [q, Q]$. When $v < Q$,

$$\sum_{u \leq p \leq v} b_p^* = \sum_{u \leq p \leq v} b_p \leq \frac{A_1}{\log u}$$

by (5.9), and when $v = Q$,

$$\sum_{u \leq p \leq Q} b_p^* = \sum_{u \leq p < Q} b_p + b_Q^* < \frac{A_1}{\log u},$$

since $b_Q^* < 0$.

To bound the sum from below, note that

$$0 \leq \sum_{q \leq p \leq v} b_p^* = \sum_{q \leq p \leq u-1} b_p + \sum_{u \leq p \leq v} b_p^* \leq \frac{A_1}{\log q} + \sum_{u \leq p \leq v} b_p^*,$$

so that

$$\sum_{u \leq p \leq v} b_p^* \geq -\frac{A_1}{\log q} > \frac{-2A_1}{\log u},$$

since $u \leq v \leq Q < q^2$.

So far we have defined a block $B_I = \{b_p^* : q \leq p \leq Q\}$ of terms whose sum is 0 and which satisfies

$$b_p^* \geq b_p \quad \text{for} \quad q \leq p \leq Q,$$

where $Q < q^2$ and such that

$$\frac{-2A_1}{\log u} \leq \sum_{u \leq p \leq v} b_p^* \leq \frac{A_1}{\log u}$$

whenever $[u, v] \subset [q, Q]$. We shall call such a B_I a *short block*.

Case II. $Q > q^2$ (possibly $Q = \infty$). We define $b_p^* = b_p$, $q \leq p < q^2$, and refer to $\{b_p^* : q \leq p < q^2\}$ as a *long block*. The sum of the elements in a long block is not 0, but we know by our construction and (5.9) that

$$0 < \sum_{q \leq p < q^2} b_p^* \leq \frac{A_1}{\log q}.$$

Suppose that $[u, v] \subset [q, q^2]$. Again, we have

$$\sum_{u \leq p \leq v} b_p^* = \sum_{u \leq p \leq v} b_p \leq \frac{A_1}{\log u};$$

and, arguing as in Case I,

$$\sum_{u \leq p \leq v} b_p^* > - \sum_{q \leq p \leq u-1} b_p \geq \frac{-A_1}{\log q} \geq \frac{-2A_1}{\log u},$$

since $u \leq v \leq q^2$. Thus, for a long block $B_I = \{b_p^* : q \leq p < q^2\}$, we have $b_p^* = b_p$ and

$$\frac{-2A_1}{\log u} \leq \sum_{u \leq p \leq v} b_p^* \leq \frac{A_1}{\log u}$$

whenever $[u, v] \subset [q, q^2]$.

We have now constructed the first block—which may be short or long—and we proceed inductively: If all elements b_p beyond the first block are non-positive, we replace them each with $b_p^* = 0$ as before; if $b_{q'}$ is the first element beyond the first block that is positive, then take $b_p^* = 0$ for all primes beyond the first block but smaller than q'. We form a second block, starting with q', and proceed as before.

We are now ready to complete the proof. Consider the sum

(5.14) $$\sum_{u \leq p \leq v} b_p^*, \quad u < v;$$

remember that the elements b_p^* that do not lie in a block are 0 and that otherwise the elements b_p^* lie in nonoverlapping blocks. We suppose, without loss of generality, that u and v are primes and that b_u^* lies in the block $\{b_p^* : r \leq p \leq r'\}$ and b_v^* lies in a later block $\{b_p^* : s \leq p \leq s'\}$. Since the sum of b_p^* over a complete block is nonnegative, as is any partial sum starting at the beginning of the block, we have, as before,

$$\sum_{u \leq p \leq r'} b_p^* \geq \frac{-2A_1}{\log u}, \quad \sum_{s \leq p \leq v} b_p^* \geq 0.$$

Thus

$$\sum_{u \leq p \leq v} b_p^* \geq -\frac{2A_1}{\log u}, \quad u < v.$$

Also, the sum in (5.14) is at most

$$\frac{A_1}{\log u} + A_1 \sum_q \frac{1}{\log q} + \frac{A_1}{\log s},$$

where $\sum_q$ extends over intervening blocks (if any). In fact, we need consider only intervening long blocks, since the sum over any (complete)

5.1 A lower bound for $G(\xi, z)$

short block is 0 by construction. If there exist no intervening long blocks, then since $u < s$, the sum is at most

$$\frac{A_1}{\log u} + \frac{A_1}{\log s} < \frac{2A_1}{\log u}.$$

On the other hand, if there are intervening long blocks, then $q > u$, and if $b_{q_1}^*$, $b_{q_2}^*$ are initial terms in successive blocks with $q_1 < q_2$ (possibly $q_2 = s$), then $q_2 > q_1^2$ and

$$\frac{1}{\log q_2} < \frac{1}{2 \log q_1}.$$

Thus the sum in (5.14) is at most

$$\frac{A_1}{\log u} + \frac{A_1}{\log u}\left(1 + \frac{1}{2} + \frac{1}{4} + \cdots\right) = \frac{3A_1}{\log u}.$$

This proves the lemma, since by (5.11), $g^*(p) \geq g(p) \geq 0$ follows from $b_p^* \geq b_p$. □

Formula (5.10) with $u = v = p$ yields $b_p^* \ll 1/\log p$. From (5.11),

$$1 + g^*(p) = \left(1 - \frac{1}{p}\right)^{-\kappa} \exp b_p^* \leq \left(1 + O\left(\frac{1}{p}\right)\right)\exp\left(\frac{c}{\log p}\right) = 1 + O\left(\frac{1}{\log p}\right),$$

whence

(5.15) $$g^*(p) \ll 1/\log p.$$

It is natural to introduce here the analogue of $\omega(\cdot)$ by defining

$$\omega^*(p) = \frac{p g^*(p)}{1 + g^*(p)},$$

so that $0 \leq \omega^*(p)/p < 1$,

$$1 + g^*(p) = (1 - \omega^*(p)/p)^{-1},$$

and

(5.16) $$g^*(p) = \frac{\omega^*(p)}{p} + \frac{\omega^*(p)}{p} g^*(p) = \frac{\omega^*(p)}{p}\left(1 + O\left(\frac{1}{\log p}\right)\right)$$

by (5.15). We may restate $\Omega^*(\kappa)$ in the form

(5.17) $$\prod_{w_1 \leq p < w}(1 + g^*(p)) = \prod_{w_1 \leq p < w}\left(1 - \frac{\omega^*(p)}{p}\right)^{-1}$$

$$= \left(\frac{\log w}{\log w_1}\right)^\kappa \exp\left\{O\left(\frac{1}{\log w_1}\right)\right\}, \quad 2 \leq w_1 < w,$$

or as
$$\sum_{w_1 \le p < w} \log(1 + g^*(p)) = \sum_{w_1 \le p < w} \log\left(1 - \frac{\omega^*(p)}{p}\right)^{-1}$$
$$= \kappa \log \frac{\log w}{\log w_1} + O\left(\frac{1}{\log w_1}\right), \quad 2 \le w_1 < w.$$

It follows that
$$\sum_{w_1 \le p < w} \frac{\omega^*(p)}{p} \le \kappa \log \frac{\log w}{\log w_1} + O\left(\frac{1}{\log w_1}\right) \le \sum_{w_1 \le p < w} g^*(p), \quad 2 \le w_1 < w,$$

and, by (5.16),
$$\sum_{w_1 \le p < w} g^*(p) \le \sum_{w_1 \le p < w} \frac{\omega^*(p)}{p} + O\left(\sum_{w_1 \le p < w} \frac{\omega^*(p)}{p \log p}\right);$$

estimating the last term by the argument leading to (1.9), we obtain

(5.18) $$\sum_{w_1 \le p < w} g^*(p) = \kappa \log \frac{\log w}{\log w_1} + O\left(\frac{1}{\log w_1}\right), \quad 2 \le w_1 < w.$$

From here, a straightforward partial summation argument (cf. the proof of (1.8)) yields
$$\sum_{w_1 \le p < w} g^*(p) \log p = \kappa \log \frac{w}{w_1} + O\left\{\log\left(e\frac{\log w}{\log w_1}\right) + \frac{1}{\log w_1}\right\}, \quad 2 \le w_1 < w;$$

in particular, if we take $w_1 = 2$ and write

(5.19) $$\ell(w) := \log(e \log(ew)) \ge 1, \quad w \ge 1,$$

we have

(5.20) $$\sum_{p < w} g^*(p) \log p = \kappa \log w + O(\ell(w)), \quad w \ge 1.$$

We note in passing the similarity of (5.18) to Mertens' formula (1.6),
$$\sum_{w_1 \le p < w} \frac{1}{p} = \log \frac{\log w}{\log w_1} + O\left(\frac{1}{\log w_1}\right), \quad 2 \le w_1 < w.$$

Lemma 5.3. *The infinite product*
$$\prod_p \left(1 + \frac{g^*(p)}{p^{s-1}}\right)\left(1 - \frac{1}{p^s}\right)^\kappa$$

5.1 A lower bound for $G(\xi, z)$

converges uniformly for $1 < s < 2$. Moreover

$$\lim_{s \to 1+} \prod_p \left(1 + \frac{g^*(p)}{p^{s-1}}\right)\left(1 - \frac{1}{p^s}\right)^\kappa = \prod_p \left(1 + g^*(p)\right)\left(1 - \frac{1}{p}\right)^\kappa.$$

Proof. Using Cauchy's criterion for uniform convergence, it is enough to consider the finite product

$$\prod_{w \leq p < w_1} \left(1 + \frac{g^*(p)}{p^{s-1}}\right)\left(1 - \frac{1}{p^s}\right)^\kappa$$

$$= \exp\left\{\sum_{w \leq p < w_1} \log\left(1 + \frac{g^*(p)}{p^{s-1}}\right)\left(1 - \frac{1}{p^s}\right)^\kappa\right\}$$

$$= \exp\left\{\sum_{w \leq p < w_1} \left(g^*(p) - \frac{\kappa}{p}\right)\frac{1}{p^{s-1}} + O\left(g^*(p)^2 + \frac{1}{p^2}\right)\right\},$$

with $2 \leq w < w_1$ and $1 < s < 2$. By (5.15), $g^*(p) \ll (\log p)^{-1}$, and therefore

$$\sum_{w \leq p < w_1} g^*(p)^2 \ll \sum_{w \leq p < w_1} g^*(p)/\log p \ll \frac{1}{\log w}$$

by $\Omega^*(\kappa)$ and a summation by parts of the kind used in Lemma 1.4. Hence

$$\sum_{w \leq p < w_1} \left(g^*(p)^2 + p^{-2}\right) \ll \frac{1}{\log w}, \quad 2 \leq w < w_1.$$

Next, by (5.18) and Mertens' formula (1.6), we have

$$\delta(t) := \sum_{w \leq p < t} \left(g^*(p) - \frac{\kappa}{p}\right) \ll \frac{1}{\log w}, \quad 2 \leq w < t,$$

and then, uniformly in $1 < s < 2$,

$$\sum_{w \leq p < w_1} \left(g^*(p) - \frac{\kappa}{p}\right)\frac{1}{p^{s-1}} = \int_w^{w_1} \frac{1}{t^{s-1}} d\delta(t)$$

$$= \frac{\delta(w_1)}{w_1^{s-1}} + (s-1)\int_w^{w_1} \frac{\delta(t)}{t^s} dt$$

$$\ll \frac{1}{\log w}\left(1 + (s-1)\int_w^\infty t^{-s} dt\right)$$

$$= \frac{1}{\log w}(1 + w^{1-s}) \ll \frac{1}{\log w}.$$

The second assertion of the lemma follows at once. $\square$

5.2 Asymptotics for $G^*(\xi, z)$

We now proceed to determine asymptotics for $G^*(\xi, z)$, defined in (5.7), on the basis of (5.20). This is a long argument, culminating in formula (5.40). We start with the Chebyshev-type identity

$$G^*(\xi, z) \log \xi = \sum_{\substack{d<\xi \\ d|P(z)}} g^*(d)(\log \frac{\xi}{d} + \log d)$$

$$= \int_1^\xi G^*(t, z)\frac{dt}{t} + \sum_{\substack{d<\xi \\ d|P(z)}} g^*(d) \sum_{p|d} \log p.$$

When we write $d = mp$ in the sum on the right, we obtain

$$G^*(\xi, z) \log \xi - \int_1^\xi G^*(t, z)\frac{dt}{t} = \sum_{\substack{m<\xi \\ m|P(z)}} g^*(m) \sum_{\substack{p<\min(\xi/m,z) \\ p\nmid m,\, p|P(z)}} g^*(p) \log p.$$

It is convenient to analyze the simpler expression

$$\Delta G^*(\xi, z) := G^*(\xi, z) \log \xi - \int_1^\xi G^*(t, z)\frac{dt}{t}$$
$$- \sum_{\substack{m<\xi \\ m|P(z)}} g^*(m) \sum_{p<\min(\xi/m,z)} g^*(p) \log p.$$

By (5.15) and (5.18) (with $w_1 = 2$), we get

(5.21) $\quad \Delta G^*(\xi, z) = - \sum_{\substack{\ell p^2 < \xi \\ \ell p | P(z)}} g^*(\ell) g^*(p)^2 \log p \ll \sum_{\substack{\ell p^2 < \xi \\ \ell p | P(z)}} g^*(\ell) g^*(p)$

$$\leq \sum_{\substack{p<\min(\sqrt{\xi},z) \\ p|P(z)}} g^*(p) G^*(\xi/p^2, z)$$

$$\leq G^*(\xi, z) \sum_{\substack{p<\min(\sqrt{\xi},z) \\ p|P(z)}} g^*(p)$$

$$\ll G^*(\xi, z)\, \ell\big(\min(\sqrt{\xi}, z)\big).$$

When $1 \leq \xi \leq z$, the condition $d \mid P(z)$ in the definition of $G^*(\xi, z)$ is superfluous and in its place we are left with

(5.22) $\quad G^*(\xi) := \sum_{d<\xi} \mu^2(d) g^*(d) < \prod_{p<\xi}(1 + g^*(p)) \ll (\log \xi)^\kappa$

5.2 Asymptotics for $G^*(\xi, z)$

by (5.17); moreover, in this case, $\ell(\min(\sqrt{\xi}, z)) = \ell(\sqrt{\xi}) \le \ell(\xi)$. Hence

(5.23) $\quad G^*(\xi)\log\xi - \int_1^\xi G^*(t)\dfrac{dt}{t} - \sum_{m<\xi} g^*(m) \sum_{p<\xi/m} g^*(p)\log p$

$$\ll G^*(\xi)\ell(\xi).$$

We are now in a position to improve (5.22) to an asymptotic.

Proposition 5.4. *Suppose that g^* is a multiplicative function satisfying (5.20). Then*

$$G^*(\xi) = C_\kappa (\log\xi)^\kappa + O((\log\xi)^{\kappa-1}\ell(\xi)), \quad \xi > 1,$$

where

$$C_\kappa := \dfrac{1}{\Gamma(\kappa+1)} \prod_p (1 + g^*(p))\left(1 - \dfrac{1}{p}\right)^\kappa \quad (\Gamma \text{ denotes Euler's function}).$$

Proof. By (5.23) and (5.20),

$$G^*(\xi)\log\xi = \int_1^\xi G^*(t)\dfrac{dt}{t} + \sum_{m<\xi} g^*(m)\left\{\kappa\log\dfrac{\xi}{m} + O\left(\ell\left(\dfrac{\xi}{m}\right)\right)\right\}$$

$$+ O(G^*(\xi)\ell(\xi))$$

$$= (1+\kappa)\int_1^\xi G^*(t)\dfrac{dt}{t} + O(G^*(\xi)\ell(\xi))$$

$$= (1+\kappa)T(\xi) + G^*(\xi)\,\epsilon(\xi)\log\xi,$$

where

$$T(\xi) := \int_1^\xi G^*(t)\dfrac{dt}{t}$$

and

(5.24) $\quad\quad\quad\quad\quad \epsilon(\xi) \ll \ell(\xi)/\log\xi.$

If we pretend for a moment that G^* is differentiable and that

$$G^*(\xi)\log\xi = (1+\kappa)T(\xi),$$

then we would obtain, on differentiating with respect to ξ, that

$$\dfrac{d}{d\xi}G^*(\xi) = \dfrac{\kappa}{\xi\log\xi}G^*(\xi)$$

or, after integration, that $G^*(\xi) = C(\log\xi)^\kappa$ for some constant C.

To develop a valid procedure along this line we proceed as follows: for sufficiently large ξ, for $\xi \geq \xi_0$ say, we can assert by (5.24) (and (5.19)) that $|\epsilon(\xi)| \leq 1/2$, so that

$$G^*(\xi) = \frac{1}{1-\epsilon(\xi)} \frac{1+\kappa}{\log \xi} T(\xi), \quad \xi \geq \xi_0.$$

Then, writing

$$E(t) := \log\left(\frac{1+\kappa}{(\log t)^{1+\kappa}} T(t)\right),$$

we obtain, after differentiating with respect to t and simplifying,

$$\frac{d}{dt} E(t) = \frac{1+\kappa}{t \log t} \frac{\epsilon(t)}{1-\epsilon(t)} \ll \frac{\ell(t)}{t \log^2 t}, \quad t \geq \xi_0.$$

Hence the integral

$$E_0 = \int_1^\infty \frac{d}{dt} E(t)\, dt$$

converges absolutely and

$$E_0 - E(\xi) = \int_\xi^\infty \frac{d}{dt} E(t)\, dt, \quad \xi \geq \xi_0.$$

Writing $C = \exp E_0$, we have, for $\xi \geq \xi_0$,

$$\frac{1+\kappa}{(\log \xi)^{1+\kappa}} T(\xi) = \exp E(\xi) = C \exp\left\{-\int_\xi^\infty \frac{d}{dt} E(t)\, dt\right\}$$

$$= C \exp\left\{O\left(\int_\xi^\infty \frac{\ell(t)}{t \log^2 t}\, dt\right)\right\} = C\left(1 + O\left(\frac{\ell(\xi)}{\log \xi}\right)\right),$$

whence

$$T(\xi) = \frac{C}{1+\kappa} (\log \xi)^{1+\kappa} \left\{1 + O\left(\frac{\ell(\xi)}{\log \xi}\right)\right\}, \quad \xi \geq \xi_0,$$

and therefore

(5.25) $$G^*(\xi) = C(\log \xi)^\kappa \left\{1 + O\left(\frac{\ell(\xi)}{\log \xi}\right)\right\}.$$

It remains to determine the constant C, and this we shall do by means of an Abelian argument. If $s > 1$ we have

$$\prod_p \left(1 + \frac{g^*(p)}{p^{s-1}}\right) = \sum_{d=1}^\infty \mu^2(d) g^*(d) d^{1-s} = (s-1) \int_1^\infty G^*(t) t^{-s}\, dt$$

$$= (s-1) \int_1^\infty \left\{C(\log t)^\kappa + O\left(\ell(t)(\log t)^{\kappa-1}\right)\right\} t^{-s}\, dt$$

5.2 Asymptotics for $G^*(\xi, z)$

by (5.25). We make the substitution $t = \exp(x/(s-1))$ in the integral on the right and obtain

$$(5.26) \quad (s-1)^\kappa \prod_p \left(1 + \frac{g^*(p)}{p^{s-1}}\right)$$

$$= C\Gamma(\kappa+1) + O\left((s-1)\int_0^\infty x^{\kappa-1}\ell(\exp\{x/(s-1)\})e^{-x}dx\right).$$

By (5.19),

$$\ell\left(\exp\left(\frac{x}{s-1}\right)\right) = \log\left(e\log\left\{\exp\left(1+\frac{x}{s-1}\right)\right\}\right)$$

$$= 1 + \log\frac{1}{s-1} + \log(s-1+x)$$

so that the error term on the right is, when $1 < s < 2$,

$$O\left\{(s-1)\left(1 + \log\frac{1}{s-1}\right)\Gamma(\kappa) + (s-1)\int_0^\infty x^{\kappa-1}\log(1+x)\cdot e^{-x}dx\right\};$$

hence the expression on the right side of (5.26) tends to $C\Gamma(\kappa+1)$ as $s \to 1+0$. On the left side of (5.26) we use the observation that

$$1 = \lim_{s\to 1+0}(s-1)\zeta(s) = \lim_{s\to 1+0}(s-1)\prod_p(1-p^{-s})^{-1}$$

and Lemma 5.3 to deduce that

$$(s-1)^\kappa \prod_p\left(1+\frac{g^*(p)}{p^{s-1}}\right) = \{(s-1)\zeta(s)\}^\kappa \prod_p\left(1+\frac{g^*(p)}{p^{s-1}}\right)\left(1-\frac{1}{p^s}\right)^\kappa$$

$$\to \prod_p(1+g^*(p))\left(1-\frac{1}{p}\right)^\kappa \quad \text{as } s \to 1+0. \quad \square$$

We assume from now on that $\xi > z$ and write

$$u := \frac{\log \xi}{\log z} > 1.$$

Returning to (5.21), we observe that now the double sum on the left is

$$\sum_{\substack{m<\xi/z \\ m|P(z)}} g^*(m) \sum_{p<z} g^*(p)\log p + \sum_{\substack{\xi/z \leq m < \xi \\ m|P(z)}} g^*(m) \sum_{p<\xi/m} g^*(p)\log p.$$

By (5.20), this sum is

$$\sum_{\substack{m<\xi/z\\m|P(z)}} g^*(m)\{\kappa\log z + O(\ell(z))\} + \sum_{\substack{\xi/z\leq m<\xi\\m|P(z)}} g^*(m)\left\{\kappa\log\frac{\xi}{m} + O\!\left(\ell\!\left(\frac{\xi}{m}\right)\right)\right\}$$

$$= \kappa \sum_{\substack{m<\xi\\m|P(z)}} g^*(m)\log\frac{\xi}{m} - \kappa \sum_{\substack{m<\xi/z\\m|P(z)}} g^*(m)\log\frac{\xi/z}{m} + O(G^*(\xi,z)\ell(z))$$

$$= \kappa \int_1^\xi G^*(t,z)\frac{dt}{t} - \kappa \int_1^{\xi/z} G^*(t,z)\frac{dt}{t} + O(G^*(\xi,z)\ell(z)).$$

On substituting in (5.21), we arrive at

(5.27) $G^*(\xi,z)\log\xi = (\kappa+1)T(\xi,z) - \kappa T(\xi/z, z) + O(G^*(\xi,z)\ell(z)),$

where

(5.28) $$T(\xi, z) := \int_1^\xi G^*(t,z)\frac{dt}{t}.$$

Since $G^*(\xi,z) \leq \prod_{p<z}(1+g^*(p))$, the order of magnitude of the error term on the right side of (5.27) is

$$\ell(z)\prod_{p<z}(1+g^*(p)) =: \ell(z)V^*(z),$$

say. By (5.17), $(\log z)^\kappa \ll V^*(z) \ll (\log z)^\kappa$.

5.3 The j and σ functions

To proceed, it is convenient to introduce a function that will have an important role here and throughout this monograph. Let $\kappa \geq 1$ and take $j(\cdot) := j_\kappa(\cdot)$ to be the continuous solution of the differential delay equation

(5.29) $uj'_\kappa(u) = \kappa j_\kappa(u) - \kappa j_\kappa(u-1), \quad u > 1,$

that is defined for other real values of u by

(5.30) $$j_\kappa(u) = \begin{cases} 0, & u \leq 0, \\ e^{-\gamma\kappa}u^\kappa/\Gamma(\kappa+1), & 0 < u \leq 1, \end{cases}$$

where γ is Euler's constant and Γ Euler's function. In Chapters 6 and 14 and afterward we shall have occasion to use the related function $\sigma_\kappa(\cdot)$

5.3 The j and σ functions

given by $\sigma_\kappa(u) = j_\kappa(u/2)$, in terms of which some of our sieve formulas appear slightly more tidy.

For $1 < \xi \le z$, we can restate Proposition 5.4 in terms of j. We have
$$G^*(\xi, z) = G^*(\xi) = C_\kappa (\log \xi)^\kappa \{1 + O(\ell(\xi)/\log \xi)\}$$
with
$$C_\kappa \Gamma(\kappa + 1) = \prod_p (1 + g^*(p)) \left(1 - \frac{1}{p}\right)^\kappa$$
$$= V^*(z) \prod_{p<z} \left(1 - \frac{1}{p}\right)^\kappa \prod_{p \ge z} (1 + g^*(p)) \left(1 - \frac{1}{p}\right)^\kappa$$
$$= V^*(z) \frac{e^{-\gamma\kappa}}{(\log z)^\kappa} \left(1 + O\left(\frac{1}{\log z}\right)\right)$$

by Mertens' formula and Lemma 5.3. Hence, for $u := \log \xi / \log z \le 1$,

(5.31) $$G^*(\xi, z) = \frac{V^*(z)}{\Gamma(\kappa+1)} \frac{e^{-\gamma\kappa}}{(\log z)^\kappa} (\log \xi)^\kappa \left\{1 + O\left(\frac{\ell(\xi)}{\log \xi}\right)\right\}$$
$$= V^*(z) j(u) \left\{1 + O\left(\frac{\ell(z)}{u \log z}\right)\right\}$$
$$= V^*(z) \left\{j(u) + O\left(u^{\kappa-1} \frac{\ell(z)}{\log z}\right)\right\}$$
$$= V^*(z) \left\{j(u) + O\left(\frac{\ell(z)}{\log z}\right)\right\}, \quad 0 < u \le 1,$$

since $\kappa \ge 1$.

To estimate $G^*(\xi, z)$ for $\xi > z$, we introduce
$$j_\kappa^{(-1)}(u) := \int_0^u j_\kappa(v) dv,$$
so that, as we see from properties of j,

(5.32) $$\begin{cases} j_\kappa^{(-1)}(u) = 0, & u \le 0, \\ j_\kappa^{(-1)}(u) = e^{-\gamma\kappa} u^{\kappa+1} / \Gamma(\kappa+2), & 0 < u \le 1, \\ u j_\kappa(u) = (\kappa+1) j_\kappa^{(-1)}(u) - \kappa j_\kappa^{(-1)}(u-1), & u > 1, \end{cases}$$

and, in particular, that

(5.33) $$\frac{j_\kappa^{(-1)}(\tau)}{\tau^{\kappa+1}} = \frac{j_\kappa^{(-1)}(\rho)}{\rho^{\kappa+1}} - \kappa \int_\rho^\tau \frac{j_\kappa^{(-1)}(v-1)}{v^{\kappa+2}} dv, \quad 0 < \rho < \tau.$$

We remark that the function $j_\kappa(u)$ is, for $u > 0$, a positive, strictly increasing function that converges exponentially to 1 as $u \to \infty$. All this will be proved later on, in Chapter 14. We shall determine asymptotics linking $T(\xi, z)$ and $j_\kappa^{(-1)}(u)V^*(z)\log z$ (with $u := (\log \xi)/\log z$), in the following manner: by (5.27)

$$\frac{d}{d\xi}\left(\frac{T(\xi, z)}{(\log \xi)^{\kappa+1}}\right) = \frac{1}{\xi(\log \xi)^{\kappa+2}}\{G^*(\xi, z)\log \xi - (\kappa+1)T(\xi, z)\}$$

$$= -\kappa \frac{T(\xi/z, z)}{\xi(\log \xi)^{\kappa+2}} + O\left(\frac{V^*(z)\ell(z)}{\xi(\log \xi)^{\kappa+2}}\right),$$

and when we integrate this relation (written with y in place of ξ) with respect to y from η to ξ, we obtain the approximate integral equation

(5.34)
$$\frac{T(\xi, z)}{(\log \xi)^{\kappa+1}} = \frac{T(\eta, z)}{(\log \eta)^{\kappa+1}}$$
$$- \kappa \int_\eta^\xi \frac{T(y/z, z)}{y(\log y)^{\kappa+2}} dy + O\left(\frac{V^*(z)\ell(z)}{(\log \eta)^{\kappa+1}}\right), \quad z \leq \eta \leq \xi.$$

Define

(5.35)
$$U(\xi, z) := \frac{1}{V^*(z)\ell(z)}\{T(\xi, z) - j_\kappa^{(-1)}(u)V^*(z)\log z\}, \quad u = \frac{\log \xi}{\log z},$$

and substitute in (5.34) to obtain, with the aid of (5.33),

(5.36) $$\frac{U(\xi, z)}{(\log \xi)^{\kappa+1}} = \frac{U(\eta, z)}{(\log \eta)^{\kappa+1}} - \kappa \int_\eta^\xi \frac{U(y/z, z)}{y(\log y)^{\kappa+2}} dy + O\left(\frac{1}{(\log \eta)^{\kappa+1}}\right)$$

for $z \leq \eta \leq \xi$. This is an approximate integral equation which we shall use to derive an order of magnitude estimate of $U(\xi, z)$.

For $\xi \leq z$, i.e., for $u \leq 1$, we have upon integrating (5.31)

$$T(\xi, z) := \int_1^\xi G^*(t)\frac{dt}{t} = V^*(z)\int_1^\xi \left\{j\left(\frac{\log t}{\log z}\right) + O\left(\frac{(\log t)^{\kappa-1}}{(\log z)^\kappa}\ell(z)\right)\right\}\frac{dt}{t}$$

$$= V^*(z)\log z \int_0^u \left\{j(t) + O\left(\frac{t^{\kappa-1}\ell(z)}{\log z}\right)\right\} dt$$

$$= V^*(z)\log z \left\{j_\kappa^{(-1)}(u) + O\left(u^\kappa \frac{\ell(z)}{\log z}\right)\right\}, \quad u \leq 1.$$

It follows from (5.35) that

(5.37) $$|U(\xi, z)| \leq Du^\kappa, \quad u \leq 1,$$

where D is a sufficiently large constant, to be specified. For $u > 1$ we estimate U by an inductive argument.

5.3 The j and σ functions

Lemma 5.5. *Let $U(\xi, z)$ be as given in (5.35) and let $u = (\log \xi)/\log z$. Let D be sufficiently large to satisfy (5.37) and $D > (4/3)B$, where B is the O-constant in (5.36). Then*

(5.38) $$|U(\xi, z)| \leq 2Du^{2\kappa+1}, \quad u \geq 1.$$

Proof. Suppose first that $z < \xi \leq z^2$, i.e., $1 < u \leq 2$. We employ (5.36) with $\eta = z$, so $U(\eta, z) = U(z, z)$ and in the integral, $y/z \leq \xi/z \leq z$. Thus the bound (5.37) applies in this range and we obtain

$$\frac{U(\xi,z)}{(\log \xi)^{\kappa+1}} \leq D\left\{\frac{1}{(\log z)^{\kappa+1}} + \kappa \int_z^{z^2} \left(\frac{\log(y/z)}{\log z}\right)^{\kappa} \frac{dy/y}{(\log y)^{\kappa+2}} + \frac{B/D}{(\log z)^{\kappa+1}}\right\}$$

$$= \frac{D}{(\log z)^{\kappa+1}}\left\{1 + \kappa \int_{1/2}^1 (1-t)^{\kappa} dt + \frac{3}{4}\right\}$$

$$= \frac{D}{(\log z)^{\kappa+1}}\left\{\frac{7}{4} + \frac{\kappa}{\kappa+1} 2^{-\kappa-1}\right\} < \frac{2D}{(\log z)^{\kappa+1}},$$

i.e.,

$$|U(\xi, z)| \leq 2Du^{\kappa+1}, \quad z < \xi \leq z^2.$$

To continue, we proceed inductively. It is convenient to establish a slightly stronger result. We show for $z^{\nu} < \xi \leq z^{\nu+1}$ that

(5.39) $$|U(\xi, z)| \leq 2Du^{\kappa+1}\nu^{\kappa},$$

which implies (5.38). This inequality was just established for $\nu = 1$, and we assume that $|U(\xi, z)| \leq 2Du^{\kappa+1}(\nu - 1)^{\kappa}$ holds for $z^{\nu-1} < \xi \leq z^{\nu}$ for some $\nu \geq 2$. When we apply the induction hypothesis to each term on the right side of (5.36) (with $\eta = z^{\nu}$), we obtain

$$\frac{|U(\xi,z)|}{(\log \xi)^{\kappa+1}} \leq \frac{2D}{(\log z)^{\kappa+1}}$$

$$\times \left\{(\nu-1)^{\kappa} + \kappa(\nu-1)^{\kappa} \int_{z^{\nu}}^{z^{\nu+1}} \frac{(\log(y/z))^{\kappa+1}}{y(\log y)^{\kappa+2}} dy + \frac{B/(2D)}{\nu^{\kappa+1}}\right\}.$$

The integral on the right (with $t = \log z/\log y$) is equal to

$$\int_{1/(\nu+1)}^{1/\nu} (1-t)^{\kappa+1} \frac{dt}{t} \leq \left(1 - \frac{1}{\nu+1}\right) \log \frac{\nu+1}{\nu} < \frac{\nu}{\nu+1} \cdot \frac{1}{\nu} = \frac{1}{\nu+1}.$$

Hence

$$\frac{|U(\xi,z)|}{(\log \xi)^{\kappa+1}} \leq \frac{2D}{(\log z)^{\kappa+1}}\left\{\left(1 + \frac{\kappa}{\nu+1}\right)(\nu-1)^{\kappa} + \frac{3/8}{\nu^{\kappa+1}}\right\},$$

and

$$\nu^\kappa - \left(1 + \frac{\kappa}{\nu+1}\right)(\nu-1)^\kappa = (\nu-1)^\kappa\left\{\left(1 + \frac{1}{\nu-1}\right)^\kappa - 1 - \frac{\kappa}{\nu+1}\right\}$$
$$\geq (\nu-1)^\kappa\left\{1 + \frac{\kappa}{\nu-1} - 1 - \frac{\kappa}{\nu+1}\right\}$$
$$= \frac{2\kappa(\nu-1)^{\kappa-1}}{\nu+1} \geq \frac{2\kappa}{\nu+1} > \frac{3/8}{\nu^{\kappa+1}}$$

for $\kappa \geq 1$ and $\nu \geq 2$. Thus

$$|U(\xi,z)| \leq 2D\left(\frac{\log\xi}{\log z}\right)^{\kappa+1}\nu^\kappa, \quad z^\nu < \xi \leq z^{\nu+1},$$

i.e., (5.39) holds. □

When we substitute (5.38) from Lemma 5.5 in (5.35), we arrive at

$$T(\xi,z) = \left\{j_\kappa^{(-1)}(u) + O\left(u^{2\kappa+1}\frac{\ell(z)}{\log z}\right)\right\}V^*(z)\log z, \quad u > 1.$$

We apply this formula in (5.27) with $u > 1$ to obtain

$$G^*(\xi,z)\log\xi$$
$$= V^*(z)(\log z)\left\{(\kappa+1)j_\kappa^{(-1)}(u) - \kappa j_\kappa^{(-1)}(u-1) + O\left(u^{2\kappa+1}\frac{\ell(z)}{\log z}\right)\right\}$$

or, by (5.32), since $\log\xi = u\log z$,

(5.40) $$G^*(\xi,z) = V^*(z)\left\{j_\kappa(u) + O\left(u^{2\kappa}\frac{\ell(z)}{\log z}\right)\right\}, \quad u > 1.$$

When we apply the Topping-Up Corollary 5.2, we conclude that

(5.41) $$G(\xi,z)V(z) = G(\xi,z)\prod_{p<z}(1+g(p))^{-1}$$
$$\geq j_\kappa(u) + O\left(u^{2\kappa}\frac{\ell(z)}{\log z}\right), \quad u > 1.$$

It is possible to refine this argument for $u > 1$ so as to eliminate the factor $u^{2\kappa}$ (see [Sng01, Sng02]), and even the ℓ factor ([Ten01]), but for our purposes these improvements are not necessary.

Since $j_\kappa(u)$ increases with u (Chapter 14), $j_\kappa(u) \geq j_\kappa(1) \gg 1$ and we may rewrite (5.41) as

(5.42) $$G(\xi,z)V(z) \geq j_\kappa(u)\left\{1 + O\left(u^{2\kappa}\frac{\ell(z)}{\log z}\right)\right\}, \quad u > 1.$$

5.3 The j and σ functions

By (4.4) and (4.5), we know that
$$G(\xi, z)V(z) \gg 1.$$
Thus we may assume that z is sufficiently large relative to ξ so that
$$(\log z)/\ell(z) > Bu^{2\kappa},$$
where B is a constant large enough to permit us to restate (5.42) as

(5.43) $$\frac{1}{G(\xi, z)} \leq V(z)\left\{\frac{1}{j_\kappa(u)} + O\left(u^{2\kappa}\frac{\ell(z)}{\log z}\right)\right\}, \quad u > 1.$$

We complement (5.43) with a similar result when $u \leq 1$. By the Topping-Up Corollary 5.2, we have
$$G(\xi, z)V(z) = G(\xi)V(\xi) \cdot \frac{V(z)}{V(\xi)} \geq \frac{G^*(\xi)}{V^*(\xi)} \cdot \frac{V(z)}{V(\xi)}$$
$$= \frac{G^*(\xi)}{V^*(z)} \cdot \frac{V^*(z)}{V^*(\xi)} \cdot \frac{V(z)}{V(\xi)} \geq \frac{G^*(\xi)}{V^*(z)}\left(1 + O\left(\frac{1}{\log \xi}\right)\right)$$
using (5.17) and $\Omega(\kappa)$ (Definition 1.3) at the last step. Hence, by (5.31),
$$G(\xi, z)V(z) \geq j(u)\left\{1 + O\left(\frac{\ell(\xi)}{\log \xi}\right)\right\}.$$
It follows that when $u \leq 1$,

(5.44) $$\frac{1}{G(\xi, z)} \leq \frac{V(z)}{j(u)}\left\{1 + O\left(\frac{\ell(\xi)}{\log \xi}\right)\right\}$$
$$= V(z)\left\{\frac{1}{j(u)} + O\left(u^{-\kappa}\frac{\ell(\xi)}{\log \xi}\right)\right\}$$
$$= V(z)\left\{\frac{1}{j(u)} + O\left(u^{-\kappa-1}\frac{\ell(z)}{\log z}\right)\right\}.$$

We are now in a position to state and establish the main result of this chapter: a Selberg theorem in a form that we shall require later. Inserting the last inequality into (4.1), we obtain

Theorem 5.6. *Subject only to the conditions $\Omega(\kappa)$ and $(q, P(z)) = 1$, and with an arbitrary parameter $u > 0$, we have*

$$S(\mathcal{A}_q, \mathcal{P}, z) \leq \frac{\omega(q)}{q} XV(z)\left\{\frac{1}{j_\kappa(u)} + O\left((u^{2\kappa} + u^{-\kappa-1})\frac{\ell(z)}{\log z}\right)\right\}$$
$$+ \sum_{\substack{n < z^{2u} \\ n \mid P(z)}} 3^{\nu(n)}|r_\mathcal{A}(qn)|.$$

Here $\ell(z) := \log\{e\log(ez)\}$ and the O-constant may depend on κ and A.

From our point of view, Theorem 5.6 will have its most important application in Chapter 9, where we shall combine it with a refinement of the Rosser–Iwaniec method to derive sharper upper *and lower* bounds for $S(\mathcal{A}, \mathcal{P}, z)$ when $\kappa > 1$ in Theorem 9.1.

Of course, Theorem 5.6 itself leads rapidly to a first lower bound (better than the Fundamental Lemma) for $S(\mathcal{A}, \mathcal{P}, z)$ by use of the Fundamental Lemma (Theorem 4.1) and a form of Buchstab's identity (3.5): we have

$$S(\mathcal{A}, \mathcal{P}, z) = S(\mathcal{A}, \mathcal{P}, z_1) - \sum_{\substack{z_1 \leq p < z \\ p \in \mathcal{P}}} S(\mathcal{A}_p, \mathcal{P}, p), \quad 2 \leq z_1 < z,$$

and we take z_1 small enough so that Theorem 4.1 applies to the first term on the right. Then, by applying Theorem 5.6 to each term in the sum on the right, we arrive after some computation at a lower bound for $S(\mathcal{A}, \mathcal{P}, z)$. Given the simplicity of this argument, the result is remarkably good (especially for small κ). We end this chapter with a simple consequence of Theorem 5.6 and some illustrations of its use.

Corollary 5.7. *Assume* $\Omega(\kappa)$ *and that* $|r_\mathcal{A}(d)| \leq \omega(d)$ *for* $d \mid P(z)$, *i.e., that* (4.9) *holds. Then, for* $2 \leq z \leq X^{1/2}$, *we have*

$$S(\mathcal{A}, \mathcal{P}, z) \leq \Gamma(\kappa+1) \prod_p \left(1 - \frac{\omega(p)}{p}\right)\left(1 - \frac{1}{p}\right)^{-\kappa} \frac{X}{\log^\kappa z} \left\{1 + O\left(\frac{\log\log 3z}{\log z}\right)\right\}.$$

Proof. Take $q = 1$ and $u = 1$ in Theorem 5.6 so that $\xi = z$, to obtain by (5.30)

$$S(\mathcal{A}, \mathcal{P}, z) \leq e^{\gamma \kappa} \Gamma(\kappa+1) X V(z) \left\{1 + O\left(\frac{\ell(z)}{\log z}\right)\right\} + \sum_{\substack{n < z^2 \\ n \mid P(z)}} 3^{\nu(n)} |r_\mathcal{A}(n)|,$$

where $\ell(\cdot)$ is defined in Theorem 5.6. By Lemma 4.3 with $K = 3$, the remainder sum is at most $z^2 V(z)^{-3}$; also $V(z)^{-1} \ll \log^\kappa z$ by $\Omega(\kappa)$. Thus

$$S(\mathcal{A}, \mathcal{P}, z) \leq e^{\gamma \kappa} \Gamma(\kappa+1) X V(z) \left\{1 + O\left(\frac{\ell(z)}{\log z}\right) + O\left(\frac{z^2 \log^{4\kappa} z}{X}\right)\right\}.$$

If we have

$$z \leq X^{1/2}/(\log X)^{2\kappa + 1/2} =: z_0,$$

then the last O-term is smaller than the preceding one and

(5.45) $$S(\mathcal{A}, \mathcal{P}, z) \leq e^{\gamma \kappa} \Gamma(\kappa+1) X V(z) \left\{1 + O\left(\frac{\ell(z)}{\log z}\right)\right\}.$$

5.3 The j and σ functions

The gap between z_0 and $X^{1/2}$ is easy to bridge, for if $z_0 \leq z \leq X^{1/2}$, then

$$S(\mathcal{A}, \mathcal{P}, z) \leq S(\mathcal{A}, \mathcal{P}, z_0)$$

and

$$\frac{1}{\log z_0} = \frac{1}{\log z}\left\{1 + \frac{\log(z/z_0)}{\log z_0}\right\} = \frac{1}{\log z}\left\{1 + O\left(\frac{\log \log z}{\log z}\right)\right\},$$

and hence by another application of $\Omega(\kappa)$,

$$V(z_0) = V(z)\frac{V(z_0)}{V(z)}$$
$$\leq V(z)\left(\frac{\log z}{\log z_0}\right)^\kappa \left(1 + \frac{A}{\log z_0}\right) \leq V(z)\left(1 + O\left(\frac{\log \log z}{\log z}\right)\right).$$

Thus (5.45) holds in all cases.

Finally, we elucidate $V(z)$. By (4.14),

$$e^{\gamma\kappa}V(z) = e^{\gamma\kappa}\prod_{p<z}\left(1 - \frac{\omega(p)}{p}\right)\left(1 - \frac{1}{p}\right)^{-\kappa}\prod_{p<z}\left(1 - \frac{1}{p}\right)^\kappa$$
$$= \prod_{p<z}\left(1 - \frac{\omega(p)}{p}\right)\left(1 - \frac{1}{p}\right)^{-\kappa}\frac{1}{\log^\kappa z}\left\{1 + O\left(\frac{1}{\log z}\right)\right\}.$$

For any $t > z$ we have

$$\prod_{p<z}\left(1 - \frac{\omega(p)}{p}\right)\left(1 - \frac{1}{p}\right)^{-\kappa}$$
$$= \prod_{p<t}\left(1 - \frac{\omega(p)}{p}\right)\left(1 - \frac{1}{p}\right)^{-\kappa}\prod_{z\leq p<t}\left(1 - \frac{\omega(p)}{p}\right)^{-1}\left(1 - \frac{1}{p}\right)^\kappa.$$

Now

$$\prod_{z\leq p<t}\left(1 - \frac{\omega(p)}{p}\right)^{-1}\left(1 - \frac{1}{p}\right)^\kappa \leq 1 + O(1/\log z)$$

uniformly for $t > z$ by $\Omega(\kappa)$ and the Mertens' product estimate (4.14). While the last estimate does not provide a Cauchy condition, it does preclude oscillation of

$$\prod_{p<z}\left(1 - \frac{\omega(p)}{p}\right)\left(1 - \frac{1}{p}\right)^{-\kappa}$$

between distinct values as $z \to \infty$; the product must either converge to a positive number or to $+\infty$. In any case, we have

$$e^{\gamma\kappa}V(z) \leq \prod_p \left(1 - \frac{\omega(p)}{p}\right)\left(1 - \frac{1}{p}\right)^{-\kappa}\frac{1}{\log^\kappa z}\left\{1 + O\left(\frac{1}{\log z}\right)\right\}. \quad \square$$

As an application of Corollary 5.7, we give an example that generalizes Example 1.2 (with $H(n)$ in place of $L(n)$).

5.4 Prime values of polynomials

Example 5.8. Let $h_1(n), \ldots, h_g(n)$ be distinct irreducible polynomials each with integer coefficients and positive leading coefficient, and write

$$H(n) := h_1(n) \cdots h_g(n).$$

Let G denote the degree of H, and let $\rho(p)$ denote the number of incongruent solutions of $H(n) \equiv 0 \bmod p$. Suppose that

(5.46) $\qquad\qquad \rho(p) < p$ for all p,

so that $H(n)$ has no fixed prime divisors. Let y and x be real numbers satisfying $1 < y \leq x$. Then

(5.47) $\quad |\{n : x - y < n \leq x : h_i(n) \text{ prime for } i = 1, \ldots, g\}|$

$$\leq 2^g\, g!\, C(H) \frac{y}{(\log y)^g}\left\{1 + O_H\left(\frac{\log\log 3y}{\log y}\right)\right\},$$

where

$$C(H) := \prod_p \left(1 - \frac{\rho(p)}{p}\right)\left(1 - \frac{1}{p}\right)^{-g} = \prod_p \left(1 - \frac{\rho(p) - 1}{p - 1}\right)\left(1 - \frac{1}{p}\right)^{-g+1}.$$

The O-constant is, of course, independent of x and y, but may depend on g and on the coefficients (as well as the degrees) of the constituent polynomials h_i ($i = 1, \ldots, g$).

Remark 5.9. If some polynomial h_i had zero degree, then (5.46) would imply that it is identically equal to 1; in that case the expression on the left side of (5.47) would be 0 and the result trivial. Thus we may as well assume that all the polynomials h_i have positive degree.

Proof. Let

$$\mathcal{A} := \{H(n) : x - y < n \leq x\}$$

and take $\mathcal{P}$ to be the set of all primes. The setup here is much like that of Example 4.7. The function ρ is multiplicative, so that, if d is squarefree, $\rho(d) = \prod_{p|d} \rho(p)$; moreover, $\rho(p)$ (which is always less than p by hypothesis) satisfies $\rho(p) \leq G$. For d squarefree, we have

$$|\mathcal{A}_d| := |\{n\colon x-y < n \leq x,\ H(n) \equiv 0 \bmod d\}| = \rho(d)(y/d+\theta),\quad |\theta| < 1,$$

so that $X = y$, $\omega(d) = \rho(d)$, and

$$r_{\mathcal{A}}(d) = |\mathcal{A}_d| - (\rho(d)/d)\,y,\quad |r_{\mathcal{A}}(d)| \leq \rho(d).$$

Also we have

$$\sum_{p \leq x} \frac{\rho(p)}{p} \log p = g \log x + O_H(1),$$

and therefore $\mathcal{A}$ satisfies $\Omega(g)$. Now Corollary 5.7 applies with $\kappa = g$ and $z = y^{1/2}$, and so the desired inequality holds for $S(\mathcal{A}, \mathcal{P}, y^{1/2})$.

Finally, suppose that n is counted on the left side of (5.47) but not in $S(\mathcal{A}, \mathcal{P}, y^{1/2})$. Since all the $h_i(n)$ counted on the left side of (5.47) are primes, at least one, $h_I(n)$, say, must be less than $y^{1/2}$. This means that any such argument n satisfies $n \ll y^{1/2}$, whence

$$|\{n\colon x-y < n \leq x,\ h_i(n) \text{prime for } i = 1,\ldots,g\}|$$
$$\leq S(\mathcal{A}, \mathcal{P}, y^{1/2}) + O_H(y^{1/2}). \qquad \square$$

Example 5.10. Prime values of polynomials at prime arguments. To extend the last example to treat prime arguments, we add two further requirements:

(i) none of the polynomials $h_i(n)$ is equal to n,

(ii) $\rho(p) < p - 1$ if $p \nmid H(0)$.

If we then apply the result of Example 5.8 to the polynomial $nH(n)$, with $g+1$ in place of g, we can readily show that

$$|\{n\colon x-y < p \leq x\colon h_i(p) \text{ prime for } i = 1,\ldots,g\}|$$
$$\ll 2^{2g+1}\, g!\, C'(H) \frac{y}{(\log y)^{g+1}} \left\{1 + O_H\!\left(\frac{\log\log 3y}{\log y}\right)\right\},$$

where

$$C'(H) = \prod_{p>2}\left(1 - \frac{1}{(p-1)^2}\right) \prod_{2 < p \nmid H(0)} \left(1 - \frac{\rho(p)-1}{p-2}\right)\left(1-\frac{1}{p}\right)^{-g+1}$$
$$\times \prod_{2 < p | H(0)} \left(1 - \frac{\rho(p)-2}{p-2}\right)\left(1-\frac{1}{p}\right)^{-g+1}.$$

Only the apparently complicated form of $C''(H)$ needs comment. Let $\rho'(p)$ denote the number of incongruent solutions of $nH(n) \equiv 0 \bmod p$, so that the product to be interpreted here (remember that $g+1$ stands in place of g) is

$$\prod_p \left(1 - \frac{\rho'(p)-1}{p-1}\right)\left(1-\frac{1}{p}\right)^{-g+1}$$

$$= 2^g \prod_{p>2}\left(1 - \frac{1}{(p-1)^2}\right) \prod_{p>2}\left(1 - \frac{\rho'(p)-2}{p-2}\right)\left(1-\frac{1}{p}\right)^{-g+1}.$$

But we have

$$\rho'(p) = \begin{cases} \rho(p)+1 & \text{if } p \nmid H(0), \\ \rho(p) & \text{if } p \mid H(0), \end{cases}$$

and from this the result follows at once.

A final remark: condition (ii) serves only to exclude a trivial case. We have $\rho(p) \leq p-1$ for all p, and if there were a prime p_0 that did not divide $H(0)$ such that $\rho(p_0) = p_0 - 1$, then $H(n) \equiv 0 \bmod p_0$ would hold for all $n \not\equiv 0 \bmod p_0$ and so we would have $p_0 \mid H(p)$ for all $p \neq p_0$.

5.5 Notes on Chapter 5

Lemma 5.1 and Corollary 5.2 were suggested by W. B. Jurkat in lectures given at the University of Illinois in Spring, 1973. Proofs of these results appear also in [Raw80], [Raw82], and [DHR88].

For a different treatment of $G(\xi, z)$, see [Grv01]. There is also an account of this sum in [HR74], but on the basis of a two-sided condition in place of $\Omega(\kappa)$, which served as a model for the evaluation of $G^*(\xi, z)$ (see Proposition 5.4). The one-sided condition $\Omega(\kappa)$ first occurs in [Iwa80].

For details on how to derive a first lower bound from Corollary 5.7, see [AO65] or [HR74], Chapter 6.5. The latter includes many applications.

6
Combinatorial foundations (continued)

6.1 Statement of the main analytic theorem

In the last chapter we saw that the solution $\sigma(\cdot)$ of a differential delay equation provides a smooth analogue of $V \cdot G$, where V is the usual product factor and G arises in Selberg's upper sieve estimate. In the Rosser–Iwaniec method and our extension of it there arise related boundary value problems that similarly reflect the combinatorial structure. We state the problem that models our situation and its solution as Theorem 6.1 below. The proof of this result is complicated and is deferred to Part II, but it is convenient to have the statement available here to help clarify several steps in our combinatorial arguments.

In the remainder of this chapter we set out the combinatorial formulas we use in our development of the Rosser–Iwaniec method and discuss the ideas underlying these formulas and their remarkable interplay with the functions of Theorem 6.1.

Theorem 6.1. (MAIN ANALYTIC THEOREM) *For each number $\kappa \geq 1$ for which $2\kappa \in \mathbb{N}$ there exist numbers $\alpha = \alpha_\kappa$ and $\beta = \beta_\kappa$ satisfying*

$$\alpha_1 = \beta_1 = 2 \ \ and \ \ \alpha_\kappa > \beta_\kappa > 2 \ \ for \ \ \kappa > 1,$$

such that the system consisting of initial conditions

(6.1) $\qquad F(u) = 1/\sigma_\kappa(u), \quad 0 < u \leq \alpha,$
(6.2) $\qquad f(u) = 0, \quad 0 < u \leq \beta,$

the simultaneous difference differential equations

(6.3) $\qquad (u^\kappa F(u))' = \kappa u^{\kappa-1} f(u-1), \quad \alpha < u,$
(6.4) $\qquad (u^\kappa f(u))' = \kappa u^{\kappa-1} F(u-1), \quad \beta < u,$

and boundary conditions

(6.5) $$F(u) = 1 + O(e^{-u}), \quad f(u) = 1 + O(e^{-u})$$

has continuous solutions $F = F_\kappa$, $f = f_\kappa$ with the properties that $F(u)$ decreases monotonically and $f(u)$ increases monotonically on $(0, \infty)$.

The function $\sigma_\kappa(u)$ in (6.1) is given by $\sigma_\kappa(u) = j_\kappa(u/2)$, where j_κ was defined in (5.29) and (5.30).

In particular, when $\kappa = 1$, the system (6.1)–(6.5) with $\alpha_1 = \beta_1 = 2$ has solutions $F = F_1$, $f = f_1$ of the kind described; indeed, here

$$F_1(u) = 1/\sigma_1(u) = 2e^\gamma/u, \quad 0 < u \leq 3,$$
$$f_1(u) = 0, \quad 0 < u \leq 2,$$
$$(uF_1(u))' = f_1(u-1), \quad u > 0,$$
$$(uf_1(u))' = F_1(u-1), \quad u > 2,$$

and

(6.6) $$uf_1(u) = \int_2^u \frac{2e^\gamma}{t-1}\, dt = 2e^\gamma \log(u-1), \quad 2 < u \leq 4.$$

For the sake of completeness, we record here that $\sigma = \sigma_\kappa$ is the continuous solution of the system

(6.7) $$u^{-\kappa}\sigma(u) = (2e^\gamma)^{-\kappa}/\Gamma(\kappa+1), \quad 0 < u \leq 2,$$
(6.8) $$(u^{-\kappa}\sigma(u))' = -\kappa u^{-\kappa-1}\sigma(u-2), \quad u > 2.$$

The last formula is equivalent to

(6.8') $$u\sigma'(u) = \kappa(\sigma(u) - \sigma(u-2)) = \kappa \int_{u-2}^u \sigma'(t)dt, \quad u > 2,$$

and this relation remains true for all real u once we define $\sigma(u) = 0$ for $u \leq 0$. Moreover, $\sigma(u)$ is positive and strictly increasing in u for all $u > 0$. In Part II, we shall prove the preceding claims and such facts as $\sigma(u) \to 1$ at a faster than exponential rate as $u \to \infty$.

The functions F_κ and f_κ of the theorem will occur in the next chapter and subsequently in upper and lower bounds respectively of $S(\mathcal{A}, \mathcal{P}, z)$.

We add here two easy consequences of Theorem 6.1 that describe further the functions F_κ and f_κ.

Lemma 6.2. Let $\kappa \geq 1$ and suppose that $1 \leq u_1 < u_2$. Then

$$0 \leq F_\kappa(u_1) - F_\kappa(u_2) \leq \frac{u_2 - u_1}{u_1} \cdot \frac{\kappa}{\sigma_\kappa(1)}$$

and
$$0 \le f_\kappa(u_2) - f_\kappa(u_1) \le \frac{u_2 - u_1}{u_1} \cdot \frac{\kappa}{\sigma_\kappa(1)}.$$

Proof. By the mean value theorem,
$$F_\kappa(u_1) - F_\kappa(u_2) = (u_2 - u_1)(-F'_\kappa(u^*)) \text{ for some } u^* \in (u_1, u_2),$$
$$f_\kappa(u_2) - f_\kappa(u_1) = (u_2 - u_1)f'_\kappa(\bar{u}) \text{ for some } \bar{u} \in (u_1, u_2),$$

provided that F and f each has a derivative throughout (u_1, u_2). If $u_1 > \alpha_\kappa$, by (6.3),
$$-F'_\kappa(u^*) = \frac{\kappa}{u^*}(F_\kappa(u^*) - f_\kappa(u^* - 1)) \le \frac{\kappa}{u^*}F_\kappa(u^*) < \frac{\kappa}{u_1}F_\kappa(1) = \frac{\kappa}{u_1 \sigma_\kappa(1)}.$$

If $1 \le u_1 < u_2 \le \alpha_\kappa$, then $F_\kappa(u) = 1/\sigma_\kappa(u)$ and therefore
$$-F'_\kappa(u) = \sigma'_\kappa(u)/\sigma_\kappa(u)^2.$$

It follows from (6.8) that
$$u\sigma'_\kappa(u) = \kappa(\sigma_\kappa(u) - \sigma_\kappa(u - 2)) \le \kappa\sigma_\kappa(u)$$
and hence that
$$-F'_\kappa(u^*) \le \frac{\kappa}{u^* \sigma_\kappa(u^*)} \le \frac{\kappa}{u_1 \sigma_\kappa(1)}.$$

By combining the two estimates (if necessary), we obtain the first inequality of the Lemma. As for the second inequality, if $\beta_\kappa \le u_1 < u_2$, then, since (by Theorem 6.1) $\bar{u} > \beta_\kappa \ge 2$, we have, by (6.4),
$$f'_\kappa(\bar{u}) = \frac{\kappa}{\bar{u}}(F_\kappa(\bar{u} - 1) - f_\kappa(\bar{u})) \le \frac{\kappa}{\bar{u}}F_\kappa(\bar{u} - 1) \le \frac{\kappa}{u_1}F_\kappa(1) = \frac{\kappa}{u_1 \sigma_\kappa(1)}.$$

If $1 \le u_1 < u_2 \le \beta_\kappa$, then $f'_\kappa(\bar{u}) = 0$, and again combining the estimates (if necessary), we get the claimed inequality. □

Lemma 6.3. *Let $\kappa \ge 1$. We have $F''_\kappa(u) > 0$ for all $u > \alpha_\kappa$ and $f''_\kappa(u) < 0$ for all $u > \beta_\kappa$.*

Proof. By the Theorem,
$$uF'(u) + \kappa F(u) = \kappa f(u - 1).$$

Upon differentiating,
$$uF''(u) = \kappa f'(u - 1) - (\kappa + 1)F'(u) > 0,$$
since f is increasing and F is decreasing. An analogous argument shows f to be concave. □

6.2 The $S(\chi)$ functions

The combinatorial formulas we develop will take one of two forms, according to the size of κ. Our starting point in each case is equation (3.13), which we restate as

$$S(\mathcal{A}, \mathcal{P}, z) = S_1(\chi) + S_2(\chi),$$

where, for given $z > z_0 \geq 2$, we set

(6.9) $$S_1(\chi) := \sum_{d \mid P(z_0, z)} \mu(d) \chi(d) S(\mathcal{A}_d, \mathcal{P}, z_0)$$

and

(6.10) $$S_2(\chi) := \sum_{d \mid P(z_0, z)} \mu(d) \overline{\chi}(d) S(\mathcal{A}_d, \mathcal{P}, p(d)).$$

The functions χ we shall choose are clearly combinatorial, and we show below, in Lemma 6.4, that they are divisor closed. (These notions were defined near the end of Chapter 3.)

With z_0 small enough, we have the Fundamental Lemma, Theorem 4.1, available to estimate the terms $S(\mathcal{A}_d, \mathcal{P}, z_0)$ in $S_1(\chi)$ and expect to lose little by doing so. This leaves us with $S_2(\chi)$. Write

(6.11) $$S_2(\chi) = \left\{ \sum_{\substack{d \mid P(z_0, z) \\ \mu(d) = 1}} - \sum_{\substack{d \mid P(z_0, z) \\ \mu(d) = -1}} \right\} \overline{\chi}(d) S(\mathcal{A}_d, \mathcal{P}, p(d))$$

$$=: S_{21}(\chi) - S_{22}(\chi),$$

say, and note that each of these sums is non-negative, since $S(\mathcal{A}, \mathcal{P}, z)$ is a counting function and, as a combinatorial function, $\overline{\chi}$ is non-negative. Thus we have

(6.12) $$S_1(\chi^-) - S_{22}(\chi^-) \leq S(\mathcal{A}, \mathcal{P}, z) \leq S_1(\chi^+) + S_{21}(\chi^+).$$

We seek functions χ^+ and χ^- so that we can estimate $S_{22}(\chi^-)$ and $S_{21}(\chi^+)$ well. Because (6.12) provides valid upper and lower estimates for $S(\mathcal{A}, \mathcal{P}, z)$, we do not *have* to treat $S_{21}(\chi^-)$ or $S_{22}(\chi^+)$ further. However, we are going to analyze $S_{21}(\chi^-)$ and $S_{22}(\chi^+)$ to help explain how χ^+ and χ^- were chosen and also to show that, with these choices, we cannot get anything better than trivial estimates of these sums.

6.3 The "linear" case $\kappa = 1$

Here we begin with the bold choices of χ^+ and χ^- for which

(a) $S_{21}(\chi^+) = 0 = S_{22}(\chi^-)$ termwise,

and

(b) there is little to gain from keeping $S_{22}(\chi^+)$ and $S_{21}(\chi^-)$.

To secure (a), we shall make choices such that

(6.13) $\overline{\chi^+}(d) = 0$ when $\mu(d) = 1$ and $\overline{\chi^-}(d) = 0$ when $\mu(d) = -1$.

To explain how one brings about b) is more complicated, and to provide a rationale requires some hindsight.

Study of Brun's and Selberg's original methods and, particularly, the method of Ankeny–Onishi (which combines the latter with Buchstab's identity) leads one to expect that an accurate sifting procedure should lead to an upper bound for $S(\mathcal{A}, \mathcal{P}, z)$ with leading term of the form $XV(z)F(v)$, where $F(v)$ is a function that tends monotonically from above to 1 as $v \to \infty$, and a lower bound with leading term $XV(z)f(v)$, where $f(v)$ tends monotonically to 1 from below as $v \to \infty$.

Moreover, as we noted in the proof of the lower bound in Theorem 4.1, we expect to have only the trivial lower bound $S(\mathcal{A}, \mathcal{P}, z) \geq 0$ if the parameter v is too small. Thus we expect that $f(v) \geq 0$, with equality holding when v is at or below a certain threshold β, called the *sieving limit*—in other words, the lower estimate is worthless if $v \leq \beta$, i.e., if the parameter $\xi \leq z^\beta$. We shall show that the function $f_1(\cdot)$ of Theorem 6.1 provides the expected lower bound function here and that $\beta = \beta_1 = 2$.

As for the linkage between F and f, the Buchstab identity hints at the interplay between them, and the high quality of the Ankeny–Onishi method demonstrates the efficacy of combining even a single application of Buchstab's identity with Selberg's upper bound. An insightful iteration of the identity carried out by Jurkat and Richert (see [JR65], and also [HR74]) actually leads to an optimal result when $\kappa = 1$ (Theorem 7.1). The more elegant procedure of Rosser–Iwaniec that we follow in Chapter 7 achieves the same result and, by avoiding use of Selberg's method, arrives at a flexible form of the remainder term that has proved important in several major applications (see [Iwa80]). For $\kappa > 1$, we have refined the Rosser–Iwaniec approach with the help of Selberg's method; the outcome of this is Theorem 9.1 below. The interested reader may wish to study a rather different refinement due to Brüdern and Fouvry (see [BrF96]).

We resume the argument dealing with the case of $\kappa = 1$ and exploit an earlier observation to construct $\chi^{\pm}$ having property (b). If, rather than abandon $-S_{22}(\chi^+)$, we sought to estimate it, we would have to estimate $S(\mathcal{A}_d, \mathcal{P}, p(d))$ from below. To do this, we return to Theorem 4.1, writing ξ there in place of z^v, and proceed as in the proof of the lower bound. In the present application, we take ξ/d in place of ξ and $p(d)$ (the least prime divisor of d) in place of z and get

$$S(\mathcal{A}_d, \mathcal{P}, p(d)) \geq \frac{\omega(d)}{d} XV(p(d)) f\left(\frac{\log(\xi/d)}{\log p(d)}\right) - \text{Error terms}.$$

If, for $\mu(d) = -1$, we take χ^+ so that $\overline{\chi^+}(d) = 1$ only for

$$v = v_d := \frac{\log(\xi/d)}{\log p(d)} \leq 2, \text{ i.e., } p(d)^2 d \geq \xi,$$

then, because $f(2) = f_1(2) = 0$, we would get nothing better than the trivial bound

(6.14) $$S(\mathcal{A}_d, \mathcal{P}, p(d)) \geq 0.$$

In other words, this choice for $\overline{\chi^+}(d)$ yields an upper estimate for $-S_{22}(\chi^+)$ whose main term is 0. Arguing in the same way for $S_{21}(\chi^-)$, we see that discarding it would be in order if $p(d)^2 d \geq \xi$ when $\mu(d) = 1$ and $\overline{\chi^-}(d) = 1$.

We are now in a position to define the functions χ^+ and χ^-. Of course, we have noted already that $\chi^+(1) = 1 = \chi^-(1)$. For $d > 1$, write

$$d = p_1 p_2 \cdots p_r \quad (p_1 > p_2 > \cdots > p_r = p(d), \ r \geq 1)$$

and take each χ to have a quasi-multiplicative structure of type

$$\chi(d) = \eta(p_1) \eta(p_1 p_2) \cdots \eta(p_1 p_2 \cdots p_r),$$

where the arithmetical functions $\eta(\cdot)$ take only the values 0 and 1. We choose

(6.15) $$\eta^+(d) = \begin{cases} 1 & \text{when } \mu(d) = 1, \\ 1 & \text{when } \mu(d) = -1 \text{ and } p(d)^2 d < Y, \\ 0 & \text{otherwise} \end{cases}$$

(the parameter Y appears where we expected ξ; Y will be taken a little smaller than ξ for technical reasons), so that

(6.16) $$\chi^+(d) = \eta^+(p_1) \eta^+(p_1 p_2) \cdots \eta^+(p_1 p_2 \cdots p_r).$$

From this formula and the definition of $\overline{\chi}$ given in Chapter 3, we have $\overline{\chi^+}(1) = 0$ (by convention) and

$$\overline{\chi^+}(d) = \chi^+(d/p(d))(1 - \eta^+(d)), \quad d > 1.$$

Now $\eta^+(d) = 1$ when $\mu(d) = 1$, $d > 1$, and so $\overline{\chi^+}(d) = 0$ for all d. Thus

$$S_{21}(\chi^+) = 0.$$

Also, when $\mu(d) = -1$ and $\overline{\chi^+}(d) = 1$, then necessarily $\eta^+(d) = 0$ and therefore $p(d)^2 d \geq Y$. It follows that

$$S_{22}(\chi^+) = \sum_{\substack{d \mid P(z_0, z),\ \mu(d)=-1 \\ p(d)^2 d \geq Y}} S(\mathcal{A}_d, \mathcal{P}, p(d)),$$

and for each summand $S(\mathcal{A}_d, \mathcal{P}, p(d))$, we have only the trivial lower bound 0, as given in (6.14). With this choice of χ^+, we may as well omit $S_{22}(\chi^+)$ from an upper estimate of $S(\mathcal{A}, \mathcal{P}, z)$.

A similar argument leads us to

(6.17) $$\chi^-(d) = \eta^-(p_1)\eta^-(p_1 p_2) \cdots \eta^-(p_1 p_2 \cdots p_r),$$

where

(6.18) $$\eta^-(d) = \begin{cases} 1 & \text{when } \mu(d) = -1, \\ 1 & \text{when } \mu(d) = 1 \text{ and } p(d)^2 d < Y, \\ 0 & \text{otherwise.} \end{cases}$$

It is straightforward to verify that $\overline{\chi^-}(d) = 0$ when $\mu(d) = -1$, whence

$$S_{22}(\chi^-) = 0;$$

and that when $\mu(d) = 1$ and $\overline{\chi^-}(d) = 1$, then $p(d)^2 d \geq Y$. We have available only a trivial lower bound estimate for the terms of $S_{21}(\chi^-)$, so we may as well omit this sum from a lower estimate of $S(\mathcal{A}, \mathcal{P}, z)$.

These choices ensure the validity of (a) and (b), and thus, by (6.12),

(6.19) $$S_1(\chi^-) \leq S(\mathcal{A}, \mathcal{P}, z) \leq S_1(\chi^+).$$

6.4 The cases $\kappa > 1$

We shall suppress most κ-subscripts in this section to ease the notation, particularly where α_κ and β_κ appear in exponents.

The original Rosser–Iwaniec method, as shown in the case $\kappa = 1$, extends to all $\kappa > 1$ on the basis of (6.19) along with a form of Theorem 6.1, without condition (6.1) and with different sieving limits. However, the resulting upper and lower bounds for $S(\mathcal{A}, \mathcal{P}, z)$ are disappointing—they are not as good even as those of the much more easily derived Ankeny–Onishi estimates. In particular, the sieving limits are relatively large.

We can do somewhat better, at least in the matter of achieving smaller sieving limits β; we return to (6.12), deal with the sums $S_1(\chi^\pm)$ by the Fundamental Lemma as before (except, of course, that we shall be working with different choices of $\chi^\pm$), but drop the constraints (a). Instead of omitting the terms in $S_{21}(\chi^+)$ and in $S_{22}(\chi^-)$, we shall estimate them using Selberg's upper bound, Theorem 2.1. The parameters α and β from Theorem 6.1 play a crucial role: the numbers β turn out to be the sieving limits, and the numbers α will limit the extent to which Theorem 2.1 will be used in estimating $S_{21}(\chi^+)$ and $S_{22}(\chi^-)$ from above. In view of the earlier discussion, we can be a little more specific—we shall apply Theorem 2.1 to the terms in these sums only when the associated parameters satisfy

$$u = u_d := (\log(Y/d))/\log p(d) \leq \alpha,$$

i.e., so long as

$$p(d)^\alpha d \geq Y.$$

We are now ready to specify $\chi^\pm$. With

$$d = p_1 p_2 \cdots p_r \quad (p_1 > p_2 > \cdots > p_r = p(d))$$

we define

(6.20) $$\chi^\pm(d) := \prod_{j=1}^{r} \eta^\pm(p_1 \ldots p_j),$$

where

(6.21) $$\eta^+(d) = \begin{cases} 1 & \text{when } \mu(d) = 1 \text{ and } p(d)^\alpha d < Y, \\ 1 & \text{when } \mu(d) = -1 \text{ and } p(d)^\beta d < Y, \\ 0 & \text{otherwise;} \end{cases}$$

and

(6.22) $$\eta^-(d) = \begin{cases} 1 & \text{when } \mu(d) = 1 \text{ and } p(d)^\beta d < Y, \\ 1 & \text{when } \mu(d) = -1 \text{ and } p(d)^\alpha d < Y, \\ 0 & \text{otherwise.} \end{cases}$$

6.4 The cases $\kappa > 1$

More explicitly, we note that $\chi^+(d) = 1$ if and only if the inequalities

(6.23)
$$\begin{cases} p_1^{\beta+1} & < Y \\ p_2^{\alpha+1} p_1 & < Y \\ p_3^{\beta+1} p_2 p_1 & < Y \\ \cdots\cdots & \cdots \end{cases}$$

all hold, and $\chi^+(d) = 0$ otherwise; also $\chi^-(d) = 1$ if and only if

(6.24)
$$\begin{cases} p_1^{\alpha+1} & < Y \\ p_2^{\beta+1} p_1 & < Y \\ p_3^{\alpha+1} p_2 p_1 & < Y \\ \cdots\cdots & \cdots \end{cases}$$

all hold and $\chi^-(d) = 0$ otherwise. The last lines in these inequalities imply that

(6.25) $$d < Y \quad \text{whenever} \quad \chi^\pm(d) = 1.$$

At the beginning of Section 6.2 we asserted that the functions χ to be formed would be divisor closed, i.e., if $\chi(d) = 1$ and $t \mid d$, then $\chi(t) = 1$. Now we show this.

Lemma 6.4. *Let $\chi^\pm(d)$ be defined by (6.20) with $\eta^\pm(d)$ defined by (6.15) and (6.18) for $\kappa = 1$ and by (6.21) and (6.22) for $\kappa > 1$. Then $\chi^\pm(d)$ is divisor closed.*

Proof. There are eight cases to consider, according to whether $\kappa = 1$ or $\kappa > 1$; whether the sign is $+$ or $-$; and whether $\nu(d)$ is odd or even. The cases are analogous, and here we treat only the one in which $\chi^+(d)$ occurs with $\nu(d)$ even and $\kappa > 1$.

We show that if $\chi^+(d) = 1$ and $p \mid d$, then $\chi^+(d/p) = 1$. Arguing inductively (and using the corresponding result for the case when $\nu(d)$ is odd), we can then conclude that $\chi^+(d/t) = 1$ for any $t \mid d$. Write, as usual,

$$d = p_1 p_2 \cdots p_r \quad \text{with} \quad p_1 > p_2 > \cdots > p_r =: p(d).$$

If $\chi^+(d) = 1$, we have upon rewriting (6.23) in reverse order:

(i) $\qquad p_r^{\alpha+1} p_{r-1} p_{r-2} \cdots p_1 < Y$

(ii) $\qquad p_{r-1}^{\beta+1} p_{r-2} \cdots p_1 < Y$

$\qquad \cdots\cdots\cdots\cdots \quad \cdots$

(r) $\qquad p_1^{\beta+1} < Y.$

We begin with the special case $p = p(d)$. Let $d_r := d/p_r$. Then by lines (ii), ..., (r), we see that $\chi(d_r) = 1$. Now suppose that $1 \leq j \leq r-1$ and consider $d_j := d/p_j$. (For ease of exposition, we treat just the cases $2 \leq j \leq r-2$.) By comparing corresponding factors, we have

(ii′) $\qquad p_r^{\beta+1} p_{r-1} \cdots p_{j+1} p_{j-1} \cdots p_1 < p_{r-1}^{\beta+1} p_{r-2} \cdots p_1 < Y$

(iii′) $\qquad p_{r-1}^{\alpha+1} p_{r-2} \cdots p_{j+1} p_{j-1} \cdots p_1 < p_{r-2}^{\alpha+1} p_{r-3} \cdots p_1 < Y$

$\qquad \cdots\cdots\cdots\cdots\cdots\cdots\cdots$

(r-j+i′) $\qquad p_{j+1}^{\alpha+1} p_{j-1} \cdots p_1 < p_j^{\alpha+1} p_{j-1} \cdots p_1 < Y \qquad (j \text{ even})$

(r-j+i′) $\qquad p_{j+1}^{\beta+1} p_{j-1} \cdots p_1 < p_j^{\beta+1} p_{j-1} \cdots p_1 < Y \qquad (j \text{ odd}).$

All subsequent lines are just those after line (r-j+i) in the preceding group. Thus $\chi(d_j) = 1$. $\qquad \square$

The arrays (6.23) and (6.24) bring to light an important feature of this construction: the inequalities in each are independent of one another only if $\alpha < \beta + 1$; if $\alpha \geq \beta + 1$, then the truth of any α-line in (6.23) implies the truth of the succeeding β-line, and the same is true in (6.24). Indeed, suppose that

$$p_r^{\alpha+1} p_{r-1} \cdots p_1 < Y.$$

Since $p_r^\alpha \geq p_r^{\beta+1} > p_{r+1}^{\beta+1}$, we necessarily have

$$p_{r+1}^{\beta+1} p_r p_{r-1} \cdots p_1 < Y.$$

Consequently, when $\alpha \geq \beta + 1$, we may modify the array (6.23) if we wish, by omitting every β-line except the first; and in the array (6.24), we may omit every β-line. In other words, when $\alpha \geq \beta + 1$, we have $\chi^+(d) = 1$ if and only if

6.4 The cases $\kappa > 1$

(6.23')
$$\begin{cases} p_1^{\beta+1} & < Y \\ p_2^{\alpha+1} p_1 & < Y \\ p_4^{\alpha+1} p_3 p_2 p_1 & < Y \\ \cdots\cdots\cdots\cdots & \cdots \end{cases}$$

all hold and $\chi^+(d) = 0$ otherwise; also $\chi^-(d) = 1$ if and only if

(6.24')
$$\begin{cases} p_1^{\alpha+1} & < Y \\ p_3^{\alpha+1} p_2 p_1 & < Y \\ p_5^{\alpha+1} p_4 p_3 p_2 p_1 & < Y \\ \cdots\cdots\cdots\cdots & \cdots \end{cases}$$

all hold and $\chi^-(d) = 0$ otherwise.

The distinction in the relative size of α_κ and β_κ emerges in the proof of Theorem 6.1 as well. There, $\alpha_\kappa < \beta_\kappa + 1$ only for (integer and half integer) dimension $\kappa = 1$ and $\kappa = 3/2$, whereas $\alpha_\kappa \geq \beta_\kappa + 1$ holds for all such dimensions $\kappa \geq 2$ (see Chapter 16). When *all* real $\kappa \geq 1$ are in play, there is a unique number $\kappa_0 = 1.8344\ldots$ such that $\alpha_\kappa < \beta_\kappa + 1$ when $\kappa < \kappa_0$ and $\alpha_\kappa \geq \beta_\kappa + 1$ when $\kappa \geq \kappa_0$ [ILR80, DHR96].

We check the effect of our choice of $\chi^\pm$ on $S_{21}(\chi^\pm)$ and $S_{22}(\chi^\pm)$. Remember that

(6.26) $$\overline{\chi^\pm}(d) = \chi^\pm(d/p(d))(1 - \eta^\pm(d))$$

holds in these sums. We first consider χ^+. In $S_{21}(\chi^+)$ we have $\mu(d) = 1$ in each term and, by the preceding formula, $\overline{\chi^+}(d) = 1$ if and only if $\chi^+(d/p(d)) = 1$ and $\eta^+(d) = 0$. The latter implies that $p(d)^\alpha d \geq Y$ by (6.21), so that $u_d := (\log(Y/d))/\log p(d) \leq \alpha$. Applying the Selberg upper bound estimate to each term $S(\mathcal{A}_d, \mathcal{P}, p(d))$ in $S_{21}(\chi^+)$ accords with the procedure we described earlier.

We now turn to $S_{22}(\chi^+)$. We remind the reader that this sum is non-negative and has been omitted from (6.12); the purpose of looking at this quantity here is to understand how χ^+ has been chosen and to show that with this choice we cannot get a better estimate of $S_{22}(\chi^+)$. Here each term of the sum has $\mu(d) = -1$. Suppose first that $\alpha < \beta + 1$. In this case, if $\overline{\chi^+}(d) = 1$, then by (6.26), $\eta^+(d) = 0$, that is, by (6.21), $p(d)^\beta d \geq Y$ or $u_d \leq \beta$. Here only the trivial bound (6.14) is available to estimate $S(\mathcal{A}_d, \mathcal{P}, p(d))$ in each term, and we may as well discard the sum $S_{22}(\chi^+)$, as was done in (6.12).

When $\alpha \geq \beta+1$, we note that $\overline{\chi^+}(p) = 1 - \eta^+(p) = 1$ when $p^{\beta+1} \geq Y$ by (6.21) or the first line of (6.23), and so $u_p \leq \beta$; by the preceding discussion this term may be discarded. No other terms occur in $S_{22}(\chi^+)$ in this case: when $\nu(d) > 1$, we show that $\overline{\chi^+}(d) = 0$. By (6.26), if $\overline{\chi^+}(d/p(d)) = 0$, then $\overline{\chi^+}(d) = 0$. On the other hand, if $\overline{\chi^+}(d/p(d)) = 1$, then $\eta^+(d/p(d)) = 1$. By the carryover of inequalities from each α-line inequality in (6.23) to the succeeding β-line, we have $\eta^+(d) = 1$ and hence $\overline{\chi^+}(d) = 0$ in this case too.

Next, we consider χ^-. In $S_{22}(\chi^-)$ we have $\mu(d) = -1$ in each term and, by (6.26), $\overline{\chi^-}(d) = 1$ if and only if $\chi^-(d/p(d)) = 1$ and $\eta^-(d) = 0$. The latter implies that $p(d)^\alpha d \geq Y$ by (6.22), so that, as in the case of $S_{21}(\chi^+)$, we have $u_d := (\log(Y/d))/\log p(d) \leq \alpha$, and once again we can apply the Selberg upper bound estimate to each term $S(\mathcal{A}_d, \mathcal{P}, p(d))$ in $S_{22}(\chi^-)$.

Finally, we consider $S_{21}(\chi^-)$. The reasons for discussing this sum are the same as those for $S_{22}(\chi^+)$. We show that if $\alpha < \beta + 1$, with our choice of χ^-, we cannot make any essential improvement in the left-hand inequality in (6.12), and if $\alpha \geq \beta + 1$, then, interestingly, *we have equality in the first relation in* (6.12).

We have $\mu(d) = 1$ in each term of $S_{21}(\chi^-)$. Suppose first that $\alpha < \beta + 1$. If $\overline{\chi^-}(d) = 1$, then by (6.26), $\eta^-(d) = 0$, that is, by (6.22), $p(d)^\beta d \geq Y$ or $u_d \leq \beta$. Here only the trivial bound (6.14) is available to estimate $S(\mathcal{A}_d, \mathcal{P}, p(d))$ in each term, and we may as well discard $S_{21}(\chi^-)$, as is done in (6.12). When $\alpha \geq \beta + 1$, we show that

$$S_{21}(\chi^-) = 0.$$

Indeed, $\overline{\chi^-}(d) = 0$ for $d = 1$ by convention, and when $\nu(d) > 1$, we consider two cases: by (6.26), if $\overline{\chi^-}(d/p(d)) = 0$, then $\overline{\chi^-}(d) = 0$. On the other hand, if $\overline{\chi^-}(d/p(d)) = 1$, then $\eta^-(d/p(d)) = 1$. By the carryover of inequalities from each α-line inequality in (6.24) to the succeeding β-line, we have $\eta^-(d) = 1$ and hence $\overline{\chi^-}(d) = 0$ in this case too.

To sum up, we have shown that with our choices of $\chi^\pm$, when $\kappa > 1$ we can set out to estimate $S(\mathcal{A}, \mathcal{P}, z)$ via (6.12) in the manner we proposed—to apply the Fundamental Lemma (Theorem 4.1) to the terms of $S_1(\chi^\pm)$, and Selberg's upper estimate (Theorem 2.1) to the terms of $S_{21}(\chi^+)$ and $S_{22}(\chi^-)$. Also, again with these choices of $\chi^\pm$ and regardless of the relative size of α and $\beta + 1$, we cannot make any essential improvements in (6.12). For possible future reference, we place on record that when $\alpha_\kappa \geq \beta_\kappa + 1$, that is, for all $\kappa \geq 2$,

$$\overline{\chi^+}(d) = 0 \text{ when } \mu(d) = -1 \text{ and } \nu(d) > 1,$$
$$\overline{\chi^-}(d) = 0 \text{ when } \mu(d) = 1.$$

6.5 Notes on Chapter 6

The combinatorial method for $\kappa > 1$ follows from [DH85] and [DHR88], and it has been referred to in several subsequent publications by these authors as the DHR sieve method. On reflection, it seems to us now more accurate to describe it as an extension of the Rosser–Iwaniec sieve method. The distinction between the ranges $\alpha_\kappa < \beta_\kappa+1$ and $\alpha_\kappa \geq \beta_\kappa+1$ that occurs here appears in a seemingly quite different, analytic, context in [ILR80], [Raw80], and also in Part II of this monograph.

The upper function $F_\kappa(u)$ coincides with $1/\sigma_\kappa(u)$, the Ankeny–Onishi function when $0 < u \leq \alpha_\kappa$, and $F_\kappa(u) < 1/\sigma_\kappa(u)$ for $u > \alpha_\kappa$. However, as Figure 6.1(b) shows for $\kappa = 2$, the difference of the function values is not large (here $\lesssim 0.08$) and it tends rapidly to 0 as $u \to \infty$. The lower function $f_\kappa(u) > 0$ for $u > \beta_\kappa$ ($\beta_2 \approx 4.266$), the so-called *sieving limit*. Below this point $f_\kappa(u) = 0$, and Theorem 9.1 yields only the trivial lower bound $S(\mathcal{A}, \mathcal{P}, z) \geq 0$. Our sieving limit of 4.266 is smaller than that of the Ankeny–Onishi sieve (about 4.42 for $\kappa = 2$) or the Rosser–Iwaniec sieve (about 4.834 for $\kappa = 2$), so we can treat some lower bound problems to which the other sieves do not apply.

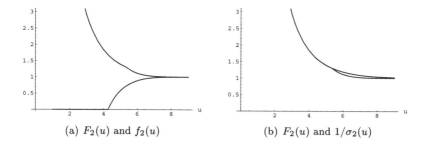

(a) $F_2(u)$ and $f_2(u)$ (b) $F_2(u)$ and $1/\sigma_2(u)$

Fig. 6.1. $F_2(u)$, f_2, and $1/\sigma_2(u)$

7
The case $\kappa = 1$: the linear sieve

7.1 The theorem and first steps

The Fundamental Lemma, Theorem 4.1, provides asymptotic estimates for $S(\mathcal{A}, \mathcal{P}, z)$ in terms of a parameter v when $z = z(X)$ is small relative to X in the sense that $\log z / \log X \to 0$ as $X \to \infty$. In this chapter we give upper and lower bounds for $S(\mathcal{A}, \mathcal{P}, z)$ that are useful in the larger range where $\log z / \log X \leq \theta$ for some constant $\theta < 1$. Theorem 7.1 is especially noteworthy because Selberg [Sel91] has shown the leading terms in both bounds to be best possible (see the Notes to this chapter). We obtain these sharper bounds by using the $\chi^{\pm}$ technology of the last chapter.

Our main result is

Theorem 7.1. *Suppose condition $\Omega(1)$ holds and y, z are any numbers satisfying $X \geq y \geq z \geq 2$. Then*

$$S(\mathcal{A}, \mathcal{P}, z) \leq XV(z)\left\{F_1\left(\frac{\log y}{\log z}\right) + O\left(\frac{(\log \log y)^{3/4}}{(\log y)^{1/4}}\right)\right\} + \sum_{\substack{m|P(z) \\ m<y}} 4^{\nu(m)} |r_\mathcal{A}(m)|$$

and

$$S(\mathcal{A}, \mathcal{P}, z) \geq XV(z)\left\{f_1\left(\frac{\log y}{\log z}\right) - O\left(\frac{(\log \log y)^{3/4}}{(\log y)^{1/4}}\right)\right\} - \sum_{\substack{m|P(z) \\ m<y}} 4^{\nu(m)} |r_\mathcal{A}(m)|,$$

where the O-constants depend at most on A (coming from $\Omega(1)$) and the functions F_1, f_1 are those specified in Theorem 6.1.

We note that here (and also for $\kappa > 1$) the parameter y plays a role analogous to that of z^{2v} in Theorem 4.1. Large values of y improve the quality of the main term by bringing F_1 and f_1 each closer to 1; on the other hand the sum defining the error term increases with y. In applications, y is chosen (if possible) so the error term is small in comparison with the main term; in particular, we always take $y \leq X$. It can happen, though, that no choice of y yields a positive lower bound for $S(\mathcal{A}, \mathcal{P}, z)$.

The remainder sums lack the flexible form of those in the original Rosser–Iwaniec method because we have used Selberg's version of a Fundamental Lemma from Chapter 4 instead of the superior version from [FI78]. We have included the linear sieve for the sake of completeness and because the proof we are about to give is a useful preparation for our principal objective, treatment of integer and half integer dimensions κ exceeding 1. Nevertheless, it should be said that the above theorem leads to many interesting applications.

The remainder of this chapter is devoted to the proof of Theorem 7.1. We begin by turning to the Fundamental Lemma, Theorem 4.1, and taking $q = 1$ there. Then

$$v \log v + 3v/2 > v \log v + v \geq 2v$$

if $v \geq e$. The Lemma now reads (remember that $\kappa = 1$ here)

$$(7.1) \qquad S(\mathcal{A}, \mathcal{P}, z) = XV(z)\{1 + O(e^{-2v})\} + \theta \sum_{\substack{n < z^{2v} \\ n \mid P(z)}} 3^{\nu(n)} |r_\mathcal{A}(n)|,$$

where $|\theta| \leq 1$.

Now let y, the parameter in the statement of Theorem 7.1, be large enough so that $\log \log y \geq 2e$. We distinguish two cases; we suppose first that

$$(7.2) \qquad z \leq \exp\left(\frac{\log y}{\log \log y}\right).$$

Formula (7.1) with $v := (\log y)/(2 \log z)$ asserts that

$$S(\mathcal{A}, \mathcal{P}, z) = XV(z)\left\{1 + O\left(\frac{1}{\log y}\right)\right\} + \theta \sum_{\substack{n < y \\ n \mid P(z)}} 3^{\nu(n)} |r_\mathcal{A}(n)|,$$

and since, by (7.2),

$$\frac{\log y}{\log z} \geq \log \log y,$$

each of $F_1((\log y)/\log z)$, $f_1((\log y)/\log z)$ is equal to $1 + O(1/\log y)$

7.1 The theorem and first steps

(by (6.5)); hence the theorem is proved under the assumption of (7.2), even as an asymptotic equality. Henceforth, we assume that (7.2) does not hold.

With the inequalities (6.19) in mind, we turn to the sums $S_1(\chi^\pm)$ defined by (6.9), (6.16), and (6.17). We apply the Fundamental Lemma, Theorem 4.1, to each of the terms $S(\mathcal{A}_d, \mathcal{P}, z_0)$ in (6.9), this time with $q = d$ (so that $(d, P(z_0)) = 1$), $z = z_0$ and $v = L/2$, where L is any number at least as large as $2e$. For any $d \mid P(z_0, z)$ we obtain

$$|S(\mathcal{A}_d, \mathcal{P}, z_0) - \frac{w(d)}{d} XV(z_0)(1 + O(e^{-L}))| \leq \sum_{\substack{n < z_0^L \\ n \mid P(z_0)}} 3^{\nu(n)} |r_\mathcal{A}(dn)|.$$

We choose $L = \log \log y$ (recall that $L \geq 2e$ by the assumption that y is sufficiently large) and take

(7.3) $$z_0 = \exp\left\{\frac{(\log y)^{3/4}}{(\log \log y)^{1/4}}\right\}$$

so that

$$z_0^L = \exp\left\{((\log y) \log \log y)^{3/4}\right\}.$$

Then, by (6.9), bearing in mind that $d < Y$ when $\chi^\pm(d) = 1$,

(7.4) $$\left|S_1(\chi^\pm) - \sum_{d \mid P(z_0, z)} \mu(d)\chi^\pm(d)\frac{w(d)}{d} XV(z_0)\left(1 + O\left(\frac{1}{\log y}\right)\right)\right|$$

$$\leq \sum_{\substack{d \mid P(z_0, z) \\ d < Y}} \sum_{\substack{n < z_0^L \\ n \mid P(z_0)}} 3^{\nu(n)}|r_\mathcal{A}(dn)| \leq \sum_{\substack{m < Y z_0^L \\ m \mid P(z)}} 4^{\nu(m)}|r_\mathcal{A}(m)|;$$

the last inequality follows from the formula

$$\sum_{n \mid m} 3^{\nu(n)} = 4^{\nu(m)}, \quad m \text{ squarefree}.$$

We remarked following equation (6.15) that Y would be taken less than ξ, and now, with y taking the place of ξ, we can be specific. We take $Y z_0^L = y$, i.e.,

(7.5) $$Y := y \exp\left\{-((\log y) \log \log y)^{3/4}\right\}.$$

We estimate the contribution arising from the $O(1/\log y)$ error term in (7.4) by using condition $\Omega(1)$:

$$\sum_{d|P(z_0,z)} \frac{\omega(d)}{d} = \prod_{z_0 \leq p < z}\left(1 + \frac{\omega(p)}{p}\right) \leq \prod_{z_0 \leq p < z}\left(1 - \frac{\omega(p)}{p}\right)^{-1}$$

$$= \prod_{z_0 \leq p < z}\left(1 - \frac{\omega(p)}{p}\right)^{-2} \frac{V(z)}{V(z_0)} \ll \left(\frac{\log z}{\log z_0}\right)^2 \frac{V(z)}{V(z_0)}$$

$$\leq \left(\frac{\log y}{\log z_0}\right)^2 \frac{V(z)}{V(z_0)} = \{\log y \log\log y\}^{1/2} \frac{V(z)}{V(z_0)},$$

so that

$$\frac{XV(z_0)}{\log y} \sum_{d|P(z_0,z)} \frac{\omega(d)}{d} \ll \left(\frac{\log\log y}{\log y}\right)^{1/2} XV(z).$$

We conclude that, by (7.5)

$$(7.6) \quad \left|S_1(\chi^{\pm}) - XV(z_0)\sum_{d|P(z_0,z)} \mu(d)\chi^{\pm}(d)\frac{\omega(d)}{d}\right|$$

$$\leq O\left(\left(\frac{\log\log y}{\log y}\right)^{1/2} XV(z)\right) + \sum_{\substack{m|P(z)\\m<y}} 4^{\nu(m)}|r_A(m)|,$$

and our remaining task is to deal with the sums

$$\sum_{d|P(z_0,z)} \mu(d)\chi^{\pm}(d)\frac{\omega(d)}{d}.$$

The classical approach here is to apply the Fundamental Sieve Identity to these sums, but we shall expedite matters by use of the following remark: writing

$$\phi^{(-)^r}(\cdot) = \begin{cases} F_1(\cdot), & r \text{ even,} \\ f_1(\cdot), & r \text{ odd,} \end{cases}$$

we have, by Theorem 6.1,

$$\mu(d) \leq \mu(d)\phi^{(-)^{\nu(d)}}\left(\frac{\log Y/d}{\log z_0}\right), \quad \mu(d) \geq \mu(d)\phi^{(-)^{\nu(d)+1}}\left(\frac{\log Y/d}{\log z_0}\right).$$

Hence

$$(7.7) \quad \sum_{d|P(z_0,z)} \mu(d)\chi^+(d)\frac{\omega(d)}{d}$$

$$\leq \Sigma^+ := \sum_{d|P(z_0,z)} \mu(d)\chi^+(d)\frac{\omega(d)}{d}\phi^{(-)^{\nu(d)}}\left(\frac{\log Y/d}{\log z_0}\right)$$

and

(7.8) $$\sum_{d|P(z_0,z)} \mu(d)\chi^-(d)\frac{\omega(d)}{d}$$

$$\geq \Sigma^- := \sum_{d|P(z_0,z)} \mu(d)\chi^-(d)\frac{\omega(d)}{d}\phi(-)^{\nu(d)+1}\left(\frac{\log Y/d}{\log z_0}\right).$$

7.2 Bounds for $V\Sigma^\pm$: the set-up

We shall prove that, up to error terms that occur, $V(z_0)\Sigma^+$ is at most $V(z)F((\log Y)/\log z)$ and $V(z_0)\Sigma^-$ is at least $V(z)f((\log Y)/\log z)$. A certain amount of technical preparation is necessary, some of which will be used also later and is formulated therefore for all $\kappa \geq 1$. We begin with the remark that, by (4.7) and subtraction

(7.9) $$V(z_0) - V(w) = \sum_{z_0 \leq p < w} \frac{\omega(p)}{p} V(p), \quad z_0 \leq w.$$

Lemma 7.2. *Suppose that $\Omega(\kappa)$ holds, that $z_0 \leq w$, and that $B(t)$ is a non-negative, continuous, and increasing function on $[z_0, w]$. Then*

(7.10) $$\sum_{z_0 \leq p < w} \frac{\omega(p)}{p} V(p) B(p)$$

$$\leq V(w)(\log w)^\kappa \left\{ \kappa \int_{z_0}^w \frac{B(t)}{t(\log t)^{\kappa+1}} dt + \frac{AB(w)}{(\log z_0)^{\kappa+1}} \right\}.$$

Proof. By (7.9) and $\Omega(\kappa)$, we have

$$\sum_{z_0 \leq p < w} \frac{\omega(p)}{p} \frac{V(p)}{V(w)} B(p)$$

$$= \sum_{z_0 \leq p < w} \frac{\omega(p)}{p} \frac{V(p)}{V(w)} \left\{ B(z_0) + \int_{z_0}^p dB(t) \right\}$$

$$= B(z_0)\left(\frac{V(z_0)}{V(w)} - 1\right) + \int_{z_0}^w \left(\sum_{t \leq p < w} \frac{\omega(p)}{p} \frac{V(p)}{V(w)}\right) dB(t)$$

$$= B(z_0)\left(\frac{V(z_0)}{V(w)} - 1\right) + \int_{z_0}^w \left(\frac{V(t)}{V(w)} - 1\right) dB(t).$$

Thus

$$\sum_{z_0 \leq p < w} \frac{\omega(p)}{p} \frac{V(p)}{V(w)} B(p)$$

$$\leq B(z_0) \left\{ \left(\frac{\log w}{\log z_0} \right)^\kappa \left(1 + \frac{A}{\log z_0} \right) - 1 \right\}$$

$$+ \int_{z_0}^w \left\{ \left(\frac{\log w}{\log t} \right)^\kappa \left(1 + \frac{A}{\log t} \right) - 1 \right\} dB(t)$$

$$= B(w) \frac{A}{\log w} - \int_{z_0}^w B(t) d \left\{ \left(\frac{\log w}{\log t} \right)^\kappa \left(1 + \frac{A}{\log t} \right) \right\}$$

$$\leq \kappa (\log w)^\kappa \int_{z_0}^w \frac{B(t)}{t (\log t)^{\kappa+1}} dt + A \frac{B(w)(\log w)^\kappa}{(\log z_0)^{\kappa+1}}. \qquad \Box$$

We now apply the preceding lemma to

(7.11) $\qquad B(t) := (-1)^\nu \left(1 - \phi^{(-)^{\nu+1}} \left(\frac{\log(x/t)}{\log t} \right) \right), \quad z_0 \leq t < x,$

with $\nu = 0$ and 1 in turn, as we may do since, by Theorem 6.1, each of

$$1 - f_\kappa \left(\frac{\log x}{\log t} - 1 \right) \quad \text{and} \quad F_\kappa \left(\frac{\log x}{\log t} - 1 \right) - 1$$

is non-negative, continuous and non-decreasing in t. Thus we obtain

Proposition 7.3. *Suppose that $\kappa \geq 1$ and $z_0 \leq w$. Also, recall from Theorem 6.1 that $\alpha_1 = \beta_1 = 2$ and $\alpha > \beta > 2$ when $\kappa > 1$. Then, if $x \geq w^{\beta_\kappa}$,*

(7.12) $\displaystyle\sum_{z_0 \leq p < w} \frac{\omega(p)}{p} V(p) F_\kappa \left(\frac{\log(x/p)}{\log p} \right) \leq V(z_0) f_\kappa \left(\frac{\log x}{\log z_0} \right)$

$$- V(w) f_\kappa \left(\frac{\log x}{\log w} \right) + \frac{A}{\sigma_\kappa(1)} \frac{V(w)}{\log z_0} \left(\frac{\log w}{\log z_0} \right)^\kappa,$$

and if $x \geq w^{\alpha_\kappa}$ for $\kappa > 1$ (resp. $x \geq w$ when $\kappa = 1$), we have

(7.13) $\displaystyle\sum_{z_0 \leq p < w} \frac{\omega(p)}{p} V(p) f_\kappa \left(\frac{\log(x/p)}{\log p} \right) \geq V(z_0) F_\kappa \left(\frac{\log x}{\log z_0} \right)$

$$- V(w) F_\kappa \left(\frac{\log x}{\log w} \right) - \frac{A}{\sigma_\kappa(1)} \frac{V(w)}{\log z_0} \left(\frac{\log w}{\log z_0} \right)^\kappa.$$

Proof. Despite its length, the argument below amounts simply to a combination of the preceding lemma and the properties of F_κ and f_κ set out

7.2 Bounds for $V\Sigma^{\pm}$: the set-up

in Theorem 6.1. Thus, with $B(t)$ now given by (7.11) with $\nu = 0$ or 1, the integral on the right side of (7.10) is

$$(-1)^{\nu}\kappa \int_{z_0}^{w} \left\{1 - \phi^{(-)^{\nu+1}}\left(\frac{\log(x/t)}{\log t}\right)\right\} \frac{dt}{t(\log t)^{\kappa+1}},$$

and on substituting $t = x^{1/\zeta}$ this becomes

$$\frac{(-1)^{\nu}\kappa}{(\log x)^{\kappa}} \int_{(\log x)/\log w}^{(\log x)/\log z_0} \{1 - \phi^{(-)^{\nu+1}}(\zeta - 1)\} \zeta^{\kappa-1} d\zeta$$

$$= (-1)^{\nu} \left\{\frac{1}{(\log z_0)^{\kappa}}\left(1 - \phi^{(-)^{\nu}}\left(\frac{\log x}{\log z_0}\right)\right) - \frac{1}{(\log w)^{\kappa}}\left(1 - \phi^{(-)^{\nu}}\left(\frac{\log x}{\log w}\right)\right)\right\}$$

by (6.3) and (6.4); when $\nu = 0$ we have invoked (6.3), as we may do, since $\zeta \geq (\log x)/\log w \geq \alpha_{\kappa}$ (resp. ≥ 1 when $\kappa = 1$), and when $\nu = 1$ we have applied (6.4) because here $\zeta \geq (\log x)/\log w \geq \beta_{\kappa}$.

Hence, by (7.9) and the preceding lemma, subject to the specified constraints on $(\log x)/\log w$,

$$(-1)^{\nu}\left\{V(z_0) - V(w) - \sum_{z_0 \leq p < w} \frac{w(p)}{p} V(p) \phi^{(-)^{\nu+1}}\left(\frac{\log(x/p)}{\log p}\right)\right\}$$

$$\leq (-1)^{\nu} V(w)\left\{\left(\frac{\log w}{\log z_0}\right)^{\kappa} - 1\right.$$

$$- \left(\frac{\log w}{\log z_0}\right)^{\kappa} \phi^{(-)^{\nu}}\left(\frac{\log x}{\log z_0}\right) + \phi^{(-)^{\nu}}\left(\frac{\log x}{\log w}\right)$$

$$\left. + \frac{A}{\log z_0}\left(\frac{\log w}{\log z_0}\right)^{\kappa}\left(1 - \phi^{(-)^{\nu+1}}\left(\frac{\log(x/w)}{\log w}\right)\right)\right\}.$$

After rearrangement this becomes

$$(-1)^{\nu}\left\{V(z_0)\phi^{(-)^{\nu}}\left(\frac{\log x}{\log z_0}\right) - V(w)\phi^{(-)^{\nu}}\left(\frac{\log x}{\log w}\right)\right.$$

$$\left. - \sum_{z_0 \leq p < w} \frac{w(p)}{p} V(p) \phi^{(-)^{\nu+1}}\left(\frac{\log(x/p)}{\log p}\right)\right\}$$

$$\leq (-1)^{\nu} V(w)\left(\frac{\log w}{\log z_0}\right)^{\kappa}\left\{\frac{V(z_0)}{V(w)}\left(\frac{\log z_0}{\log w}\right)^{\kappa}\left(\phi^{(-)^{\nu}}\left(\frac{\log x}{\log z_0}\right) - 1\right)\right.$$

$$\left. - \left(\phi^{(-)^{\nu}}\left(\frac{\log x}{\log z_0}\right) - 1\right) + \frac{A}{\log z_0}\left(1 - \phi^{(-)^{\nu+1}}\left(\frac{\log(x/w)}{\log w}\right)\right)\right\}.$$

The last formula can be restated as

$$V(w)\left(\frac{\log w}{\log z_0}\right)^\kappa (-1)^\nu \left\{ \left(\frac{V(z_0)}{V(w)}\left(\frac{\log z_0}{\log w}\right)^\kappa - 1\right)\left(\phi^{(-)^\nu}\left(\frac{\log x}{\log z_0}\right) - 1\right) \right.$$
$$\left. + \frac{A}{\log z_0}\left(1 - \phi^{(-)^{\nu+1}}\left(\frac{\log(x/w)}{\log w}\right)\right)\right\}$$
$$\leq A\frac{V(w)}{\log z_0}\left(\frac{\log w}{\log z_0}\right)^\kappa (-1)^\nu \left\{ \phi^{(-)^\nu}\left(\frac{\log x}{\log z_0}\right) - \phi^{(-)^{\nu+1}}\left(\frac{\log x}{\log w} - 1\right)\right\}$$
$$\leq A\frac{V(w)}{\log z_0}\left(\frac{\log w}{\log z_0}\right)^\kappa F_\kappa(\beta_\kappa - 1),$$

the last since

$$(-1)^\nu \left\{ \phi^{(-)^\nu}\left(\frac{\log x}{\log z_0}\right) - \phi^{(-)^{\nu+1}}\left(\frac{\log x}{\log w} - 1\right)\right\}$$
$$= \begin{cases} F_\kappa\left(\frac{\log x}{\log z_0}\right) - f_\kappa\left(\frac{\log x}{\log w} - 1\right), & \nu = 0, \\ F_\kappa\left(\frac{\log x}{\log w} - 1\right) - f_\kappa\left(\frac{\log x}{\log z_0}\right), & \nu = 1, \end{cases}$$
$$\leq \begin{cases} F_\kappa\left(\frac{\log x}{\log z_0}\right), & \nu = 0, \\ F_\kappa\left(\frac{\log x}{\log w} - 1\right), & \nu = 1, \end{cases}$$
$$\leq F_\kappa(\beta_\kappa - 1)$$

in both cases, because F_κ is decreasing and

$$\frac{\log x}{\log z_0} \geq \frac{\log x}{\log w} > \frac{\log x}{\log w} - 1 \geq \beta_\kappa - 1.$$

Finally, since $\beta_\kappa - 1 \geq 1$, $F_\kappa(\beta_\kappa - 1) \leq F_\kappa(1) = 1/\sigma_\kappa(1)$ by (6.1). □

The two inequalities of Proposition 7.3 constitute the principal technical tools in the proofs of Theorem 7.1 and of Theorem 9.1. We shall demonstrate their use next in the case of $\kappa = 1$, in the course of proving Proposition 7.4. Theorem 9.1, which deals with all integer and half integer dimensions $\kappa > 1$, will be established by an argument similar to that used in Proposition 7.4; while the details will be more complicated, the ideas are the same.

7.3 Bounds for $V\Sigma^\pm$: conclusion

We now revert to the case $\kappa = 1$ (where $\alpha_1 = \beta_1 = 2$) and are ready to estimate $V(z_0)\Sigma^+$, defined in (7.7), from above, and $V(z_0)\Sigma^-$, defined in (7.8), from below.

7.3 Bounds for $V\Sigma^{\pm}$: conclusion

We shall prove

Proposition 7.4. *Suppose that $\Omega(1)$ holds and that $\nu = 0$ or 1. Then, for $2 \leq z_0 \leq z \leq Y^{1/(\nu+1)}$, we have*

$$(-1)^{\nu}\left\{V(z_0)\Sigma^{(-)^{\nu}} - V(z)\phi^{(-)^{\nu}}\left(\frac{\log Y}{\log z}\right)\right\} \leq A\exp\left(\frac{2A}{\log z_0}\right)\frac{V(z)(\log z)^2}{\sigma_1(1)(\log z_0)^3}.$$

Proof. With $\nu = 0$ or 1 and $r = 1, 2, 3 \ldots$, introduce the expressions

$$E_r^{(-)^{\nu}} := (-1)^{\nu}\Bigg\{V(z_0)\sum_{\substack{d \mid P(z_0, z) \\ \nu(d) < r}}\mu(d)\chi^{(-)^{\nu}}(d)\frac{\omega(d)}{d}\phi^{(-)^{\nu(d)+\nu}}\left(\frac{\log(Y/d)}{\log z_0}\right)$$

$$- V(z)\phi^{(-)^{\nu}}\left(\frac{\log Y}{\log z}\right)$$

$$+ \sum_{\substack{d \mid P(z_0, z) \\ \nu(d) = r}}\mu(d)\chi^{(-)^{\nu}}(d)\frac{\omega(d)}{d}V(p(d))\,\phi^{(-)^{\nu(d)+\nu}}\left(\frac{\log(Y/d)}{\log p(d)}\right)\Bigg\}.$$

The expression on the left side of the inequality of the Proposition is

$$E_{\infty}^{(-)^{\nu}} := \lim_{r \to \infty} E_r^{(-)^{\nu}}.$$

This limit exists, since for each pair $\{z_0, z\}$, $E_r^{\pm}$ is in fact constant for all sufficiently large values of r. (We hope the reader will not find confusing our use of $\nu(d)$ as the counter of distinct prime factors of d and $\nu = 0$ or 1 as a parity index.) Observe that

$$(7.14) \qquad E_{\infty}^{(-)^{\nu}} = E_1^{(-)^{\nu}} + \sum_{r=1}^{\infty}\left(E_{r+1}^{(-)^{\nu}} - E_r^{(-)^{\nu}}\right).$$

While the expressions $E_r^{(-)^{\nu}}$ look unwieldy, differences of consecutive Es, when suitably arranged, can be estimated readily by one or more applications of Proposition 7.3 in conjunction with the properties of the functions $\chi^{(-)^{\nu}}$.

Consider $E_1^{(-)^{\nu}}$ first. The first sum for this quantity has only one term, corresponding to $d = 1$, and the second sum extends over primes only. Thus

$$E_1^{(-)^{\nu}} = (-1)^{\nu}\Bigg\{V(z_0)\,\phi^{(-)^{\nu}}\left(\frac{\log Y}{\log z_0}\right) - V(z)\,\phi^{(-)^{\nu}}\left(\frac{\log Y}{\log z}\right)$$

$$- \sum_{z_0 \leq p < z}\chi^{(-)^{\nu}}(p)\frac{\omega(p)}{p}V(p)\,\phi^{(-)^{1+\nu}}\left(\frac{\log(Y/p)}{\log p}\right)\Bigg\}.$$

When $\nu = 0$,
$$\phi^{(-)^{\nu+1}}\left(\frac{\log(Y/p)}{\log p}\right) = f_1\left(\frac{\log(Y/p)}{\log p}\right) = 0 \quad \text{if} \quad \log(Y/p) \le 2\log p,$$

i.e., if $p^3 \ge Y$; thus we may assume that the last sum is further restricted to primes $p < Y^{1/3}$. For such primes, $\chi^{(-)^{\nu}}(p) = \chi^{+}(p) = \eta^{+}(p) = 1$ (see (6.15), (6.16)). When $\nu = 1$, we have $\chi^{(-)^{\nu}}(p) = \chi^{-}(p) = 1$ for all p by (6.17) and (6.18). Hence, whether $\nu = 0$ or 1, the term $\chi^{(-)^{\nu}}(p)$ in the sum may be replaced by 1. Now Proposition 7.3, with z in place of w and Y in place of x, tells us (remember that $\kappa = 1$ here) that

(7.15) $$E_1^{(-)^{\nu}} \le \frac{A}{\sigma_1(1)} \frac{\log z}{\log^2 z_0} V(z), \quad z^{1+\nu} \le Y.$$

Next consider
$$E_{r+1}^{(-)^{\nu}} - E_r^{(-)^{\nu}} = (-1)^{\nu}\Bigg\{V(z_0) \sum_{\substack{d|P(z_0,z) \\ \nu(d)=r}} \mu(d)\chi^{(-)^{\nu}}(d)\frac{\omega(d)}{d}\phi^{(-)^{\nu}}\left(\frac{\log(Y/d)}{\log z_0}\right)$$
$$+ \sum_{\substack{d|P(z_0,z) \\ \nu(d)=r+1}} \mu(d)\chi^{(-)^{\nu}}(d)\frac{\omega(d)}{d}V(p(d))\,\phi^{(-)^{r+\nu+1}}\left(\frac{\log(Y/d)}{\log p(d)}\right)$$
$$- \sum_{\substack{d|P(z_0,z) \\ \nu(d)=r}} \mu(d)\chi^{(-1)^{\nu}}(d)\frac{\omega(d)}{d}V(p(d))\,\phi^{(-)^{r+\nu}}\left(\frac{\log(Y/d)}{\log p(d)}\right)\Bigg\}.$$

Here it is natural to write d in the second sum as $d = pt$, $p = p(d)$, so that $p(t) > p$, $\chi^{(\nu)}(d) = \chi^{(\nu)}(t)\eta^{(\nu)}(pt)$, and then replace t by d so that all three sums extend over $d \mid P(z_0, z)$ and $\nu(d) = r$. We obtain

$$E_{r+1}^{(-)^{\nu}} - E_r^{(-)^{\nu}} = (-1)^{\nu+r}\Bigg\{\sum_{\substack{d|P(z_0,z) \\ \nu(d)=r}} \chi^{(-)^{\nu}}(d)\frac{\omega(d)}{d}\Bigg\{V(z_0)\phi^{(-)^{r+\nu}}\left(\frac{\log(Y/d)}{\log z_0}\right)$$
$$- V(p(d))\,\phi^{(-)^{r+\nu}}\left(\frac{\log(Y/d)}{\log p(d)}\right)$$
$$- \sum_{z_0 \le p < p(d)} \eta^{(-)^{\nu}}(dp)\frac{\omega(p)}{p}V(p)\,\phi^{(-)^{r+\nu+1}}\left(\frac{\log(Y/dp)}{\log p}\right)\Bigg\}.$$

It is necessary now to elucidate the role of $\eta^{(-)^{\nu}}(dp)$ in the inner sum over p. Take $\nu = 0$ first, so that $\eta^{(-)^{\nu}}(dp) = \eta^{+}(dp) = 1$ always when $\nu(d) = r$ is odd. When r is even, $\eta^{+}(dp) = 1$ if $p^3d < Y$. But if $p^3d \ge Y$,

7.3 Bounds for $V\Sigma^\pm$: conclusion

that is, if $\log\{Y/(pd)\} \leq 2\log p$, and r is even,

(7.16) $$\phi^{(-)^{r+\nu+1}}\left(\frac{\log(Y/pd)}{\log p}\right) = f_1\left(\frac{\log(Y/pd)}{\log p}\right) = 0$$

so that the sum over p is unchanged when $\eta^{(-)^\nu}(dp)$ is replaced by 1.

Next, suppose that $\nu = 1$, so that $\eta^{(-)^\nu}(dp) = \eta^-(dp) = 1$ always when $\nu(d) = r$ is even. When r is odd, $\eta^-(dp) = 1$ if $p^3 d < Y$ and is 0 otherwise. Now the argument goes as before: $r + \nu + 1$ is odd and (7.16) applies, and once again the factor $\eta^-(dp)$ in the inner sum over p may be replaced by 1. Thus

$$E_{r+1}^{(-)^\nu} - E_r^{(-)^\nu} = \sum_{\substack{d \mid P(z_0, z) \\ \nu(d) = r}} \chi^{(-)^\nu}(d) \frac{\omega(d)}{d} (-1)^{r+\nu} \Big\{ V(z_0) \phi^{(-)^{r+\nu}}\left(\frac{\log(Y/d)}{\log z_0}\right)$$
$$- V(p(d)) \phi^{(-)^{r+\nu}}\left(\frac{\log(Y/d)}{\log p(d)}\right)$$
$$- \sum_{z_0 \leq p < p(d)} \frac{\omega(p)}{p} V(p) \phi^{(-)^{r+\nu+1}}\left(\frac{\log(Y/dp)}{\log p}\right) \Big\}.$$

Proposition 7.3 applies to the last sum on the right, this time with $p(d)$ in place of w and Y/d in place of x, provided that $p(d)^2 < Y/d$ when $r+\nu$ is odd and $p(d) < Y/d$ when $r+\nu$ is even. We find under these conditions that

$$(-1)^{r+\nu}\{\cdots\} \leq \frac{A}{\sigma_1(1)} \frac{\log p(d)}{\log^2 z_0} V(p(d)).$$

We turn to the factor $\chi^{(-)^\nu}(d)$, subject to $\nu(d) = r$, to check these conditions. Suppose that $\nu = 0$. Then, by (6.16), $\chi^+(d) = 1$ implies that $\eta^+(d) = 1$ when r is odd and $\eta^+(d/p(d)) = 1$ when r is even. But "$\eta^+(d) = 1$ with $\nu(d) = r$ odd" is precisely the statement (see (6.15)) that $p(d)^2 d < Y$; when $\nu(d) = r$ is even and we have

$$d = p_1 p_2 \cdots p_r \ (p_1 > \cdots > p_r = p(d)),$$

then $\eta^+(d/p(d)) = 1$ implies that $p_{r-1}^3 p_{r-2} \cdots p_1 < Y$, and consequently that $p(d)d < Y$, since $p_{r-1}^3 > p(d) p_r p_{r-1}$. A parallel argument deals with $\nu = 1$. By (6.17), $\chi^-(d) = 1$ implies that $\eta^-(d) = 1$ when r is even and that $\eta^-(d/p(d)) = 1$ when r is odd; and then, by (6.18), $p(d)^2 d < Y$

when r is even and $p(d)d < Y$ when r is odd. Hence

$$E_{r+1}^{(-)^\nu} - E_r^{(-)^\nu} \leq \sum_{\substack{d|P(z_0,z) \\ \nu(d)=r}} \frac{\omega(d)}{d} \frac{A}{\sigma_1(1)} \frac{\log p(d)}{\log^2 z_0} V(p(d)), \quad r = 1, 2, 3, \ldots,$$

and, by (7.14), (7.15), the last relation and condition $\mathbf{\Omega}(1)$,

$$E_\infty^{(-)^\nu} = E_1^{(-)^\nu} + \sum_{r=1}^\infty \left(E_{r+1}^{(-)^\nu} - E_r^{(-)^\nu} \right)$$

$$\leq \frac{AV(z)}{\sigma_1(1)\log^2 z_0} \left\{ \log z + \sum_{r=1}^\infty \sum_{\substack{d|P(z_0,z) \\ \nu(d)=r}} \frac{\omega(d)}{d} \log p(d) \frac{V(p(d))}{V(z)} \right\}$$

$$\leq \frac{AV(z)}{\sigma_1(1)\log^2 z_0} \left\{ \log z + \sum_{\substack{d|P(z_0,z) \\ d>1}} \frac{\omega(d)}{d} \log z \left(1 + \frac{A}{\log p(d)} \right) \right\}$$

$$\leq \frac{AV(z)\log z}{\sigma_1(1)\log^2 z_0} \left(1 + \frac{A}{\log z_0} \right) \sum_{d|P(z_0,z)} \frac{\omega(d)}{d}$$

$$\leq \frac{AV(z)\log z}{\sigma_1(1)\log^2 z_0} \left(1 + \frac{A}{\log z_0} \right) \prod_{z_0 \leq p < z} \left(1 - \frac{\omega(p)}{p} \right)^{-1}$$

$$\leq \frac{A}{\sigma_1(1)} \left(1 + \frac{A}{\log z_0} \right)^2 \frac{\log^2 z}{\log^3 z_0} V(z). \qquad \square$$

7.4 Completion of the proof of Theorem 7.1

By (6.19), (7.6), and (7.7), and then by the case $\nu = 0$ of Proposition 7.3, we get

$$S(\mathcal{A},\mathcal{P},z) \leq S_1(\chi^+) \leq XV(z_0)\Sigma^+ + O\left\{ \left(\frac{\log\log y}{\log y} \right)^{1/2} XV(z) \right\}$$

$$+ \sum_{\substack{m|P(z) \\ m<y}} 4^{\nu(m)} |r_\mathcal{A}(m)|$$

$$\leq XV(z) \left\{ F_1\left(\frac{\log Y}{\log z} \right) + O\left(\frac{\log^2 z}{\log^3 z_0} + \left(\frac{\log\log y}{\log y} \right)^{1/2} \right) \right\}$$

$$+ \sum_{\substack{m|P(z) \\ m<y}} 4^{\nu(m)} |r_\mathcal{A}(m)|$$

for $2 \leq z_0 \leq z \leq Y$.

7.4 Completion of the proof of Theorem 7.1

Similarly, by (6.19), (7.6), (7.8), and the case $\nu = 1$ of Proposition 7.3,

$$S(\mathcal{A}, \mathcal{P}, z) \geq S_1(\chi^-) \geq XV(z)\Sigma^- + O\left\{\left(\frac{\log \log y}{\log y}\right)^{1/2} XV(z)\right\}$$
$$- \sum_{\substack{m \mid P(z) \\ m < y}} 4^{\nu(m)} |r_{\mathcal{A}}(m)|$$
$$\geq XV(z)\left\{f_1\left(\frac{\log Y}{\log z}\right) - O\left(\frac{\log^2 z}{\log^3 z_0} + \left(\frac{\log \log y}{\log y}\right)^{\frac{1}{2}}\right)\right\}$$
$$- \sum_{\substack{m \mid P(z) \\ m < y}} 4^{\nu(m)} |r_{\mathcal{A}}(m)|, \quad 2 \leq z_0 < z \leq Y^{1/2}.$$

The choice (7.3) of z_0 leads to

(7.17) $\quad S(\mathcal{A}, \mathcal{P}, z) \leq XV(z)\left\{F_1\left(\frac{\log Y}{\log z}\right) + O\left(\frac{(\log \log y)^{3/4}}{(\log y)^{1/4}}\right)\right\}$
$$+ \sum_{\substack{m < y \\ m \mid P(z)}} 4^{\nu(m)} |r_{\mathcal{A}}(m)|, \quad 2 \leq z_0 \leq z \leq Y,$$

and

(7.18) $\quad S(\mathcal{A}, \mathcal{P}, z) \geq XV(z)\left\{f_1\left(\frac{\log Y}{\log z}\right) - O\left(\frac{(\log \log y)^{3/4}}{(\log y)^{1/4}}\right)\right\}$
$$- \sum_{\substack{m < y \\ m \mid P(z)}} 4^{\nu(m)} |r_{\mathcal{A}}(m)|, \quad 2 \leq z_0 \leq z \leq Y^{1/2}.$$

In (7.5), we set $\log Y = \log y - ((\log y) \log \log y)^{3/4}$. With this choice,

$$F_1\left(\frac{\log Y}{\log z}\right) - F_1\left(\frac{\log y}{\log z}\right) \quad \text{and} \quad f_1\left(\frac{\log y}{\log z}\right) - f_1\left(\frac{\log Y}{\log z}\right)$$

are each non-negative and small. The non-negativity follows from the monotonicity of F_1 and f_1; and by Lemma 6.2 with $\kappa = 1$ each of these differences is

$$\ll \frac{\log y - \log Y}{\log Y} \ll \frac{(\log)^{3/4}(\log \log y)^{3/4}}{\log y} = \frac{(\log \log y)^{3/4}}{(\log y)^{1/4}}.$$

Thus we may replace $F_1(\log Y/\log z)$ by $F_1(\log y/\log z)$ on the right side of (7.17) and $f_1(\log Y/\log z)$ by $f_1(\log y/\log z)$ on the right side of (7.18).

We are now very close to finishing the proof of Theorem 7.1. The upper bound for $S(\mathcal{A}, \mathcal{P}, z)$ differs from the altered version of (7.17) only in that it is asserted to hold for the range $z \leq y$ in place of $z \leq Y$. But if $Y < z \leq y$,

$$S(\mathcal{A}, \mathcal{P}, z) \leq S(\mathcal{A}, \mathcal{P}, Y) \leq XV(Y)\left\{F_1\left(\frac{\log y}{\log Y}\right) + O\left(\frac{(\log \log y)^{3/4}}{(\log y)^{1/4}}\right)\right\}$$
$$+ \sum_{\substack{m \mid P(Y) \\ m < y}} 4^{\nu(m)} |r_\mathcal{A}(m)|$$

by (7.17). Since $Y < z$, the inequality remains all the more true when $m \mid P(Y)$ is replaced by $m \mid P(z)$. Next, by $\mathbf{\Omega(1)}$ and (7.3)

$$V(Y) = V(z) \prod_{Y \leq p < z} \left(1 - \frac{\omega(p)}{p}\right)^{-1} \leq V(z) \frac{\log z}{\log Y}\left(1 + \frac{A}{\log Y}\right)$$
$$\leq V(z) \frac{\log y}{\log Y}\left(1 + \frac{A}{\log y} \cdot \frac{\log y}{\log Y}\right)$$
$$\leq V(z)(1+\delta)\left(1 + \frac{A}{\log y}(1+\delta)\right), \quad \delta = 2\frac{(\log \log y)^{3/4}}{(\log y)^{1/4}},$$
$$= V(z)\left\{1 + O\left(\frac{(\log \log y)^{3/4}}{(\log y)^{1/4}}\right)\right\}.$$

Finally, since F_1 is decreasing,

$$F_1\left(\frac{\log y}{\log Y}\right) < F_1\left(\frac{\log y}{\log z}\right),$$

and this completes the first part of the theorem.

As for the lower bound of Theorem 7.1, it is true but worthless when $\log y / \log z \leq 2$, for here $f_1 = 0$. Now suppose that $Y^{1/2} < z < y^{1/2}$. By Lemma 6.2,

$$0 \leq f_1\left(\frac{\log y}{\log z}\right) - f_1(2) \ll \frac{\log y}{\log z} - 2 \leq 2\frac{\log y}{\log Y} - 2$$
$$= 2\frac{(\log y - \log Y)}{\log Y} \ll \frac{(\log \log y)^{3/4}}{(\log y)^{1/4}}.$$

It follows that for suitable constants B, B',

$$S(\mathcal{A}, \mathcal{P}, z) \geq 0 \geq XV(z)\left\{f_1(2) - B\frac{(\log \log y)^{3/4}}{(\log y)^{1/4}}\right\} - \sum_{\substack{m < y \\ m \mid P(z)}} 4^{\nu(m)} |r_\mathcal{A}(m)|$$
$$\geq XV(z)\left\{f_1\left(\frac{\log y}{\log z}\right) - \frac{B'(\log \log y)^{3/4}}{(\log y)^{1/4}}\right\} - \sum_{\substack{m < y \\ m \mid P(z)}} 4^{\nu(m)} |r_\mathcal{A}(m)|.$$

This concludes the proof of Theorem 7.1.

7.5 Notes on Chapter 7

Theorem 7.1 was first proved by Jurkat and Richert ([JR65]; see also [HR74], Chapter 8). Their proof used Selberg's upper sieve method (cf. Theorem 5.6), as we shall do in Chapter 9; but then it was combined with many skillfully chosen iterations of Buchstab's identity.

The method of proof here derives from the Rosser–Iwaniec approach in [Iwa80], a seminal memoir in sieve theory. We follow a somewhat different path, however, via the inequalities (7.7) and (7.8). While this costs some precision in the remainder terms it has the merit of bringing the functions F_1 and f_1 into play from the start. (The remainder term in the Rosser–Iwaniec method for $\kappa = 1$ can be given in a highly flexible form to which deeper methods can be applied—character sum estimates, for example.)

Rosser had discovered his sieve method in the late 1930s but published no details; the only evidence in print is to be found in two research announcements, in volumes 43 (p. 173) and 47 (p. 383) of the AMS Bulletin (the latter jointly with W. J. Harrington) and in an article of W. J. LeVeque [Lev49], where a version of the Fundamental Lemma is presented. Iwaniec's work paralleled but was developed independently of Rosser's earlier studies. There is a comprehensive account of the Rosser–Iwaniec linear sieve method in [Grv01].

The upper and lower bounds of Theorem 7.1 are optimal, as shown by the following example of Selberg [Sel91]: for

$$\mathcal{A} = \mathcal{B}_\nu = \{n \colon 1 \leq n \leq x,\ \Omega(n) \equiv \nu \bmod 2\}, \quad \nu = 1, 2,$$

the upper bound of the theorem holds with equality when $\nu = 1$ and the lower bound holds with equality when $\nu = 2$. Apropos of this example, he remarked famously that the sieve method (as enshrined in Theorem 7.1) "cannot distinguish between numbers with an odd or an even number of prime factors." Circumventing this parity "problem" by combining a sieve method with other information is a feature of most recent advances in the application of sieve ideas ([Sel91], p. 204, [Grv01], p. 171.)

We could not prove Lemma 7.2 by appealing to Lemma 1.4, because $V(\cdot)B(\cdot)$ need not be monotonic.

8
An application of the linear sieve

8.1 Toward the twin prime conjecture

Before considering the sieve method for $\kappa > 1$ we pause to demonstrate—and test—the quality of Theorem 7.1. Specifically, we ask: How close can it bring us to the famous twin prime conjecture? We show, as an approximation to the conjecture, that there are an infinite number of pairs $(p, p+2)$, where p is prime and $p+2$ has only a few prime factors.

Let $\mathcal{A} = \{p+2 \colon p \leq x\}$ and take $\mathcal{P}$ to be the set of odd primes. For d a positive integer and a an integer, set

$$\pi(x, d, a) := |\{p \leq x \colon p \equiv a \bmod d\}|.$$

When d is odd and squarefree, we have

$$|\mathcal{A}_d| = \pi(x, d, -2) = \frac{\operatorname{li} x}{\varphi(d)} + r_\mathcal{A}(d),$$

so that $X = \operatorname{li} x$ and $\omega(p) = p/(p-1)$ for $p \mid d$. If we write

$$\mathcal{E}(x, d) := \max_{\substack{1 \leq m \leq d \\ (m,d)=1}} \left| \pi(x, d, m) - \frac{\operatorname{li} x}{\varphi(d)} \right|,$$

then the Bombieri–Vinogradov theorem ([Bom65]; [Dav00], Chapter 28) applies, and states: *given an arbitrary constant $A > 0$ there exists a number $B = B(A) > 0$ (it is known that $B = A + 5$ is a valid choice) such that*

(8.1) $$\sum_{d \leq x^{1/2} (\log x)^{-B}} \mathcal{E}(x, d) \ll \frac{x}{(\log x)^A}.$$

We want the following variant upon this estimate:

Lemma 8.1. *With $\mathcal{E}$, $\mathcal{A}$, and B as above and for any positive constant K, we have*

$$\sum_{d \leq x^{1/2}(\log x)^{-B}} \mu^2(d) K^{\nu(d)} \mathcal{E}(x,d) \ll x(\log x)^{-(A-K^2)/2}.$$

Proof. It is easy to see that $\mathcal{E}(x,d) \ll x/d$. Hence the sum in question, by an application of Cauchy's inequality, is at most of order

$$\sum_{d \leq x^{1/2}(\log x)^{-B}} \mu^2(d) K^{\nu(d)} \left(\frac{x}{d}\right)^{1/2} \mathcal{E}(x,d)^{1/2}$$

$$\leq x^{1/2} \left(\sum_{d \leq x^{1/2}} \mu^2(d) \frac{K^{2\nu(d)}}{d}\right)^{1/2} \left(\sum_{d \leq x^{1/2}(\log x)^{-B}} \mathcal{E}(x,d)\right)^{1/2}$$

$$\ll \frac{x}{(\log x)^{A/2}} \prod_{2 < p \leq x^{1/2}} \left(1 + \frac{K^2}{p}\right)^{1/2}$$

$$\ll \frac{x}{(\log x)^{A/2}} \prod_{2 < p \leq x^{1/2}} \left(1 + \frac{1}{p}\right)^{K^2/2} \ll x(\log x)^{-(A-K^2)/2}. \quad \Box$$

When we choose $K = 4$, $A = 36$, and $B = B(36)$ in the lemma and note that $|r_\mathcal{A}(d)| = |\pi(x,d,-2) - \operatorname{li} x/\varphi(d)| \leq \mathcal{E}(x,d)$, we obtain

$$(8.2) \qquad \sum_{d \leq x^{1/2}(\log x)^{-B}} \mu^2(d) 4^{\nu(d)} |r_\mathcal{A}(d)| \ll \frac{x}{(\log x)^{10}}.$$

Now we apply the *lower bound* estimate of Theorem 7.1 with

$$z = x^{1/(4+\delta)} \quad (0 < \delta < 1), \quad y = x^{1/2}(\log x)^{-B(36)},$$

and we obtain

$$|\{p+2 : p \leq x, (p+2, P(x^{1/(4+\delta)})) = 1\}| = S(\mathcal{A}, \mathcal{P}, x^{1/(4+\delta)})$$

$$\geq \operatorname{li} x \prod_{2 < p < x^{1/(4+\delta)}} \left(1 - \frac{1}{p-1}\right) \left\{f_1\left(2 + \frac{\delta}{2}\right) + O\left(\frac{(\log \log x)^{3/4}}{(\log x)^{1/4}}\right)\right\}$$

$$- O\left(\frac{x}{(\log x)^{10}}\right);$$

noting that we have used Lemma 6.2 to show that

$$0 \leq f_1(2 + \delta/2) - f_1((\log y)/\log z) \ll (\log \log x)/\log x.$$

8.1 Toward the twin prime conjecture

But

$$\prod_{2<p<z}\left(1-\frac{1}{p-1}\right) = \prod_{2<p<z}\left(1-\frac{1}{p}\right)\left(1-\frac{1}{(p-1)^2}\right)$$

$$> H\prod_{p<z}\left(1-\frac{1}{p}\right) = H\frac{e^{-\gamma}}{\log z}\left(1+O\left(\frac{1}{\log z}\right)\right),$$

where

$$H := 2\prod_{p>2}\left(1-(p-1)^{-2}\right).$$

Also, since $2 + \delta/2 < 3$, it follows from (6.4) and (6.1) that

$$\left(2+\frac{\delta}{2}\right)f_1\left(2+\frac{\delta}{2}\right) = \left(2+\frac{\delta}{2}\right)f_1\left(2+\frac{\delta}{2}\right) - 2f_1(2) = \int_2^{2+\delta/2} F_1(t-1)dt$$

$$= \int_2^{2+\delta/2} \frac{2e^\gamma}{t-1}dt = 2e^\gamma \log\left(1+\frac{\delta}{2}\right)$$

and therefore

$$f_1(2+\delta/2) = e^\gamma \frac{\log(1+\delta/2)}{1+\delta/4} > \frac{1}{4}e^\gamma \delta.$$

The last inequality holds for $0 < \delta < 1$, since

$$\log\left(1+\frac{\delta}{2}\right) > \frac{\delta}{2} - \frac{1}{2}\left(\frac{\delta}{2}\right)^2 = \frac{\delta}{2}\left(1-\frac{\delta}{4}\right) > \frac{\delta}{4}\left(1+\frac{\delta}{4}\right).$$

Hence

$$\left|\{p+2: p \leq x, (p+2, P(x^{1/(4+\delta)}))=1\}\right| > (4+\delta)H\left(\frac{\delta}{4}+o(1)\right)\frac{x}{(\log x)^2}$$

$$> \delta H\frac{x}{\log^2 x}, \quad x \geq x_0.$$

This means that any number $p+2$ counted on the left is the product of at most *four* prime factors; for a product of five or more primes, each at least as large as $x^{1/(4+\delta)}$, would exceed or equal $x^{5/(4+\delta)} > x$, by our choice of δ. Writing P_r for any positive integer that is the product of at most r primes, we have proved that

(8.3) there are at least $\delta H x(\log x)^{-2}$ primes $p \leq x$ with $p+2$ a P_4.

The reader may well find the preceding result rather disappointing; given the effort expended in proving Theorem 7.1, one might have expected a better approximation to the twin prime problem to emerge from it. We note here how we could do better *if* (8.1), *the Bombieri–Vinogradov theorem, is valid over a longer range of d, say over* $d \le x^\tau$ *for some* $\tau > 1/2$.

For example, assume (8.1) holds with $\tau = 1/2+\epsilon$ for some $\epsilon \in (0, 1/6)$, and we wish to use this estimate to improve upon (8.3). From the preceding argument, we see that a reasonable choice of parameters is

$$z = x^{1/(3+\delta)} \quad (\delta \in (0,1)), \quad y = x^\tau.$$

With $\delta = 1 - 6\epsilon \in (0, 1)$, we have $(3+\delta)\tau = 2 + \delta\epsilon$, and by (6.6)

$$f_1\left(\frac{\log y}{\log z}\right) = f_1(2+\delta\epsilon) = 2e^\gamma \frac{\log(1+\delta\epsilon)}{2+\delta\epsilon} > \delta\epsilon,$$

the last because

$$\log(1+t) > t - \frac{t^2}{2} > \frac{t(2+t)}{2e^\gamma}$$

for $0 < t < 1/2$ (say). This estimate implies that (8.3) holds with P_3 in place of P_4.

There is, however, an ingenious argument that leads to the same conclusion *without* appealing to an unproved conjecture: we shall prove that the number of primes p counted in (8.3) which are such that $p + 2 = p_1 p_2 p_3 p_4$ (with $p_j > x^{1/(4+\delta)}$, $1 \le j \le 4$), is relatively negligible compared with the estimate of (8.3), and hence *there are infinitely many ps such that* $p+2 = P_3$. We begin with the observation that we may as well assume that the four primes p_1, p_2, p_3, p_4 are distinct, for the number of integers $\le x$ that are divisible by the square of a prime of size at least $x^{1/5}$ is at most

$$\sum_{p>x^{1/5}} \left[\frac{x}{p^2}\right] \le x \sum_{p>x^{1/5}} \frac{1}{p^2} \le x \sum_{n>x^{1/5}} \frac{1}{n^2} \ll x^{4/5}.$$

We may assume therefore that we need to count the number of primes $p \le x$ such that

$$p+2 = p_1 p_2 p_3 p_4 \quad (p_1 > p_2 > p_3 > p_4 \ge x^{1/(4+\delta)}).$$

First note that

$$x+2 \ge p+2 > x^{3/(4+\delta)} p_1$$

so that

$$p_1 < 2x^{1-3/(4+\delta)} = 2x^{(1+\delta)/(4+\delta)} < x^{(1+\delta)/4} \quad (x \ge x_0).$$

8.1 Toward the twin prime conjecture

Next, let
$$Q := \{q = p_2 p_3 p_4 < (x+2)x^{-1/(4+\delta)} : x^{1/(4+\delta)} < p_4 < p_3 < p_2 < x^{(1+\delta)/4}\},$$

and, writing p in place of p_1, estimate the quantity

(8.4) $$\sum_{q \in Q} \left|\{p : x^{1/(4+\delta)} < p < (x+2)/q,\ qp - 2 = p',\ \text{a prime}\}\right|$$

from above. We apply the *upper bound* of Theorem 7.1 to each term of this sum, namely to the set

$$\mathcal{B}(q) := \{qp - 2 : x^{1/(4+\delta)} < p < (x+2)/q\}, \quad q \in Q,$$

with $X = \mathrm{li}((x+2)/q)$, $\omega(p) = p/(p-1)$ as before, $y = x^{1/2}(\log x)^{-B(36)}$ (so that the remainder sum can be estimated by (8.2)), $V(z) \ll 1/\log z$, and $z = x^{1/100}$ say. We obtain

$$S(\mathcal{B}(q), \mathcal{P}, x^{1/100}) \ll \frac{x}{q}(\log x)^{-2},$$

where the implied constant is absolute. Hence

(8.5) $$\sum_{q \in Q} S(\mathcal{B}(q), \mathcal{P}, x^{1/100}) \ll \frac{x}{(\log x)^2} \sum_{q \in Q} \frac{1}{q}.$$

But

$$\sum_{q \in Q} \frac{1}{q} \ll \Big(\sum_{x^{1/(4+\delta)} < p < x^{(1+\delta)/4}} \frac{1}{p} \Big)^3 \ll \Big(\log \frac{(1+\delta)/4}{1/(4+\delta)} \Big)^3$$
$$= \Big(\log(1+\delta) + \log(1+\delta/4) \Big)^3 \ll \delta^3,$$

so that the sum in (8.4), and also in (8.5), is at most of order

$$\delta^3 x / \log^2 x,$$

a quantity that is plainly dwarfed by the lower bound (8.3) if δ is sufficiently small. We conclude that $p+2$ is *infinitely often a* P_3.

There is a more complicated but not dissimilar procedure by which the P_3 in this result can be replaced by a P_2, the famous theorem of J. R. Chen [Chn73], the closest approach that has been achieved to date to the twin prime conjecture. We remark that Chen's result would follow in a more straightforward way, using Corollary 11.2 below, if the Bombieri–Vinogradov estimate (8.1) holds for a sum over a range $d \leq x^\tau$ for $\tau = 0.5453$.

8.2 Notes on Chapter 8

The deduction from Theorem 7.1 that $p + 2 = P_3$ infinitely often was shown to one of us a long time ago by R. C. Vaughan. His argument is a suggestive preliminary to Chen's famous theorem [Chn73]. Other accounts of Chen's result are given in [HR74], Chapter 11, and in [Ros75].

9
A sieve method for $\kappa > 1$

9.1 The main theorem and start of the proof

Now we turn to the central result of this monograph. We shall prove Theorem 9.1, the analogue for integer and half integer dimensions $\kappa > 1$ of Theorem 7.1. We have prepared the ground in Chapter 6 for the method to be presented here, and we resume the discussion by recalling from (6.12), (6.9), (6.11), and (6.20) that, with $2 \leq z_0 < z$,

(9.1) $\qquad S_1(\chi^-) - S_{22}(\chi^-) \leq S(\mathcal{A}, \mathcal{P}, z) \leq S_1(\chi^+) + S_{21}(\chi^+),$

where

(9.2) $\qquad S_1(\chi^\pm) = \sum_{d | P(z_0, z)} \mu(d) \chi^\pm(d) S(\mathcal{A}_d, \mathcal{P}, z_0),$

(9.3) $\qquad S_{21}(\chi^+) = \sum_{\substack{d | P(z_0, z) \\ \mu(d) = 1}} \overline{\chi^+}(d) S(\mathcal{A}_d, \mathcal{P}, p(d)),$

(9.4) $\qquad S_{22}(\chi^-) = \sum_{\substack{d | P(z_0, z) \\ \mu(d) = -1}} \overline{\chi^-}(d) S(\mathcal{A}_d, \mathcal{P}, p(d)),$

and

(9.5) $\qquad \chi^\pm(d) = \prod_{j=1}^{r} \eta^\pm(p_1 \cdots p_j), \quad d = p_1 \cdots p_r \quad (p_1 > \cdots > p_r).$

The functions $\eta^\pm(\cdot)$ were defined in (6.21) and (6.22), but we restate these definitions here for convenience of reference. We have

(9.6) $$\eta^+(d) = \begin{cases} 1 & \text{when } \mu(d) = 1 \text{ and } p(d)^{\alpha_\kappa}d < Y, \\ 1 & \text{when } \mu(d) = -1 \text{ and } p(d)^{\beta_\kappa}d < Y, \\ 0 & \text{otherwise,} \end{cases}$$

(9.7) $$\eta^-(d) = \begin{cases} 1 & \text{when } \mu(d) = 1 \text{ and } p(d)^{\beta_\kappa}d < Y, \\ 1 & \text{when } \mu(d) = -1 \text{ and } p(d)^{\alpha_\kappa}d < Y, \\ 0 & \text{otherwise,} \end{cases}$$

and we have to remember that $\overline{\chi^\pm}(1) = 0$ and

(9.8) $$\overline{\chi^\pm}(d) = \chi^\pm\left(\frac{d}{p(d)}\right)\left(1 - \eta^\pm(d)\right), \quad d > 1.$$

We shall estimate $S_1(\chi^\pm)$ by applying Theorem 4.1 (the Fundamental Lemma) to the functions $S(\mathcal{A}_d, \mathcal{P}, z_0)$, just as we did in Chapter 7 when dealing with the linear sieve, and we shall estimate each of $S_{21}(\chi^+)$, $S_{22}(\chi^-)$ by applying Theorem 5.6 (Selberg's upper bound sieve method) to the functions $S(\mathcal{A}_d, \mathcal{P}, p(d))$.

Our objective is to prove

Theorem 9.1. *Suppose that $\kappa \geq 1$ and that 2κ is an integer. If $\Omega(\kappa)$ holds and y is a parameter such that $2 \leq z \leq y$, then we have*

(9.9) $$S(\mathcal{A}, \mathcal{P}, z) \leq XV(z)\left\{F_\kappa\left(\frac{\log y}{\log z}\right) + O\left(\frac{(\log\log y)^2}{(\log y)^{1/(2\kappa+2)}}\right)\right\}$$
$$+ 2\sum_{\substack{m|P(z)\\m<y}} 4^{\nu(m)}|r_\mathcal{A}(m)|,$$

and

(9.10) $$S(\mathcal{A}, \mathcal{P}, z) \geq XV(z)\left\{f_\kappa\left(\frac{\log y}{\log z}\right) - O\left(\frac{(\log\log y)^2}{(\log y)^{1/(2\kappa+2)}}\right)\right\}$$
$$- 2\sum_{\substack{m|P(z)\\m<y}} 4^{\nu(m)}|r_\mathcal{A}(m)|,$$

where F_κ and f_κ are the functions in Theorem 6.1, and the constants implied by the O-notation depend at most on κ and A.

Proof. The case $\kappa = 1$ has been dealt with in Theorem 7.1. The result given there has slightly better error terms. We henceforth assume $\kappa > 1$.

9.1 The main theorem and start of the proof

Next, we point out, as we did in Chapter 7, that the theorem follows from Theorem 4.1 (with z^{2v} replaced by y) when $\log z$ is very small compared with $\log y$. Specifically (see (7.1)), if $\log \log y \geq 2e$ and

$$z \leq \exp\left(\frac{\log y}{\log \log y}\right),$$

then

$$S(\mathcal{A}, \mathcal{P}, z) = XV(z)\left\{1 + O\left(\frac{1}{\log y}\right)\right\} + \theta \sum_{\substack{n \mid P(z) \\ n < y}} 3^{\nu(n)} |r_{\mathcal{A}}(n)|, \quad |\theta| \leq 1.$$

Since $(\log y)/\log z \geq \log \log y$, it follows from (6.5) that

$$F_\kappa\left(\frac{\log y}{\log z}\right) = 1 + O\left(\frac{1}{\log y}\right) = f_\kappa\left(\frac{\log y}{\log z}\right),$$

so that the theorem reduces to Theorem 4.1.

Henceforth we take

(9.11) $$\log z_0 := \left(\frac{\log^{2\kappa+1} y}{\log \log y}\right)^{1/(2\kappa+2)}$$

and assume (until near the end of the argument) that

(9.12) $$z_0 \leq z \leq Y,$$

where, with $L := \log \log y \geq 2e$, we now choose

(9.13) $$Y := y z_0^{-L} = y \exp\left(-\{(\log y) \log \log y\}^{(2\kappa+1)/(2\kappa+2)}\right).$$

We follow as closely as possible the procedure in the case of $\kappa = 1$, especially when dealing with the sums $S_1(\chi^\pm)$, our first task. We apply Theorem 4.1 to each $S(\mathcal{A}_d, \mathcal{P}, z_0)$ on the right side of (9.2), this time with $q = d$, z_0 as in (9.11), and z^{2v} replaced by z_0^L, and obtain

$$\left| S_1(\chi^\pm) - \sum_{d \mid P(z_0, z)} \mu(d) \chi^\pm(d) \frac{\omega(d)}{d} XV(z_0)\left(1 + O\left(\frac{1}{\log y}\right)\right) \right|$$

$$\leq \sum_{\substack{m \mid P(z) \\ m < y}} 4^{\nu(m)} |r_{\mathcal{A}}(m)|.$$

The contribution arising from the O-term in the sum on the left is at most of order

$$\frac{X}{\log y} V(z_0) \sum_{d \mid P(z_0, z)} \frac{\omega(d)}{d} = \frac{X}{\log y} V(z_0) \prod_{z_0 \leq p < z}\left(1 + \frac{\omega(p)}{p}\right).$$

Now

$$V(z_0) \prod_{z_0 \leq p < z} \left(1 + \frac{\omega(p)}{p}\right) \leq V(z_0) \prod_{z_0 \leq p < z} \left(1 - \frac{\omega(p)}{p}\right)^{-1}$$
$$= V(z) \prod_{z_0 \leq p < z} \left(1 - \frac{\omega(p)}{p}\right)^{-2} \ll V(z) \left(\frac{\log z}{\log z_0}\right)^{2\kappa}$$
$$\leq V(z)\big((\log y)(\log \log y)\big)^{\kappa/(\kappa+1)}$$

by $\Omega(\kappa)$, (9.11), and because $z \leq Y < y$; hence

(9.14) $\quad \left| S_1(\chi^\pm) - XV(z_0) \sum_{d \mid P(z_0,z)} \mu(d)\chi^\pm(d)\frac{\omega(d)}{d} \right|$
$$\leq O\left(XV(z)\frac{\log\log y}{(\log y)^{1/(\kappa+1)}}\right) + \sum_{\substack{m \mid P(z) \\ m < y}} 4^{\nu(m)}|r_{\mathcal{A}}(m)|.$$

We are not finished with $S_1(\chi^\pm)$; we write

(9.15) $\quad\quad \phi_\kappa^+(u) := F_\kappa(u) \quad \text{and} \quad \phi_\kappa^-(u) := f_\kappa(u),$

where F_κ and f_κ are the functions defined in Theorem 6.1 and note, since $f_\kappa(u) < 1 < F_\kappa(u)$, that when d is squarefree

$$\mu(d)\phi_\kappa^{(-)^{\nu(d)+1}}(u) < \mu(d) < \mu(d)\phi_\kappa^{(-)^{\nu(d)}}(u), \quad u > 0.$$

On writing

(9.16) $\quad \Sigma_1^{(-)^\nu} := \sum_{d \mid P(z_0,z)} \mu(d)\chi^{(-)^\nu}(d)\frac{\omega(d)}{d}\phi_\kappa^{(-)^{\nu(d)+\nu}}\left(\frac{\log(Y/d)}{\log z_0}\right),$

for $\nu = 0, 1$, we derive from (9.14)

(9.17) $\quad S_1(\chi^+) \leq XV(z_0)\Sigma_1^+ + O(XV(z)\delta_1(y)) + \sum_{\substack{m \mid P(z) \\ m < y}} 4^{\nu(m)}|r_{\mathcal{A}}(m)|$

and

(9.18) $\quad S_1(\chi^-) \geq XV(z_0)\Sigma_1^- - O(XV(z)\delta_1(y)) - \sum_{\substack{m \mid P(z) \\ m < y}} 4^{\nu(m)}|r_{\mathcal{A}}(m)|,$

where we have set

(9.19) $\quad\quad \delta_1(y) := (\log\log y)(\log y)^{-1/(\kappa+1)}.$

9.2 The S_{21} and S_{22} sums

So far we have followed the development in Chapter 7, but now we turn to the sums $S_{21}(\chi^+)$ from (9.3) and $S_{22}(\chi^-)$ from (9.4), which have no counterpart in that chapter. We need to estimate each of these sums from above, and to this end we apply Theorem 5.6 to the expressions $S(\mathcal{A}_d, \mathcal{P}, p(d))$ that appear in both. In that theorem we replace q by d, z by $p(d)$, z^{2u} by Y/d, and u by

$$(9.20) \qquad u_d := \frac{\log(Y/d)}{2\log p(d)}, \quad d > 1,$$

so that $j_\kappa(u_d) = \sigma_\kappa(2u_d)$. In $S_{21}(\chi^+)$ we have $\mu(d) = 1$ in each term. Also $\overline{\chi^+}(d) = 1$ if and only if $\chi^+(d/p(d)) = 1$ and $\eta^+(d) = 0$, by (9.8) and (9.6). We claim that

$$(9.21) \qquad 1 < \beta_\kappa - 1 < 2u_d := \frac{\log(Y/d)}{\log p(d)} \leq \alpha_\kappa.$$

The right-hand inequality is the familiar consequence of $\eta^+(d) = 0$. For the middle inequality, $\chi^+(d/p(d)) = 1$ implies that $\eta^+(d/p(d)) = 1$; writing

$$d = p_1 p_2 \cdots p_r \quad \text{with} \quad p_1 > p_2 > \ldots > p_r$$

and r even, it follows that $d/p(d) = p_1 p_2 \cdots p_{r-1}$ with $r-1$ odd. Thus $\eta^+(d/p(d)) = 1$ implies that

$$(9.22) \qquad p_{r-1}^{\beta+1} \cdots p_1 < Y$$

and consequently

$$p_r^{\beta-1}(p_r p_{r-1} \cdots p_1) < Y,$$

i.e., $p(d)^{\beta-1} d < Y$, which is equivalent to the middle inequality in (9.21). The left-hand inequality in (9.21) is a consequence of the first assertion of Theorem 6.1.

Since u_d is bounded from above and below, we have $u_d^{2\kappa} + u_d^{-\kappa-1} \ll 1$ and, by (9.3),

$$S_{21}(\chi^+) \leq \sum_{\substack{d \mid P(z_0, z) \\ \mu(d)=1}} \overline{\chi^+}(d) \Big\{ \frac{\omega(d)}{d} XV(p(d)) \Big(\frac{1}{\sigma_\kappa(2u_d)} + O\Big(\frac{\log\log p(d)}{\log p(d)}\Big) \Big)$$

$$+ \sum_{\substack{n \mid P(p(d)) \\ n < Y/d}} 3^{\nu(n)} |r_A(dn)| \Big\}.$$

The remainder sums on the right add up to at most
$$\sum_{\substack{m|P(z) \\ m<Y}} 4^{\nu(m)}|r_{\mathcal{A}}(m)|,$$

and the O-terms to an expression of order

$$XV(z) \sum_{d|P(z_0,z)} \frac{1}{\log p(d)} \frac{V(p(d))}{V(z)} \frac{\omega(d)}{d} \ell(p(d)) \ll XV(z) \frac{(\log \log y)^2}{(\log y)^{1/(2\kappa+2)}}$$

after two applications of $\Omega(\kappa)$ and by (9.11), using the simple argument that preceded (9.14) (recall $\ell(w) := \log\{e \log(ew)\}$). Thus, if we write

(9.23) $$\delta_2(y) = \frac{(\log \log y)^2}{(\log y)^{1/(2\kappa+2)}},$$

we obtain

(9.24) $$S_{21}(\chi^+) \leq X \sum_{\substack{d|P(z_0,z) \\ \mu(d)=1}} \overline{\chi^+}(d) \frac{\omega(d)}{d} \frac{V(p(d))}{\sigma_\kappa(2u_d)}$$

$$+ O(XV(z)\delta_2(y)) + \sum_{\substack{m|P(z) \\ m<Y}} 4^{\nu(m)}|r_{\mathcal{A}}(m)|,$$

with $2u_d := \log(Y/d)/\log p(d)$, as given in (9.20).

We turn to $S_{22}(\chi^-)$, defined in (9.4), and proceed in much the same way on the basis of Theorem 5.6. In all the terms, $\mu(d) = -1$. Now

$$\overline{\chi^-}(d) = \chi^-(d/p(d))(1 - \eta^-(d)) = 1$$

provided that $\chi^-(d/p(d)) = 1$ and $\eta^-(d) = 0$. By (9.7), $\eta^-(d) = 0$ precisely when $p(d)^{\alpha_\kappa} d \geq Y$, i.e., when $2u_d \leq \alpha_\kappa$. For a lower estimate on u_d we distinguish the cases $\nu(d) > 1$ and $\nu(d) = 1$. For $\nu(d) = r > 1$, write as usual $d = p_1 p_2 \cdots p_r$ with $p_1 > p_2 > \ldots > p_r = p(d)$. The condition $\chi^-(d/p(d)) = 1$ implies that $\eta^-(d/p_r) = 1$ and hence, by (9.7) again, that $p_{r-1}^{\beta_\kappa}(d/p_r) < Y$. It follows that $p_r^{\beta_\kappa - 1} d < Y$, i.e., that $2u_d > \beta_\kappa - 1$. For $\nu(d) = 1$, i.e., $d = p$, the condition $\chi^-(d/p(d)) = 1$ becomes $\chi^-(1) = 1$, which provides no information.

Note, however, the occurrence of $S_{22}(\chi^-)$ in the lower estimate for $S(\mathcal{A}, \mathcal{P}, z)$. To obtain a non-trivial lower bound for the latter quantity, it is necessary to have $f_\kappa(\log y/\log z) > 0$ (see (6.2)), so we henceforth assume that $\log y/\log z > \beta_\kappa$. Thus we have, uniformly for $z_0 \leq p < z$,

$$2u_p = \frac{\log Y}{\log p} - 1 > \frac{\log Y}{\log z} - 1 = \frac{\log y - o(\log y)}{\log z} - 1 > \beta_\kappa - o(1) - 1 \gg 1$$

9.2 The S_{21} and S_{22} sums

since $\beta_\kappa > 2$. It follows that
$$u_d^{2\kappa} + u_d^{-\kappa-1} \ll 1$$
uniformly for $d \mid P(z_0, z)$ if $\overline{\chi^-}(d) = 1$ and $\mu(d) = -1$.

We are now in a position to apply Theorem 5.6 to each term in the sum $S_{22}(\chi^-)$ in the same way we did in $S_{21}(\chi^+)$, and we obtain the inequality

$$(9.25) \quad S_{22}(\chi^-) \leq X \sum_{\substack{d\mid P(z_0,z) \\ \mu(d)=-1}} \overline{\chi^-}(d) \frac{\omega(d)}{d} \frac{V(p(d))}{\sigma_\kappa(2u_d)} + O(XV(z)\delta_2(y))$$

$$+ \sum_{\substack{m\mid P(z) \\ m<Y}} 4^{\nu(m)} |r_\mathcal{A}(m)|.$$

Using the notation

$$(9.26) \quad \Sigma_2^{(-)^\nu} := \sum_{\substack{d\mid P(z_0,z) \\ \mu(d)=(-1)^\nu}} \overline{\chi^{(-)^\nu}}(d) \frac{\omega(d)}{d} \frac{V(p(d))}{\sigma_\kappa(2u_d)}, \quad \nu = 0, 1,$$

and

$$(9.27) \quad R_\mathcal{A}(Y, z) := \sum_{\substack{m\mid P(z) \\ m<Y}} 4^{\nu(m)} |r_\mathcal{A}(m)|,$$

we may restate (9.24) and (9.25) in the form

$$(9.28) \quad S_{21}(\chi^+) \leq X\Sigma_2^+ + O(XV(z)\delta_2(y)) + R_\mathcal{A}(Y, z)$$

and

$$(9.29) \quad S_{22}(\chi^-) \leq X\Sigma_2^- + O(XV(z)\delta_2(y)) + R_\mathcal{A}(Y, z).$$

This is a good point at which to summarize progress. Writing

$$(9.30) \quad \Sigma^+ := V(z_0)\Sigma_1^+ + \Sigma_2^+, \quad \Sigma^- := V(z_0)\Sigma_1^- - \Sigma_2^-,$$

we combine (9.1), (9.17), (9.18), (9.28) and (9.29) to yield

$$(9.31) \quad S(\mathcal{A}, \mathcal{P}, z) \leq X\Sigma^+ + O(XV(z)\delta_2(y)) + 2R_\mathcal{A}(Y, z)$$

and

$$(9.32) \quad S(\mathcal{A}, \mathcal{P}, z) \geq X\Sigma^- - O(XV(z)\delta_2(y)) - 2R_\mathcal{A}(Y, z),$$

the latter under the assumption $z < y^{1/\beta_\kappa}$. Here we have used (9.19) and (9.23) to infer that $\delta_1(y) < \delta_2(y)$.

9.3 Bounds on $\Sigma^{\pm}$

The chief task in proving (9.31) and (9.32) is to bound Σ^+ from above and Σ^- from below (the latter subject to $\log y/\log z > \beta_\kappa$, that is, to $z < y^{1/\beta_\kappa}$), and we shall accomplish this in Proposition 9.2 below. The argument will proceed along the same lines used in Chapter 7, but the details are more complicated.

We recall here the notation introduced in (9.15),

$$\phi^{(-)^r}(\cdot) = \phi_\kappa^{(-)^r}(\cdot) = \begin{cases} F_\kappa(\cdot), & r \text{ even,} \\ f_\kappa(\cdot), & r \text{ odd,} \end{cases}$$

where $(-)^r = 1$ when r is even and -1 when r is odd, and we take

$$D_\kappa^{(-)^r}(x, w, z_0) := (-1)^r \Big\{ V(z_0)\phi^{(-)^r}\Big(\frac{\log x}{\log z_0}\Big) - V(w)\phi^{(-)^r}\Big(\frac{\log x}{\log w}\Big)$$
$$- \sum_{z_0 \le p < w} \frac{\omega(p)}{p} V(p) \phi^{(-)^{r+1}}\Big(\frac{\log(x/p)}{\log p}\Big) \Big\}.$$

Also, we should recall that Proposition 7.3 asserts in (7.13) that

$$(9.33) \qquad D_\kappa^-(x, w, z_0) \le \frac{A}{\sigma(1)} \frac{V(w)}{\log z_0} \Big(\frac{\log w}{\log z_0}\Big)^\kappa,$$

provided $x \ge w^{\beta_\kappa}$, and in (7.15), provided $x \ge w^{\alpha_\kappa}$, that

$$(9.34) \qquad D_\kappa^+(x, w, z_0) \le \frac{A}{\sigma(1)} \frac{V(w)}{\log z_0} \Big(\frac{\log w}{\log z_0}\Big)^\kappa.$$

Proposition 9.2. *Given y, let $Y(<y)$ be defined by (9.13). We have*

$$(9.35) \qquad \Sigma^+ \le V(z)\Big\{ F_\kappa\Big(\frac{\log Y}{\log z}\Big) + O(\delta_2(y)) \Big\}, \quad z_0 < z \le Y^{1/\alpha_\kappa},$$

and

$$(9.36) \qquad \Sigma^- \ge V(z)\Big\{ f_\kappa\Big(\frac{\log Y}{\log z}\Big) - O(\delta_2(y)) \Big\}, \quad z_0 < z < Y^{1/\beta_\kappa}.$$

Our application of this proposition in Theorem 9.1 will need an estimate of the type (9.35) valid for $z_0 < z \le y$; we shall deal with the gap $Y^{1/\alpha_\kappa} < z \le y$ later.

Proof. We make a start with Σ^+, given by (9.30), (9.16), and (9.26) (the last two with $\nu = 0$). Keeping in mind the restriction

$$(9.37) \qquad Y \ge z^{\alpha_\kappa},$$

we introduce the expressions

9.3 Bounds on $\Sigma^{\pm}$

$$\text{(9.38)} \quad E_r^+ := V(z_0) \sum_{\substack{d|P(z_0,z) \\ \nu(d)<r}} \mu(d)\chi^+(d) \frac{\omega(d)}{d} \phi_\kappa^{(-)^{\nu(d)}} \left(\frac{\log(Y/d)}{\log z_0}\right)$$

$$+ \sum_{\substack{d|P(z_0,z) \\ \mu(d)=1,\, \nu(d)<r}} \overline{\chi^+}(d) \frac{\omega(d)}{d} \frac{V(p(d))}{\sigma_\kappa(2u_d)}$$

$$+ (-1)^r \sum_{\substack{d|P(z_0,z) \\ \nu(d)=r}} \chi^+(d) \frac{\omega(d)}{d} V(p(d)) \phi_\kappa^{(-)^r}(2u_d)$$

$$- V(z) F_\kappa\left(\frac{\log Y}{\log z}\right), \quad r = 1, 2, 3, \ldots$$

Our aim is to deal with Σ^+ by means of the representation

$$\text{(9.39)} \quad \Sigma^+ = V(z) F_\kappa\left(\frac{\log Y}{\log z}\right) + \lim_{k \to \infty} E_{2k+1}^+$$

$$= V(z) F_\kappa\left(\frac{\log Y}{\log z}\right) + \lim_{k \to \infty} \left\{E_1^+ + \sum_{s=1}^{k} \left(E_{2s+1}^+ - E_{2s-1}^+\right)\right\}.$$

We observe at the outset that $\lim_{r \to \infty} E_r^+$ exists for each fixed pair z_0, z. For r sufficiently large, each of the first two sums in (9.38) is constant, the third is empty, and the last term in (9.38) is independent of r.

We estimate E_1^+ and the differences $E_{2s+1}^+ - E_{2s-1}^+$ ($s = 1, \ldots, k$) by repeated appeals to (9.33) and (9.34) and the properties of χ^+, as we did in Chapter 7; however, the "new" terms from the sum Σ_2^+ (see (9.28) and (9.26) with $\nu = 0$) introduce an extra degree of complication. The trick is to express E_1^+ and each of the k differences, perhaps after some rearrangement, in terms of the expressions $D_\kappa^\pm(\cdot, \cdot, z_0)$ in such a way that (9.33) or (9.34) applies. Each time it turns out that the properties of χ^+ make such applications possible.

We turn to E_1^+ first; note that the second sum on the right side of (9.38) is empty when $r = 1$. Thus

$$\text{(9.40)} \quad E_1^+ = V(z_0) F_\kappa\left(\frac{\log Y}{\log z_0}\right) - V(z) F_\kappa\left(\frac{\log Y}{\log z}\right)$$

$$- \sum_{z_0 \leq p < z} \chi^+(p) \frac{\omega(p)}{p} V(p) f_\kappa\left(\frac{\log(Y/p)}{\log p}\right).$$

We have $E_1^+ = D_\kappa^+(Y, z, z_0)$ when $\chi^+(p) = \eta^+(p) = 1$, that is, when $p < Y^{1/(\beta_\kappa+1)}$ by (9.6). To apply (9.34) with $x = Y$ and $w = z$, we require that $Y \geq z^{\alpha_\kappa}$, which we have assumed in (9.37); and when $\alpha_\kappa \geq \beta_\kappa + 1$, then, indeed, $Y \geq z^{\beta_\kappa+1}$. Of course, we may apply (9.34) even when $\alpha_\kappa < \beta_\kappa + 1$ provided that $z \le Y^{1/(\beta_\kappa+1)}$; in either case

(9.41) $$E_1^+ \le \frac{A}{\sigma(1)} \frac{V(z_0)}{\log z_0} \left(\frac{\log z}{\log z_0}\right)^\kappa.$$

It remains to deal with the case when $\alpha_\kappa < \beta_\kappa + 1$ and

$$Y^{1/(\beta_\kappa+1)} < z \le Y^{1/\alpha_\kappa}.$$

Here the sum in E_1^+,

$$\sum_{z_0 \le p < z} \chi^+(p) \frac{\omega(p)}{p} V(p) f_\kappa\left(\frac{\log(Y/p)}{\log p}\right),$$

extends only over the primes $p \in [z_0, Y^{1/(\beta_\kappa+1)})$, since $\chi^+(p) = 0$ for $p \ge Y^{1/(\beta_\kappa+1)}$ by (9.6). It follows by a small calculation that

(9.42) $$E_1^+ = V(Y^{1/(\beta_\kappa+1)}) F_\kappa(\beta_\kappa+1) - V(z) F_\kappa\left(\frac{\log Y}{\log z}\right)$$
$$+ D_\kappa^+(Y, Y^{1/(\beta_\kappa+1)}, z_0).$$

The contribution of the first two terms on the right side is, by $\Omega(\kappa)$,

$$V(z)\left\{\frac{V(Y^{1/(\beta_\kappa+1)})}{V(z)} F_\kappa(\beta_\kappa+1) - F_\kappa\left(\frac{\log Y}{\log z}\right)\right\}$$
$$\le V(z)\left\{\left(\frac{\log z}{\log Y}\right)^\kappa (\beta_\kappa+1)^\kappa \left(1 + \frac{A(\beta_\kappa+1)}{\log Y}\right) F_\kappa(\beta_\kappa+1) - F_\kappa\left(\frac{\log Y}{\log z}\right)\right\}$$
$$= V(z)\left(\frac{\log z}{\log Y}\right)^\kappa \left\{(\beta_\kappa+1)^\kappa F_\kappa(\beta_\kappa+1) - \left(\frac{\log Y}{\log z}\right)^\kappa F_\kappa\left(\frac{\log Y}{\log z}\right)\right\}$$
$$+ A(\beta_\kappa+1)^{\kappa+1} F_\kappa(\beta_\kappa+1) \frac{V(z)}{\log Y} \left(\frac{\log z}{\log Y}\right)^\kappa.$$

Since here $\alpha_\kappa < (\log Y)/\log z < \beta_\kappa + 1$, the expression in the last set of curly brackets is 0 by (6.3) and (6.2). In the last expression on the right, $\log Y > (\beta_\kappa + 1) \log z_0$ for large y by (9.11) and (9.13), and therefore that expression is at most

$$A F_\kappa(\beta_\kappa+1) \frac{V(z)}{\log z_0} \left(\frac{\log z}{\log z_0}\right)^\kappa.$$

We can estimate the last term of (9.42) by using (9.34) with $x = Y$ and $w = Y^{1/(\beta_\kappa+1)}$, because $Y > (Y^{1/(\beta_\kappa+1)})^{\alpha_\kappa}$. Then, making another application of $\Omega(\kappa)$, we obtain

$$D_\kappa^+(Y, Y^{1/(\beta_\kappa+1)}, z_0) \leq \frac{A}{\sigma(1)} \frac{V(Y^{1/(\beta_\kappa+1)})}{\log z_0} \left(\frac{\log Y^{1/(\beta_\kappa+1)}}{\log z_0}\right)^\kappa$$

$$\leq \frac{A}{\sigma_\kappa(1)}\left(1 + \frac{A}{\log z_0}\right) \frac{V(z)}{\log z_0} \left(\frac{\log z}{\log z_0}\right)^\kappa.$$

Combining these estimates in (9.42), we find that

$$E_1^+ \leq \left\{\frac{A}{\sigma_\kappa(1)}\left(1 + \frac{A}{\log z_0}\right) + AF_\kappa(\beta_\kappa + 1)\right\} \frac{V(z)}{\log z_0} \left(\frac{\log z}{\log z_0}\right)^\kappa$$

when $\alpha_\kappa < \beta_\kappa + 1$ and $Y^{1/(\beta_\kappa+1)} < z < Y^{1/\alpha_\kappa}$. In these circumstances

$$F_\kappa(\beta_\kappa + 1) < F_\kappa(\alpha_\kappa) < F_\kappa(1) = 1/\sigma_\kappa(1)$$

by Theorem 6.1, so that in this case

$$E_1^+ \leq \frac{A(2+A)}{\sigma_\kappa(1)} \frac{V(z)}{\log z_0} \left(\frac{\log z}{\log z_0}\right)^\kappa.$$

This estimate and (9.41) along with the assumption that $A > 1$ in $\Omega(\kappa)$ lets us conclude that in any case

(9.43) $$E_1^+ \leq \frac{3A^2}{\sigma_\kappa(1)} \frac{V(z)}{\log z_0} \left(\frac{\log z}{\log z_0}\right)^\kappa, \quad Y \geq z^{\alpha_\kappa}.$$

Recalling that equation (9.38) defines E_r^+, we consider next, for any integer $s \geq 1$, the difference

(9.44) $$E_{2s+1}^+ - E_{2s-1}^+ = V(z_0) \sum_{\substack{d \mid P(z_0, z) \\ \nu(d) = 2s}} \chi^+(d) \frac{\omega(d)}{d} F_\kappa\left(\frac{\log(Y/d)}{\log z_0}\right)$$

$$- V(z_0) \sum_{\substack{d \mid P(z_0, z) \\ \nu(d) = 2s-1}} \chi^+(d) \frac{\omega(d)}{d} f_\kappa\left(\frac{\log(Y/d)}{\log z_0}\right)$$

$$+ \sum_{\substack{d \mid P(z_0, z) \\ \nu(d) = 2s}} \overline{\chi^+}(d) \frac{\omega(d)}{d} \frac{V(p(d))}{\sigma_\kappa(2u_d)}$$

$$- \sum_{\substack{d \mid P(z_0, z) \\ \nu(d) = 2s+1}} \chi^+(d) \frac{\omega(d)}{d} V(p(d)) f_\kappa(2u_d)$$

$$+ \sum_{\substack{d = P(z_0, z) \\ \nu(d) = 2s-1}} \chi^+(d) \frac{\omega(d)}{d} V(p(d)) f_\kappa(2u_d)$$

$$=: H_1 - H_2 + H_3 - H_4 + H_5.$$

Our principal tools in the sequel are formulas (9.33) and (9.34), and to be in a position to apply them we introduce the sum

$$(9.45) \qquad H_6 := \sum_{\substack{d \mid P(z_0, z) \\ \nu(d) = 2s}} \chi^+(d) \frac{\omega(d)}{d} V(p(d)) F_\kappa(2u_d).$$

We must remember, here and below, that $2u_d = (\log(Y/d))/\log p(d)$.

With the aid of H_6 we rewrite (9.44) in the form

$$(9.46) \qquad E^+_{2s+1} - E^+_{2s-1} = (H_1 - H_6 - H_4) \\ + (H_5 - H_2 + H_6 + H_3).$$

Then

$$H_1 - H_6 - H_4$$
$$= \sum_{\substack{d \mid P(z_0, z) \\ \nu(d) = 2s}} \chi^+(d) \frac{\omega(d)}{d} \left\{ V(z_0) F_\kappa\left(\frac{\log(Y/d)}{\log z_0}\right) - V(p(d)) F_\kappa\left(\frac{\log(Y/d)}{\log p(d)}\right) \right.$$
$$\left. - \sum_{z_0 \le p < p(d)} \eta^+(dp) \frac{\omega(p)}{p} V(p) f_\kappa\left(\frac{\log(Y/pd)}{\log p}\right) \right\}$$
$$= \sum_{\substack{d \mid P(z_0, z) \\ \nu(d) = 2s}} \frac{\omega(d)}{d} D^+_\kappa\left(\frac{Y}{d}, p(d), z_0\right)$$

provided that $\chi^+(d) = 1$ in each term, and also when $\eta^+(dp) = 1$ in the inner sum. We can assume that these conditions hold. Indeed, since $\mu(dp) = -1$, $\eta^+(dp) = 1$ when $p^{\beta_\kappa+1}d < Y$ and is otherwise 0, by (9.6); but when $p^{\beta_\kappa+1}d \ge Y$, it follows that $u := (\log(Y/dp))/\log p \le \beta_\kappa$ and then that $f_\kappa(u) = 0$, by equation (6.2). Thus we may take $\eta^+(dp) = 1$ throughout the inner sum. Also, since $\chi^+(d) = 1$ and $\mu(d) = 1$ together imply that $Y/d > p(d)^{\alpha_\kappa}$, (9.34) with $x = Y/d$ and $w = p(d)$ applies in each term, and we arrive at

$$(9.47) \qquad H_1 - H_6 - H_4 \le \frac{A}{\sigma_\kappa(1)} \sum_{\substack{d \mid P(z_0, z) \\ \nu(d) = 2s}} \frac{\omega(d)}{d} \frac{V(p(d))}{\log z_0} \left(\frac{\log p(d)}{\log z_0}\right)^\kappa.$$

9.3 Bounds on $\Sigma^{\pm}$

Similarly, after applying (9.8) to H_3,

$$H_5 - H_2 + H_6 + H_3 = \sum_{\substack{d \mid P(z_0, z) \\ \nu(d) = 2s-1}} \chi^+(d) \frac{\omega(d)}{d} \Big\{ V(p(d)) f_\kappa \left(\frac{\log(Y/d)}{\log p(d)} \right)$$

$$- V(z_0) f_\kappa \left(\frac{\log(Y/d)}{\log z_0} \right)$$

$$+ \sum_{z_0 \leq p < p(d)} \left(\eta^+(dp) \frac{\omega(p)}{p} V(p) F_\kappa \left(\frac{\log(Y/dp)}{\log p} \right) \right.$$

$$\left. + (1 - \eta^+(dp)) \frac{\omega(p)}{p} \frac{V(p)}{\sigma_\kappa(\log(Y/dp)/\log p)} \right) \Big\}.$$

In this awkward-looking sum we may assume that $\chi^+(d) = 1$ and $\mu(d) = -1$ in each term, and these imply by (9.6) that $Y/d > p(d)^{\beta_\kappa}$, an inequality that puts at our disposal (9.33) with $x = Y/d$ and $w = p(d)$. This is precisely what we need, for, despite appearances, the expression within parentheses is in fact $D_\kappa^-(Y/d, p(d), z_0)$. This relation is clear when $\eta^+(pd) = 1$; when $\eta^+(pd) = 0$, since $\mu(dp) = 1$, $Y/(dp) \leq p^{\alpha_\kappa}$ by (9.6) and

$$u = (\log(Y/dp)/\log p \leq \alpha_\kappa$$

tells us that, by (6.1), $(1/\sigma_\kappa)(u) = F_\kappa(u)$. Hence

$$(9.48) \quad H_5 - H_2 + H_6 + H_3 \leq \frac{A}{\sigma_\kappa(1)} \sum_{\substack{d \mid P(z_0, z) \\ \nu(d) = 2s-1}} \frac{\omega(d)}{d} \frac{V(p(d))}{\log z_0} \left(\frac{\log p(d)}{\log z_0} \right)^\kappa.$$

By the combination of (9.44), (9.47), and (9.48), we arrive at

$$E_{2s+1}^+ - E_{2s-1}^+ \leq \frac{A}{\sigma_\kappa(1)} \sum_{\substack{d \mid P(z_0, z) \\ \nu(d) = 2s, 2s-1}} \frac{\omega(d)}{d} \frac{V(p(d))}{\log z_0} \left(\frac{\log p(d)}{\log z_0} \right)^\kappa,$$

and hence, on summing over s,

$$E_{2s+1}^+ - E_1^+ \leq \frac{A}{\sigma_\kappa(1) \log z_0} \sum_{\substack{d \mid P(z_0, z) \\ 0 < \nu(d) \leq 2s}} \frac{\omega(d)}{d} V(p(d)) \left(\frac{\log p(d)}{\log z_0} \right)^\kappa,$$

or, by (9.43),

$$E_{2s+1}^+ \leq \frac{3A^2}{\sigma_\kappa(1) \log z_0} \frac{V(z)}{\log z_0} \left(\frac{\log z}{\log z_0} \right)^\kappa \Big\{ 1 + \sum_{\substack{d \mid P(z_0, z) \\ 0 < \nu(d) \leq 2s}} \frac{\omega(d)}{d} \frac{V(p(d))}{V(z)} \left(\frac{\log p(d)}{\log z} \right)^\kappa \Big\}$$

for $Y \geq z^{\alpha_\kappa}$. By $\Omega(\kappa)$, the last sum is at most

$$\sum_{d|P(z_0,z)} \frac{\omega(d)}{d}\left(1 + \frac{A}{\log p(d)}\right) \leq \left(1 + \frac{A}{\log z_0}\right) \prod_{z_0 \leq p < z}\left(1 + \frac{\omega(p)}{p}\right)$$

$$\leq \left(1 + \frac{A}{\log z_0}\right) \prod_{z_0 \leq p < z}\left(1 - \frac{\omega(p)}{p}\right)^{-1} \leq \left(1 + \frac{A}{\log z_0}\right)^2 \left(\frac{\log z}{\log z_0}\right)^\kappa$$

after a second application of $\Omega(\kappa)$. Hence

$$\lim_{s \to \infty} E_{2s+1}^+ \leq \frac{3A^2}{\sigma_\kappa(1)}\left(1 + \frac{A}{\log z_0}\right)^2 \frac{V(z)}{\log z_0}\left(\frac{\log z}{\log z_0}\right)^{2\kappa},$$

and therefore, by (9.39), since $z \leq y$,

$$\Sigma^+ \leq V(z)\left\{F_\kappa\left(\frac{\log Y}{\log z}\right) + O\left(\frac{1}{\log y}\left(\frac{\log y}{\log z_0}\right)^{2\kappa+1}\right)\right\}$$

$$\leq V(z)\left\{F_\kappa\left(\frac{\log Y}{\log z}\right) + O\left(\frac{\log \log y}{(\log y)^{1/(2\kappa+2)}}\right)\right\}$$

by (9.11). This proves the first inequality in Proposition 9.2.

We turn to proving the second part of the proposition and accordingly assume

(9.49) $\qquad z_0 < z \leq Y^{1/\beta_\kappa}, \quad \beta_\kappa > 2.$

We introduce (cf. (9.38)), for $r \geq 1$,

(9.50) $\quad E_r^- := V(z)f_\kappa\left(\frac{\log Y}{\log z}\right)$

$$+ \sum_{\substack{d|P(z_0,z) \\ \mu(d)=-1 \\ \nu(d)<r}} \overline{\chi^-}(d)\frac{\omega(d)}{d}\frac{V(p(d))}{\sigma_\kappa(2u_d)}$$

$$- V(z_0) \sum_{\substack{d|P(z_0,z) \\ \nu(d)<r}} \mu(d)\chi^-(d)\frac{\omega(d)}{d}\phi_\kappa^{(-)^{\nu(d)+1}}\left(\frac{\log Y/d}{\log z_0}\right)$$

$$- (-1)^r \sum_{\substack{d|P(z_0,z) \\ \nu(d)=r}} \chi^-(d)\frac{\omega(d)}{d}V(p(d))\phi_\kappa^{(-)^{r+1}}(2u_d),$$

with u_d as defined by (9.20). By (9.50), with Σ^- as specified above,

(9.51) $\qquad \Sigma^- = V(z)f_\kappa\left(\frac{\log Y}{\log z}\right) - \lim_{r \to \infty} E_r^-,$

where the limit exists by the same reasoning as was applied to E_r^+.

9.3 Bounds on $\Sigma^{\pm}$

We require an *upper* bound for E_r^-. Our procedure follows closely the argument dealing with Σ^+, and a dedicated reader might try to carry it out without looking at the details, which are given below for the sake of completeness.

It is convenient to extend the definition of $\{E_r^-\}$ to include E_0^-, in order to express

$$E_{2k}^- = \sum_{s=0}^{k-1} \left(E_{2s+2}^- - E_{2s}^- \right) + E_0^-.$$

To this end, we recall that $p(1) = \infty$ so that $u_1 = 0$. Several of the terms of E_r^-, given by (9.50), vanish when $r = 0$ because of empty sums or by the convention that $V(\infty) = 0$, and we get

$$E_0^- = V(z) f_\kappa \left(\frac{\log Y}{\log z} \right).$$

We begin by examining the E^- differences. For $s \geq 0$ we have by (9.50) (again using (9.15))

$$E_{2s+2}^- - E_{2s}^- = V(z_0) \sum_{\substack{d \mid P(z_0,z) \\ \nu(d)=2s+1}} \chi^-(d) \frac{\omega(d)}{d} F_\kappa \left(\frac{\log Y/d}{\log z_0} \right)$$

$$- V(z_0) \sum_{\substack{d \mid P(z_0,z) \\ \nu(d)=2s}} \chi^-(d) \frac{\omega(d)}{d} f_\kappa \left(\frac{\log Y/d}{\log z_0} \right)$$

$$+ \sum_{\substack{d \mid P(z_0,z) \\ \nu(d)=2s+1}} \overline{\chi^-}(d) \frac{\omega(d)}{d} \frac{V(p(d))}{\sigma_\kappa(2u_d)}$$

$$- \sum_{\substack{d \mid P(z_0,z) \\ \nu(d)=2s+2}} \chi^-(d) \frac{\omega(d)}{d} V(p(d)) f_\kappa(2u_d)$$

$$+ \sum_{\substack{d \mid P(z_0,z) \\ \nu(d)=2s}} \chi^-(d) \frac{\omega(d)}{d} V(p(d)) f_\kappa(2u_d).$$

Let us look at some of the sums that are on the right side of the last equation. By convention, the last sum is 0 for $s = 0$. In the third sum, $\overline{\chi^-}(d) = 1$ only if $\eta^-(d) = 0$ (by (9.8)) and, since $\nu(d)$ is odd here, by (9.7), $p(d)^{\alpha_\kappa} d \geq Y$, i.e.,

$$2u_d = (\log Y/d)/\log p(d) \leq \alpha_\kappa;$$

it follows that $1/\sigma_\kappa(2u_d) = F_\kappa(2u_d)$ in that sum. Still in the third sum, write

$$\overline{\chi^-}(d) = \chi^-(d/p(d)) - \chi^-(d),$$

by (9.8), and that sum then becomes

$$(9.52) \quad \sum_{\substack{d|P(z_0,z) \\ \nu(d)=2s+1}} \chi^-\left(\frac{d}{p(d)}\right)\frac{\omega(d)}{d} V(p(d)) F_\kappa\left(\frac{\log Y/d}{\log p(d)}\right)$$

$$- \sum_{\substack{d|P(z_0,z) \\ \nu(d)=2s+1}} \chi^-(d)\frac{\omega(d)}{d} V(p(d)) F_\kappa\left(\frac{\log Y/d}{\log p(d)}\right).$$

With this change, we can prepare the expression $E^-_{2s+2} - E^-_{2s}$, which now comprises six sums (for $s \geq 1$), for a two-fold application of Proposition 7.3.

We have

$$E^-_{2s+2} - E^-_{2s}$$

$$= \sum_{\substack{d|P(z_0,z) \\ \nu(d)=2s+1}} \chi^-(d)\frac{\omega(d)}{d}\left\{V(z_0)F_\kappa\left(\frac{\log Y/d}{\log z_0}\right) - V(p(d))F_\kappa\left(\frac{\log Y/d}{\log p(d)}\right)\right.$$

$$- \sum_{z_0 \leq p < p(d)} \eta^-(dp)\frac{\omega(p)}{p}V(p)f_\kappa\left(\frac{\log Y/dp}{\log p}\right)\Bigg\}$$

$$+ \sum_{\substack{d|P(z_0,z) \\ \nu(d)=2s}} \chi^-(d)\frac{\omega(d)}{d}\left\{\sum_{z_0 \leq p < p(d)} \frac{\omega(p)}{p}V(p)F_\kappa\left(\frac{\log Y/pd}{\log p}\right)\right.$$

$$- V(z_0)f_\kappa\left(\frac{\log Y/d}{\log z_0}\right) + V(p(d))f_\kappa\left(\frac{\log Y/d}{\log p(d)}\right)\Bigg\}$$

$$=: B_{2s+1} + B_{2s}, \text{ say.}$$

Take B_{2s+1} first. We may suppose that $\chi^-(d) = 1$, $\mu(d) = -1$ in each term, so that, by (9.7), $Y/d \geq p(d)^{\alpha_\kappa}$; this means that (9.34) with $x = Y/d$, $w = p(d)$ is available, and indeed $\eta^-(dp) = 1$ in each term of the inner sum when $\alpha_\kappa \geq \beta_\kappa + 1$, since $Y/d \geq p(d)^{\alpha_\kappa}$ implies each time that $Y/(pd) \geq p^{\beta_\kappa}$, which is precisely the condition for $\eta^-(dp)$ to be equal to 1. So in this case

$$B_{2s+1} = \sum_{\substack{d|P(z_0,z) \\ \nu(d)=2s+1}} \frac{\omega(d)}{d} D^+_\kappa\left(\frac{Y}{d}, p(d), z_0\right).$$

This statement is true also when $\alpha_\kappa < \beta_\kappa + 1$. To see this, note that $\eta^-(dp) = 0$ if $Y/(pd) \leq p^{\beta_\kappa}$ by (9.7), and then

$$f_\kappa(\{\log Y/(pd)\}/\log p) \leq f_\kappa(\beta_\kappa) = 0$$

by equation (6.2). It follows from (9.34) applied in each term that

$$B_{2s+1} \leq \frac{A}{\sigma_\kappa(1)} \frac{V(z)}{\log z_0} \left(\frac{\log z}{\log z_0}\right)^\kappa \sum_{\substack{d|P(z_0,z) \\ \nu(d)=2s+1}} \frac{w(d)}{d} \frac{V(p(d))}{V(z)} \left(\frac{\log p(d)}{\log z}\right)^\kappa$$

$$\leq \frac{A}{\sigma_\kappa(1)} \left(1 + \frac{A}{\log z_0}\right) \frac{V(z)}{\log z_0} \left(\frac{\log z}{\log z_0}\right)^\kappa \sum_{\substack{d|P(z_0,z) \\ \nu(d)=2s+1}} \frac{w(d)}{d}.$$

Next turn to B_{2s}, first with $s \geq 1$. In each term $\chi^-(d) = 1$, $\mu(d) = 1$ imply, by (9.7), that $Y/d \geq p(d)^{\beta_\kappa}$, and therefore, for $s \geq 1$,

$$B_{2s} = \sum_{\substack{d|P(z_0,z) \\ \nu(d)=2s}} \frac{w(d)}{d} D_\kappa^-\left(\frac{Y}{d}, p(d), z_0\right)$$

$$\leq \frac{A}{\sigma_\kappa(1)} \left(1 + \frac{A}{\log z_0}\right) \frac{V(z)}{\log z_0} \left(\frac{\log z}{\log z_0}\right)^\kappa \sum_{\substack{d|P(z_0,z) \\ \nu(d)=2s}} \frac{w(d)}{d}, \quad s \geq 1.$$

B_0 takes the simpler form

$$B_0 = \sum_{z_0 \leq p < z} \frac{w(p)}{p} V(p) F_\kappa\left(\frac{\log Y/p}{\log p}\right) - V(z_0) f_\kappa\left(\frac{\log Y}{\log z_0}\right)$$

$$= D^-(Y, z, z_0) - V(z) f_\kappa\left(\frac{\log Y}{\log z}\right).$$

Applying (9.33) with $x = Y$ and $w = z$, as we may do by (9.49), we obtain

$$B_0 \leq -V(z) f_\kappa\left(\frac{\log Y}{\log z}\right) + \frac{A}{\sigma_\kappa(1)} \frac{V(z)}{\log z_0} \left(\frac{\log z}{\log z_0}\right)^\kappa.$$

Combining the estimates, we find that

$$E_{2k}^- \leq \frac{A}{\sigma_\kappa(1)} \left(1 + \frac{A}{\log z_0}\right) \left(\frac{\log z}{\log z_0}\right)^\kappa \sum_{\substack{d|P(z_0,z) \\ 0 \leq \nu(d) \leq k-1}} \frac{w(d)}{d}.$$

Using $\Omega(\kappa)$, we can bound the sum on the right by

$$\sum_{d|P(z_0,z)} \frac{\omega(d)}{d} = \prod_{z_0 \leq p < z} \left(1 + \frac{\omega(p)}{p}\right)$$

$$\leq \prod_{z_0 \leq p < z} \left(1 - \frac{\omega(p)}{p}\right)^{-1} \leq \left(1 + \frac{A}{\log z_0}\right) \left(\frac{\log z}{\log z_0}\right)^\kappa.$$

Hence

$$\lim_{k \to \infty} E_{2k}^- \leq \frac{A}{\sigma_\kappa(1)} \left(1 + \frac{A}{\log z_0}\right)^2 \left(\frac{\log z}{\log z_0}\right)^\kappa \frac{V(z)}{\log z_0}$$

and, since $z \leq y$, we obtain by (9.51)

$$\Sigma^- \geq V(z)\left\{f_\kappa\left(\frac{\log Y}{\log z}\right) - O\left(\frac{1}{\log y}\right)\left(\frac{\log y}{\log z_0}\right)^{2\kappa+1}\right\}, \quad z_0 \leq z \leq Y^{1/\beta_\kappa}.$$

By (9.11) and (9.23), the O-term on the right is $O(\delta_2(y))$. □

9.4 Completion of the proof of Theorem 9.1

To derive Theorem 9.1 from Proposition 9.2 is now a simple matter. By (9.31), (9.23), and (9.37),

$$S(\mathcal{A}, \mathcal{P}, z) \leq XV(z)\left\{F_\kappa\left(\frac{\log Y}{\log z}\right) + O(\delta_2(y))\right\} + 2R_\mathcal{A}(Y, z)$$

for $z_0 \leq z \leq Y^{1/\alpha_\kappa}$. When $Y^{1/\alpha_\kappa} < z \leq Y$, we have

$$1 \leq (\log Y)/\log z < \alpha_\kappa$$

and therefore, by (6.1),

$$1/\sigma_\kappa(\log Y/\log z) = F_\kappa(\log Y/\log z).$$

Hence Theorem 5.6 applies with $u = \frac{1}{2}(\log Y)/\log z$, $q = 1$ and $\xi^2 = Y$; also, the range $2 \leq z \leq z_0$ was handled at the outset of the proof, and so we arrive at the inequality

(9.53) $\quad S(\mathcal{A}, \mathcal{P}, z) \leq XV(z)\{F_\kappa(\log Y/\log z) + O(\delta_2(y))\} + 2R_\mathcal{A}(Y, z)$

for $2 \leq z \leq Y$.

We are now very close to a proof of (9.9); it remains to show that,

9.4 Completion of the proof of Theorem 9.1

in (9.53), each Y can be replaced by a y. First, by Lemma 6.2 with $u_1 = (\log Y)/\log z$ and $u_2 = (\log y)/\log z$,

$$(9.54) \qquad F_\kappa\left(\frac{\log Y}{\log z}\right) - F_\kappa\left(\frac{\log y}{\log z}\right) \ll \frac{\log(y/Y)}{\log Y} \ll \delta_2(y)$$

by (9.13). Next, suppose that $Y < z \leq y$. Then, trivially, we have $S(\mathcal{A}, \mathcal{P}, z) \leq S(\mathcal{A}, \mathcal{P}, Y)$, and by (9.53) and (9.27),

$$S(\mathcal{A}, \mathcal{P}, Y) \leq XV(Y)\{F_\kappa(1) + O(\delta_2(y))\} + 2R_\mathcal{A}(y, z).$$

Finally, since $F_\kappa(u)$ is decreasing in u,

$$F_\kappa(1) \leq F_\kappa\left(\frac{\log Y}{\log z}\right) \leq F_\kappa\left(\frac{\log y}{\log z}\right) + O(\delta_2(y)), \quad Y < z \leq y,$$

by (9.54), and

$$V(Y) = V(z)\frac{V(Y)}{V(z)} \leq V(z)\left(\frac{\log z}{\log Y}\right)^\kappa\left(1 + \frac{A}{\log Y}\right)$$

$$\leq V(z)\left(\frac{\log y}{\log Y}\right)^\kappa\left(1 + \frac{A}{\log Y}\right) \leq V(z)\{1 + O(\delta_2(y))\}$$

by (9.13). It follows from these computations that the proof of (9.9), the upper bound part of Theorem 9.1, is now complete.

As for the second part of Theorem 9.1, it follows from Lemma 6.2 that in the inequality for $\sum^-$ from Proposition 9.2 that $f_\kappa((\log Y)/\log z)$ on the right may be replaced by $f_\kappa((\log y)/\log z)$ at no greater cost. To deduce (9.36), it remains to deal with the range $Y^{1/\beta_\kappa} \leq z < y^{1/\beta_\kappa}$. However, in this narrow range

$$\beta_\kappa < \frac{\log y}{\log z} < \beta_\kappa \frac{\log y}{\log Y} = \beta_\kappa\left(1 + \frac{\log(y/Y)}{\log Y}\right) = \beta_\kappa + O(\delta_2(y))$$

by (9.13). Since $f_\kappa(\beta_\kappa) = 0$, we deduce from Lemma 6.2 that on this range of values of z, $f_\kappa((\log y)/\log z) = O(\delta_2(y))$. This proves (9.36) of Proposition 9.2 and hence also (9.32), since the range $2 \leq z \leq z_0$ was handled at the outset. The proof of Theorem 9.1 is now complete. □

Later, in Chapter 11, when dealing with weighted sieves, we shall need upper bounds for the functions $S(\mathcal{A}_p, \mathcal{P}, z)$ when these are summed over a range of primes p; accordingly, we state here an extended version of (9.9). We made a similar change in Chapter 4 in passing from the Fundamental Lemma to Lemma 4.2.

Lemma 9.3. *Suppose that $\kappa \geq 1$, that 2κ is an integer, and $\Omega(\kappa)$ holds. Let q be a natural number such that $(q, P(z)) = 1$, and also that $2 \leq z \leq q < y$. Then*

$$S(\mathcal{A}_q, \mathcal{P}, z) \leq \frac{\omega(q)}{q} XV(z) \left\{ F_\kappa\left(\frac{\log(y/q)}{\log z}\right) + O\left(\frac{(\log \log y)^2}{(\log(y/p))^{1/(2\kappa+2)}}\right) \right\}$$
$$+ 2 \sum_{\substack{m|P(z) \\ m<y/q}} 4^{\nu(m)} |r_{\mathcal{A}}(qm)|,$$

where F_κ is the function in Theorem 6.1, and the constants implied by the O-notation depend at most on κ and A.

Inequality (9.10) can be modified for $\mathcal{A}_q$ in the same way, but we shall have no need for this in the sequel.

9.5 Notes on Chapter 9

Our proof of the main result, Theorem 9.1, uses the methodology of Theorem 7.1, but insofar as it rests also on Selberg's Theorem 5.6, it may be said to be the natural extension of [JR65].

We must stress that the validity of the theorem rests crucially on the existence of the functions F_κ, f_κ, and that its applicability relies on knowledge of the parameters α_κ, β_κ and of values taken by $F_\kappa(u)$, $f_\kappa(u)$ at specific values of κ and u (see Chapters 17 and 18). The chief burden of the series of DHR papers was to show that F_κ and f_κ exist for all real $\kappa > 1$.

Brüdern and Fouvry [BrF94], [BrF96] have devised what they refer to as the "vector sieve," which aims *to reduce an integer $\kappa > 1$ problem to applications of the linear sieve*. The details are complicated and the results only a little better than those coming from Theorem 9.1 (see the next two chapters), but the idea goes back to Brun's "pure" sieve (see [FH00], §5): with the notation of Example 1.2 and with P the product of all primes up to z,

$$\sum_{d|(L(n),P)} \mu(d) = \prod_{j=1}^{r} \sum_{d|(a_j n + b_j, P)} \mu(d) \leq \prod_{j=1}^{r} \sum_{d|(a_j n + b_j, P)} \mu(d) \chi^+(d)$$

on the one hand, where $\chi^+(d)$ characterizes the Rosser–Iwaniec *linear*

9.5 Notes on Chapter 9

upper bound method; and on the other,

$$\sum_{d|(L(n),P)} \mu(d) \geq \prod_{j=1}^{r} \sum_{d|(a_j n+b_j,P)} \mu(d)\,\chi^+(d)$$

$$-\sum_{\ell=1}^{r} \left(\sum_{\substack{d|(a_\ell n+b_\ell,P) \\ p(d)=p((a_\ell n+b_\ell,P))}} \overline{\chi^+}(d) \right) \prod_{\substack{j=1 \\ j\neq\ell}}^{r} \sum_{d|(a_j n+b_j,P)} \mu(d)\,\chi^+(d).$$

10
Some applications of Theorem 9.1

10.1 A Mertens-type approximation

The aims of this short chapter are, first, to establish an analogue for algebraic numbers of Mertens' prime number formula and, second, to illustrate the use of Theorem 9.1 by means of a few straightforward applications. In the next chapter we shall describe a more elaborate use of the theorem in the construction of "weighted sieve methods" which lead, in an important respect, to better results.

With the prime g-tuples conjecture described in Example 1.2 in mind, we begin with a more general configuration of which that example is a special case. Let $H(\cdot)$ be a monic polynomial of degree G in a single variable with integer coefficients and let $1 \leq y \leq x$. Consider

$$\mathcal{A} = \{H(n): x - y < n \leq x\}$$

as a sequence to be sifted by the set $\mathcal{P}$ of all primes. Let $\rho(d)$ denote the number of solutions of the congruence

$$H(n) \equiv 0 \bmod d$$

that are incongruent modulo d. We have

$$|\mathcal{A}_d| = |\{n: x-y < n \leq x, H(n) \equiv 0 \bmod d\}| = \rho(d)\left(\frac{y}{d}+\vartheta\right), \quad |\vartheta| \leq 1,$$

so that, in our notation, $X = y$ and $\omega(d) = \rho(d)$, also (see (1.7))

$$r_{\mathcal{A}}(d) = |\mathcal{A}_d| - \frac{\rho(d)}{d}y, \quad |r_{\mathcal{A}}(d)| \leq \rho(d).$$

By the Chinese Remainder Theorem (see [HW79], Theorem 121), ρ is multiplicative. Further, $\rho(p) = p$ or $\rho(p) \leq G$ by Lagrange's theorem on the number of solutions of a polynomial congruence (see [HW79],

Theorem 107). We stipulate that there are no fixed prime divisors of $\{H(n)\colon n \in \mathbb{N}\}$ by requiring that $\rho(p) < p$ for all primes p.

The average behavior of $\rho(p)$ is given (implicitly) by the Prime Ideal Theorem
$$\sum_{p \leq x} \rho(p) \sim g \frac{x}{\log x}, \quad x \to \infty,$$
where g is the number of irreducible components of H over $\mathbb{Z}$. It is easier to establish

Proposition 10.1. *The function $\rho(p)$ satisfies*

(10.1) $$\sum_{p \leq x} \frac{\rho(p)}{p} \log p = g \log x + O_H(1),$$

where the notation O_H indicates that the implicit constant may depend on the polynomial.

Remark 10.2. The preceding two formulas, which were given first by Landau [Lan03] in 1903, tell us that $\rho(p)$ is, on average, equal to g. These relations are analogues of the Prime Number Theorem and of Mertens' formula for $\sum (\log p)/p$ respectively. We make a short digression here to sketch a proof of the second one, which we need for many examples.

Proof of Proposition 10.1. We denote our polynomial by
$$H(z) = z^G + c_1 z^{G-1} + \ldots + c_G = \prod_{i=1}^{g} h_i(z),$$
where the h_i are monic irreducible polynomials over $\mathbb{Z}$. If p does not divide the discriminant of H, then no two of the h_is have a common root, and so
$$\rho(p) = \rho_H(p) = \sum_{i=1}^{g} \rho_{h_i}(p);$$
thus we may as well focus on the case of irreducible H, that is, $g = 1$.

Let θ be an algebraic integer that is a root of $H(z) = 0$, $K = \mathbb{Q}(\theta)$ the extension field of $\mathbb{Q}$ by θ, and $\mathcal{O}_K$ the integral domain of K. For p not dividing the discriminant of H, there is a striking correspondence due to Dedekind–Kummer ([Mrc77], Chapter 3) between the factorization of $H(z)$ into distinct irreducible polynomials modulo p:
$$H(z) \equiv \ell_1(z)^{e_1} \cdots \ell_\nu(z)^{e_\nu} \mod p,$$

10.1 A Mertens-type approximation

and the unique factorization of (p) into of prime ideals:

$$(p) = \mathfrak{p}_1^{e_1} \cdots \mathfrak{p}_\nu^{e_\nu},$$

where $\mathfrak{p}_i = (\ell_i(\theta), p)$. Furthermore, writing k_i for the degree of ℓ_i, we have $N\mathfrak{p}_i = p^{k_i}$. One refers to k_i as the "degree" of $\mathfrak{p}_i$. Thus

$$G = e_1 k_1 + \ldots + e_\nu k_\nu \geq \nu.$$

Note in particular that every prime factors into at most G prime ideals.

From this correspondence we see that the roots of the congruence $H(n) \equiv 0 \bmod p$ derive from the polynomials l_i that are linear or, equivalently, from the prime ideals $\mathfrak{p}_i$ that have degree 1; in other words,

$$\rho_H(p) = |\{i \colon 1 \leq i \leq \nu,\ N\mathfrak{p}_i = p\}|.$$

This identification of $\rho(p)$ suggests that we transfer attention from the sum occurring in (10.1) to

$$M(x) := \sum_{N\mathfrak{p} \leq x} \frac{\log N\mathfrak{p}}{N\mathfrak{p}}.$$

Letting $\sum'$ denote a sum restricted to ideals $\mathfrak{p}$ for which $N\mathfrak{p}$ has no common factors with the discriminant of H, we exclude at most a finite number of primes, and we have first

$$M(x) = \sum_{N\mathfrak{p} \leq x}{}' \frac{\log N\mathfrak{p}}{N\mathfrak{p}} + O(1).$$

Thus

$$M(x) = \sum_{k=1}^{G} \sum_{\substack{p^k \leq x \\ N\mathfrak{p} = p^k}}{}' \frac{\log p^k}{p^k} + O(1) = \sum_{k=1}^{G} k \sum_{p^k \leq x}{}' \rho_H^{(k)}(p) \frac{\log p}{p^k} + O(1),$$

where

$$\rho_H^{(k)}(p) := |\{i \colon 1 \leq i \leq \nu,\ N\mathfrak{p}_i = p^k\}|, \quad \rho_H^{(1)}(p) = \rho(p).$$

Next, the sum over indices $k \geq 2$ is dominated by the convergent series

$$\nu \sum_p \sum_{k=2}^{\infty} \frac{k \log p}{p^k},$$

and thus we have the desired formula

(10.2) $$\sum_{N\mathfrak{p} \leq x} \frac{\log N\mathfrak{p}}{N\mathfrak{p}} = \sum_{p \leq x} \rho(p) \frac{\log p}{p} + O(1).$$

It remains to evaluate $M(x)$, and this Landau accomplished by imitating Mertens' well-known treatment of $\sum p^{-1} \log p$ via the prime decomposition of $[x]!$. Let $I(n)$ denote the number of (integral) ideals $\mathfrak{n}$ with norm equal to n, and introduce the ideal counting function

$$C(x) := \sum_{N\mathfrak{n} \leq x} 1 = \sum_{n \leq x} I(n).$$

By a famous theorem of Weber from 1896 ([Mrc77], Chapter 6),

$$C(x) = \alpha x + O(x^{1-1/G}),$$

where α is a positive constant of the field K. We can regard $C(\cdot)$ as an analogue of the integer counting function $[\,\cdot\,]$.

Also define

$$T(x) := \log \prod_{N\mathfrak{n} \leq x} N\mathfrak{n} = \sum_{N\mathfrak{n} \leq x} \log N\mathfrak{n} = \sum_{n \leq x} I(n) \log n,$$

an analogue of $\log[x]!$. Using summation by parts and Weber's approximation, we obtain

(10.3) $$T(x) = \sum_{n \leq x} \{C(n) - C(n-1)\} \log n$$
$$= C(x) \log x + O(x) = \alpha x \log x + O(x).$$

The prime ideal decomposition

$$\prod_{N\mathfrak{n} \leq x} N\mathfrak{n} = \prod_{N\mathfrak{p} \leq x} \prod_{\ell \geq 1} (N\mathfrak{p})^{C(x/N\mathfrak{p}^\ell)}$$

yields the analogue of Chebyshev's classical identity (called by Landau the Poincaré Identity)

(10.4) $$T(x) = \sum_{N\mathfrak{p} \leq x} \left\{ \sum_{\ell \geq 1} C\left(\frac{x}{N\mathfrak{p}^\ell}\right) \right\} \log N\mathfrak{p}.$$

If we recall that the norm is a completely multiplicative function and estimate the sum over the higher prime powers by

$$O\left(\sum_{N\mathfrak{p} \leq x} \sum_{\ell \geq 2} \frac{x \log N\mathfrak{p}}{(N\mathfrak{p})^\ell} \right) = O\left(x \sum_{N\mathfrak{p} \leq x} \frac{\log N\mathfrak{p}}{(N\mathfrak{p})^2} \right) = O(x),$$

we obtain

$$T(x) = \sum_{N\mathfrak{p} \leq x} C\left(\frac{x}{N\mathfrak{p}}\right) \log N\mathfrak{p} + O(x).$$

Inserting Weber's estimate for C into the last sum, we find that

(10.5) $$T(x) = \sum_{N\mathfrak{p} \leq x} \frac{\alpha x}{N\mathfrak{p}} \log N\mathfrak{p} + \sum_{N\mathfrak{p} \leq x} O\left(\frac{x}{N\mathfrak{p}}\right)^{1-1/G} \log N\mathfrak{p} + O(x).$$

We have noted earlier that each rational prime p is associated with at most G prime ideals $\mathfrak{p}_i$ and that $p \leq N\mathfrak{p}_i$. Thus

$$\sum_{N\mathfrak{p} \leq x} (x/N\mathfrak{p})^{1-1/G} \log N\mathfrak{p} \ll \sum_{p \leq x} (x/p)^{1-1/G} \log p.$$

To estimate the last sum, we set

$$\vartheta(x) := \sum_{p \leq x} \log p$$

and recall Chebyshev's classical bound $\vartheta(x) \ll x$. We find by summation by parts and simple estimates that

$$\sum_{p \leq x} (x/p)^{1-1/G} \log p = \vartheta(x) + \sum_{1}^{x-1} \left\{ \left(\frac{x}{n}\right)^{1-1/G} - \left(\frac{x}{n+1}\right)^{1-1/G} \right\} \vartheta(n)$$

$$\ll x + x^{1-1/G} \sum_{1}^{x-1} n^{1/G-2} \cdot n$$

$$\ll x + x^{1-1/G} \int_{1}^{x} t^{1/G-1} \, dt \ll x.$$

It follows from (10.5) and the preceding estimates that

$$T(x) = \alpha x \sum_{N\mathfrak{p} \leq x} \frac{\log N\mathfrak{p}}{N\mathfrak{p}} + O(x).$$

Combining this formula for $T(x)$ with (10.3) and using (10.2) to convert to $\sum (\log p)/p$, we find that Landau's formula (10.1) holds. □

10.2 The sieve setup and examples

Now we show that $\mathcal{A}$ satisfies $\Omega(g)$. Indeed, (10.1) implies an asymptotic formula for the product on the left side of $\Omega(g)$ by the classical method used to derive Mertens' product formula (4.14), so that the "topping up" procedure described in Chapter 5 is not needed here. Also, it is worth noting that, from above,

$$\rho(d) \leq G^{\nu(d)}, \quad \mu(d) \neq 0.$$

Hence, by Lemma 4.3,

$$\sum_{\substack{m|P(z)\\m<\xi}} 4^{\nu(m)}|r_A(m)| \le \sum_{\substack{m|P(z)\\m<\xi}} 4^{\nu(m)}\rho(m) \le \sum_{\substack{m|P(z)\\m<\xi}} (4G)^{\nu(m)}$$

$$\le \xi V(z)^{-4G} \ll \xi(\log z)^{4G}.$$

By using (10.1), we can obtain even the sharper estimate $\xi(\log \xi)^{4g}$ if this is desired. Also

$$V(z) := \prod_{p<z}\left(1 - \frac{\rho(p)}{p}\right) = \exp\left\{\sum_{p<z} \log(1 - \frac{\rho(p)}{p})\right\}$$

$$\gg \exp\left\{-\sum_{p<z}\frac{\rho(p)}{p}\right\} \ge \exp\left\{-G\sum_{p<z}\frac{1}{p}\right\} \gg (\log z)^{-G};$$

again, by (10.1) or the asymptotic form of $\Omega(z)$, actually

$$V(z) \gg (\log z)^{-g}.$$

Thus, in any case,

$$\sum_{\substack{m|P(z)\\m<\xi}} 4^{\nu(m)}|r_A(m)| \ll \xi V(z)(\log z)^{5G} = o(yV(z))$$

on choosing $\xi = y(\log y)^{-5G-1}$.

We are ready to apply Theorem 9.1 (with ξ here playing the role of "y" there). The problem of estimating $S(\mathcal{A}, \mathcal{P}, z)$ is of dimension g. Let δ_0 be a small positive constant so that $f_g(\beta_g + \delta_0) > 0$. Choosing

$$z := \xi^{1/(\beta_g+\delta_0)} = y^{1/(\beta_g+\delta_0)}(\log y)^{-(5G+1)/(\beta_g+\delta_0)},$$

formula (9.10) of Theorem 9.1 tells us that

(10.6) $\qquad S(\mathcal{A}, \mathcal{P}, z) \ge \{f_g(\beta_g + \delta_0) + o(1)\}\, yV(z).$

For large enough y, the expression in curly brackets is strictly positive. Thus there are

(10.7) $\qquad\qquad\qquad \gg yV(y)$

integers n in the interval $(x - y, x]$ for which all the prime divisors of $H(n)$ are at least as large as z.

Example 10.3. Suppose H is the product of g polynomial factors h_i each irreducible over $\mathbb{Z}$ and of degree h (so that $G = hg$). Thus each $h_i(n)$ ($n \le x$) is at most as large as cx^h for some suitable constant c. For

10.2 The sieve setup and examples

any n counted in (10.7) for which the factor $h_i(n)$ has k prime divisors of size at least z, we have $z^k \ll x^h$ or

$$k \leq h(\beta_g + 2\delta_0).$$

In other words, each polynomial $h_i(n)$ has at most $[h\beta_g]$ prime factors for infinitely many n. When $g = 1$, so that $H(\cdot) = h_1(\cdot)$ and $\beta_1 = 2$, we obtain $H(n) = P_{2h}$ infinitely often. In Chapter 11 we shall obtain a much better result. Take $H(n)$ to be the product $L(n)$ of Example 1.2, so that $G = g$ and $h = 1$. Then there are very many $n \leq x$ for which each of the linear factors of $L(n)$ has at most $\beta_g + 2\delta_0$ prime factors. From Table 17.1 of values of β_κ we see that when $g = 2$, $k \leq 4$; when $g = 3$, $k \leq 6$; when $g = 4$, $k \leq 9$; etc. In general, we may say that, whenever $[\beta_g] < \beta_g$ (which is usually the case), there are (as $x \to \infty$) infinitely many n such that each linear factor of $L(n)$ is a $P_{[\beta_g]}$.

Next, consider the same polynomial $H(\cdot)$, but with prime arguments. We take $\mathcal{A} = \{H(p)\colon p \leq x\}$ and $\mathcal{P}$ as the set of all primes.

Example 10.4. Suppose $H(\cdot)$ is a polynomial with integer coefficients, of degree G and the product of g irreducible components, having no fixed prime divisors. Then

$$(10.8) \quad |\{p\colon p \leq x, (H(p), P(z)) = 1\}| \gg (\text{li}\, x) V(z), \quad z = x^{1/(2\beta_g + 3\delta_0)},$$

for any $\delta_0 > 0$ and all sufficiently large x.

In particular, when $H = L$, so that $G = g$ and the irreducible factors are linear, with the coefficients satisfying the conditions of Example 1.2, there are infinitely many primes p for which the linear factors $a_i p + b_i$, $1 \leq i \leq g$, are simultaneously $P_{[2g]}$'s.

To show this, we take

$$|\mathcal{A}_d| = |\{p\colon p \leq x, H(p) \equiv 0 \bmod d\}|$$

$$= \sum_{\substack{m=1 \\ H(m) \equiv 0 \bmod d}}^{d} |\{p\colon p \leq x, p \equiv m \bmod d\}|,$$

and we note that the sum contains $\rho(d)$ terms. If $(m, d) > 1$, the arithmetical progression $m \bmod d$ contains at most one prime. Hence

$$|\mathcal{A}_d| = \sum_{\substack{m=1,\, (m,d)=1 \\ H(m) \equiv 0 \bmod d}}^{d} \pi(x, d, m) + \vartheta \rho(d), \quad 0 \leq \vartheta \leq 1.$$

The number of terms in the sum is $\rho_1(d)$, the number of solutions of

$$H(m) \equiv 0 \bmod d, \ 1 \leq m \leq d, \ (m,d) = 1,$$

which is also a multiplicative function of d. Now $\rho(p)$ is the number of solutions of $H(m) \equiv 0 \bmod p$, $0 \leq m < p$, and $m = 0$ is a solution if and only if p divides $H(0)$. Hence

$$\rho_1(p) = \begin{cases} \rho(p), & p \nmid H(0), \\ \rho(p) - 1, & p \mid H(0). \end{cases}$$

In either case, $\rho_1(p) \leq \rho(p) \leq G$ and $\rho_1(d) \leq G^{\nu(d)}$ when $\mu(d) \neq 0$. We suppose again, as we did earlier, that $\rho(p) < p$ for every prime p, so that $H(\cdot)$ has no fixed prime divisors, and that H is the product of g irreducible components.

Let

$$\mathcal{E}(x,d) := \max_{\substack{1 \leq m \leq d \\ (m,d)=1}} \left| \pi(x,d,m) - \frac{\operatorname{li} x}{\phi(d)} \right|.$$

Writing

$$r_{\mathcal{A}}(d) = |\mathcal{A}_d| - \frac{\rho_1(d)}{\phi(d)} \operatorname{li} x,$$

we have

$$|r_{\mathcal{A}}(d)| \leq \rho(d)\mathcal{E}(x,d) + \rho(d),$$

so that

$$X = \operatorname{li} x, \quad \omega(d) = d\rho_1(d)/\phi(d)$$

in our estimate of $S(\mathcal{A}, \mathcal{P}, z)$. Note that $\omega(p)/p = \rho_1(p)/(p-1)$. Before we can apply (9.10), we need an upper bound for

$$(10.9) \quad \sum_{\substack{d \mid P(z) \\ d < y}} 4^{\nu(d)} |r_{\mathcal{A}}(d)| \leq \sum_{\substack{d \mid P(z) \\ d < y}} (4G)^{\nu(d)} \mathcal{E}(x,d) + \sum_{\substack{d \mid P(z) \\ d < y}} (4G)^{\nu(d)}.$$

The last sum is at most of order $y(\log z)^{4G}$, by Lemma 4.3. We estimate the first sum on the right side of (10.9) using Lemma 8.1. Choose $y = x^{1/2}(\log x)^{-B}$ for a suitably large value of B and $A = B + 5$ (from [Dav00], Chapter 28). Then

$$\sum_{\substack{d \mid P(z) \\ d < y}} (4G)^{\nu(d)} \mathcal{E}(x,d) \ll x(\log x)^{-(A-16G^2)/2}.$$

10.2 The sieve setup and examples

Recalling that $z \leq x$, we see that the last estimate is of greater order than $y(\log z)^{4G}$, i.e., one estimate in (10.9) dominates the other. Thus

$$\sum_{\substack{d \mid P(z) \\ d < x^{1/2}(\log x)^{-B}}} 4^{\nu(d)} |r_A(d)| \ll x(\log x)^{-A/2+8G^2}.$$

Since

$$V(z) = \prod_{p<z} \left(1 - \frac{\rho_1(p)}{p-1}\right) \geq \prod_{p<z}\left(1 - \frac{\rho(p)}{p-1}\right) \gg \exp\left(-G\sum_{p<z}\frac{1}{p}\right)$$
$$\gg (\log z)^{-G} \geq (\log x)^{-G},$$

we conclude that

$$\sum_{\substack{d \mid P(z) \\ d < x^{1/2}(\log x)^{-B}}} 4^{\nu(d)} |r_A(d)| = o(V(x)\operatorname{li} x)$$

on choosing

$$A = 16G^2 + 2G + 3.$$

Finally, we take δ_0 to be a small positive constant and

$$z = x^{1/(2\beta_g + 3\delta_0)}.$$

Then $f_g(\beta_g + \delta_0) > 0$. Since

$$\frac{\log y}{\log z} = \beta_g + \frac{3}{2}\delta_0 - (2\beta_g + 3\delta_0)c\frac{\log\log x}{\log x} > \beta_g + \delta_0$$

for all large enough x, we deduce (10.8) from (9.10).

In the next chapter we shall augment Theorem 9.1 by some combinatorial devices to improve significantly the results of this chapter.

11
A weighted sieve method

11.1 Introduction and additional conditions

We showed in Examples 10.3 and 10.4 (among other things) that, when $g = 2$, the product $L(n)$ of two linear forms is infinitely often a P_8 and $L(p)$ is infinitely often a P_{16}. We shall describe now a different and quite general approach that leads to improved results of this kind for a broad range of problems. Let $\mathcal{A}$ again denote a finite integer sequence of about X elements and $\mathcal{P}$ a set of primes having the properties (1.3), (1.4), and $\Omega(\kappa)$. The goal of this chapter is to establish Theorem 11.1 (see Section 11.3), which determines a number r for $\mathcal{A}$ such that many elements of $\mathcal{A}$ have at most r prime divisors. Also, we give an estimate of r in terms of the sieving limit β_κ that involves no numerical integration.

In place of $S(\mathcal{A}, \mathcal{P}, z)$, we consider a new type of sifting function, a "weighted" sum

$$(11.1) \qquad W(\mathcal{A}, \mathcal{P}, z, y) := \sum_{\substack{a \in \mathcal{A} \\ (a, P(z)) = 1}} w(a),$$

where the "weight" $w(a)$ is to be constructed in such a way that $w(a) > 0$ only when a has very few prime divisors lying between $z = X^{1/v}$ and a number $y = X^{1/u} > z$. For example,

$$w_0(a) := 1 - \frac{1}{b+1} \sum_{\substack{z \leq p < y \\ p \in \mathcal{P},\, p \mid a}} 1 \quad (b > 0)$$

is the simplest weight of this kind [Kuh54]: if $w_0(a) > 0$, then a (which will be taken to have no prime divisors from $\mathcal{P}$ that are less than z) has at most b prime factors coming from the "interval" $[z, y) \cap \mathcal{P}$. If the largest element of $\mathcal{A}$ is at most a power of X, e.g., $\mathcal{A}$ is the set of values assumed by a polynomial on an interval, then an element a cannot have too many prime factors exceeding y. To show that there are many elements a in $\mathcal{A}$ for which $w_0(a) > 0$, we need to prove that

$$W_0(\mathcal{A}, \mathcal{P}, z, y) := \sum_{\substack{a \in \mathcal{A} \\ (a, P(z))=1}} w_0(a) = S(\mathcal{A}, \mathcal{P}, z) - \frac{1}{b+1} \sum_{z \leq p < y} S(\mathcal{A}_p, \mathcal{P}, z)$$

is large and positive; and, to accomplish this, (9.10) and Lemma 9.3 are at our disposal.

We shall use a smoother, more efficient weight ([AO65], [Ric69]), but the underlying idea is the same. There is scope for the design of superior weight functions, as we indicate in Section 11.7 below.

Before we begin, we introduce several new conditions on $\mathcal{A}$ and $\mathcal{P}$ (in addition to (1.3), (1.4), and $\Omega(\kappa)$ (Definition 1.3)): the first requires all elements of $\mathcal{A}$ to be divisible only by primes of $\mathcal{P}$. (We could allow for a small number of exceptions which could be estimated by an error term.) Next, we require the number of elements of $\mathcal{A}$ that are divisible by the square of a prime in $\mathcal{P} \cap (z, y]$ to be relatively small: we postulate that

(11.2) $\boldsymbol{Q_0}$: $\displaystyle\sum_{\substack{z \leq p < y \\ p \in \mathcal{P}}} |\mathcal{A}_{p^2}| \ll \frac{X \log X}{z} + y, \quad 2 \leq z < y.$

In most cases of interest this condition is satisfied without much trouble.

With applications of Theorem 9.1 in mind, we formulate also a general remainder sum condition that serves as a quantitative measure of "quasi-independence": for some constants $\tau \in (0, 1]$, $A_1 \geq 1$ and $A_2 \geq 2$,

(11.3) $\boldsymbol{R_0}$: $\displaystyle\sum_{d < X^\tau (\log X)^{-A_1}} \mu^2(d) 4^{\nu(d)} |r_\mathcal{A}(d)| \leq A_2 \frac{X}{(\log X)^{\kappa+1}}.$

Finally—and this is something that we have not had to take into account before—we require a measure of the largest element of $\mathcal{A}$: let μ_0 be a positive constant such that

(11.4) $\boldsymbol{M_0}$: $\displaystyle\max_{a \in \mathcal{A}} |a| \leq X^{\tau \mu_0},$

where τ is the constant from (11.3).

We shall repeatedly need the following consequence of $\Omega(\kappa)$:

(11.5) $\qquad\qquad V(X) \gg (\log X)^{-\kappa}.$

11.2 A set of weights

We are now ready to make a start. Introduce the weighted sum

$$(11.6) \quad W(\mathcal{A}, \mathcal{P}, z, y, \lambda) := \sum_{\substack{a \in \mathcal{A} \\ (a, P(z)) = 1}} \left\{ \lambda - \sum_{\substack{p \in \mathcal{P}, p \mid a \\ z \leq p < y}} \left(1 - \frac{\log p}{\log y}\right)\right\},$$

where the positive parameter λ plays a role similar to $b+1$ in the earlier discussion. It is clear from (11.6) that

$$(11.7) \quad W(\mathcal{A}, \mathcal{P}, z, y, \lambda) = \lambda S(\mathcal{A}, \mathcal{P}, z) - \sum_{\substack{z \leq p < y \\ p \in \mathcal{P}}} \left(1 - \frac{\log p}{\log y}\right) S(\mathcal{A}_p, \mathcal{P}, z).$$

We write

$$(11.8) \quad z = X^{1/v}, \ y = X^{1/u} \text{ with } 1/\tau < u < v \text{ and } \beta_\kappa < \tau v,$$

and change notation from $W(\mathcal{A}, \mathcal{P}, z, y, \lambda)$ to $W(\mathcal{A}, \mathcal{P}, v, u, \lambda)$. Thus equation (11.7) now reads

$$(11.9) \quad W(\mathcal{A}, \mathcal{P}, v, u, \lambda) = \lambda S(\mathcal{A}, \mathcal{P}, X^{1/v})$$
$$- \sum_{\substack{X^{1/v} \leq p < X^{1/u} \\ p \in \mathcal{P}}} \left(1 - u\frac{\log p}{\log X}\right) S(\mathcal{A}_p, \mathcal{P}, X^{1/v}).$$

We apply (9.10) of Theorem 9.1 to the first expression on the right with $z = X^{1/v}$ and $y = X^\tau (\log X)^{-A_1}$, the latter condition taken in order to use (11.3). The argument of $f_\kappa(\cdot)$ is

$$\frac{\log\{X^\tau (\log X)^{-A_1}\}}{\log X^{1/v}} = \tau v - A_1 v \frac{\log \log X}{\log X},$$

and by Lemma 6.2, for v a bounded number,

$$f_\kappa\left(\tau v - A_1 v \frac{\log \log X}{\log X}\right) \geq f_\kappa(\tau v) - O\left(\frac{\log \log X}{\log X}\right).$$

If we now apply (11.3) and recall (11.5), we obtain

$$(11.10) \quad S(\mathcal{A}, \mathcal{P}, X^{1/v}) \geq XV(X^{1/v})\{f_\kappa(\tau v) - O(\log \log X/\log X)\}.$$

This is a non-trivial inequality, since $\tau v > \beta_\kappa$ by (11.8).

Next, apply Lemma 9.3 (with $z = X^{1/v}$ and $y = X^\tau (\log X)^{-A_1}$) to each S-function in the sum on the right side of (11.9); we obtain for this sum the upper bound

$$\sum_{X^{1/v}\leq p<X^{1/u}}\left(1-u\frac{\log p}{\log X}\right)\left(\frac{\omega(p)}{p}\right)XV(X^{1/v})\left\{F_\kappa\left(\frac{\log(X^\tau(\log X)^{-A_1}/p)}{(1/v)\log X}\right)\right.$$
$$\left.+O\left(\frac{(\log\log X)^2}{(\log(X^\tau/p))^{1/(2\kappa+2)}}\right)\right\}+\sum_{\substack{m|P(X^{1/v})\\m\leq X^\tau(\log X)^{-A_1}/p}}2\cdot 4^{\nu(m)}|r_A(pm)|.$$

Now $X^\tau/p > X^{\tau-1/u}$ and by (11.8), $\tau - 1/u > 0$, whence by another application of Lemma 6.2, this time to F_κ, the upper bound is at most

$$XV(X^{1/v})\left\{\sum_{X^{1/v}\leq p<X^{1/u}}\left(1-u\frac{\log p}{\log X}\right)\frac{\omega(p)}{p}F_\kappa\left(v\frac{\log(p^{-1}X^\tau)}{\log X}\right)\right.$$
$$\left.+O\left(\sum_{X^{1/v}\leq p<X^{1/u}}\frac{\omega(p)}{p}\frac{(\log\log X)^2}{(\log X)^{1/(2\kappa+2)}}\right)\right\}$$
$$+\sum_{n\leq X^\tau(\log X)^{-A_1}}\mu^2(n)4^{\nu(n)}|r_A(n)|.$$

We now apply (1.7) in the O-term to note that

$$\sum_{X^{1/v}\leq p<X^{1/u}}\frac{\omega(p)}{p}\leq\kappa\log\frac{v}{u}+\frac{Av}{\log X}\ll 1$$

and recall that $V(X^{1/v})\gg(\log X)^{-\kappa}$ by (11.5). Hence, by (11.3) and these last remarks, the sum in (11.9) is at most

(11.11) $\qquad XV(X^{1/v})\times$
$$\left\{\sum_{X^{1/v}\leq p<X^{1/u}}\left(1-u\frac{\log p}{\log X}\right)\frac{\omega(p)}{p}F_\kappa\left(v\frac{\log(p^{-1}X^\tau)}{\log X}\right)+C_1\delta(X)\right\}$$

with $\delta(X) = (\log\log X)^2/(\log X)^{1/(2\kappa+2)}$ and C_1 a suitable positive constant. Our next step is to convert the sum on the right to an integral; before we do that we summarize progress so far by combining (11.9), (11.10), and (11.11) to form the inequality

(11.12) $\quad W(\mathcal{A},\mathcal{P},v,u,\lambda)$
$$\geq XV(X^{1/v})\Big\{\lambda f_\kappa(\tau v)-C_2\delta(X)$$
$$-\sum_{X^{1/v}\leq p<X^{1/u}}\left(1-\frac{u\log p}{\log X}\right)\frac{\omega(p)}{p}F_\kappa\left(v\left(\tau-\frac{\log p}{\log X}\right)\right)\Big\}.$$

11.2 A set of weights

Proceeding, we set

$$D(w_1, w) := \sum_{w_1 \leq p < w} \frac{\omega(p)}{p}\left(1 - \frac{\log p}{\log w}\right), \quad 2 \leq w_1 < w.$$

Since $(1 - (\log t)/\log w)$ is positive, continuous, and decreasing in t for $w_1 \leq t < w$, we deduce from (1.9) of Lemma 1.4 that

(11.13)
$$D(w_1, w) \leq \frac{A}{\log w_1}\left(1 - \frac{\log w_1}{\log w}\right) + \kappa \int_{w_1}^{w}\left(1 - \frac{\log t}{\log w}\right)\frac{dt}{t \log t}$$
$$= \kappa\left\{\log\left(\frac{\log w}{\log w_1}\right) - 1 + \frac{\log w_1}{\log w}\right\} + A\left(\frac{1}{\log w_1} - \frac{1}{\log w}\right).$$

Next, let

$$\Phi(t) := F_\kappa\left(v\left(\tau - \frac{\log t}{\log X}\right)\right), \quad z = X^{1/v} \leq t \leq X^{1/u} = y,$$

and note that $\Phi(t)$ is an increasing function in t on $[z, y]$, since $F_\kappa(s)$ is decreasing in s. By the argument leading to the first part of Lemma 1.4 (with D in place of L and Φ in place of f) and by (11.13), we have

$$\sum_{z \leq p < y}\left(1 - \frac{\log p}{\log y}\right)\frac{\omega(p)}{p}\Phi(p) - \int_z^y \Phi(t)\left\{\kappa\left(\frac{1}{t\log t} - \frac{1}{t\log y}\right) + \frac{A}{t \log^2 t}\right\}dt$$
$$= -\int_z^y \Phi(t)d\left\{D(t, y) - \kappa\left(\log\frac{\log y}{\log t} - 1 + \frac{\log t}{\log y}\right) - A\left(\frac{1}{\log t} - \frac{1}{\log y}\right)\right\}$$
$$= -\Phi(t)\left\{D(t, y) - \kappa\left(\log\frac{\log y}{\log t} - 1 + \frac{\log t}{\log y}\right) - A\left(\frac{1}{\log t} - \frac{1}{\log y}\right)\right\}\Big|_z^y$$
$$+ \int_z^y \left\{D(t, y) - \kappa\left(\log\frac{\log y}{\log t} - 1 + \frac{\log t}{\log y}\right) - A\left(\frac{1}{\log t} - \frac{1}{\log y}\right)\right\}d\Phi(t)$$
$$\leq 0.$$

When we apply this calculation to the sum over p on the right side of (11.12), we obtain in its place the upper bound

$$\int_{X^{1/v}}^{X^{1/u}} F_\kappa\left(v\left(\tau - \frac{\log t}{\log X}\right)\right)\left\{\kappa\left(\frac{1}{\log t} - \frac{u}{\log X}\right) + \frac{A}{\log^2 t}\right\}\frac{dt}{t}$$
$$= \kappa \int_u^v F_\kappa\left(v\left(\tau - \frac{1}{s}\right)\right)\left(1 - \frac{u}{s}\right)\frac{ds}{s} + O(1/\log X)$$

on substituting $X^{1/s}$ for t in the integral and noting for the error term that $F_\kappa(v(\tau - 1/u)) \ll 1$, since $\tau - 1/u$ is a positive number. Hence, by (11.12),

(11.14) $\quad W(\mathcal{A}, \mathcal{P}, v, u, \lambda) \geq XV(X^{1/v}) \times$
$$\left\{ \lambda f_\kappa(\tau v) - \kappa \int_u^v F_\kappa\left(v\left(\tau - \frac{1}{s}\right)\right)\left(1 - \frac{u}{s}\right)\frac{ds}{s} - C_3 \delta(X) \right\}.$$

Remember that λ is a positive number at our disposal, τ is the parameter in the remainder condition (11.3), and u, v are the numbers in (11.8).

11.3 Arithmetic interpretation

It is time to take steps towards the arithmetic interpretation of (11.14). First, the numbers λ, τ, u, v will be chosen so that

$$\lambda f_\kappa(\tau v) - \kappa \int_u^v F_\kappa\left(v\left(\tau - \frac{1}{s}\right)\right)\left(1 - \frac{u}{s}\right)\frac{ds}{s} > 0.$$

Having ensured that the right side of (11.14) is positive (for sufficiently large X), we turn to supplying a useful upper bound for $W(\mathcal{A}, \mathcal{P}, v, u, \lambda)$. Recall that any a counted in W has no prime divisors from $\mathcal{P}$ that are less than $X^{1/v}$. (For X so large that $X > 2^v$, any element 0 in $\mathcal{A}$ is not counted by W—for if 0 lies in $\mathcal{A}$, then g.c.d. $(0, P(X^{1/v})) \geq 2 > 1$.)

Next, by (11.2) (the $\mathcal{Q}_0$ condition), the number of elements of $\mathcal{A}$ that are divisible by p^2 for some p in $[X^{1/v}, X^{1/u}) \cap \mathcal{P}$ is negligible, since

$$\sum_{\substack{X^{1/v} \leq p < X^{1/u} \\ p \in \mathcal{P}}} |\mathcal{A}_{p^2}| \ll X^{1-1/v} \log X + X^{1/u} \ll XV(X^{1/v})/\log X$$

by (11.5) and the condition $u > 1$ from (11.8). Henceforth we consider

$$\mathcal{A}' = \mathcal{A} \setminus \bigcup_{\substack{X^{1/v} \leq p < X^{1/u} \\ p \in \mathcal{P}}} \mathcal{A}_{p^2}$$

and absorb the discarded elements of $\mathcal{A}$ into our error estimates.

Let a' denote a generic element of $\mathcal{A}'$ with $(a', P(X^{1/v})) = 1$ and recall that $(a', \mathcal{P}^c) = 1$, where $\mathcal{P}^c$ denotes the primes not in $\mathcal{P}$. If a' contains a repeated prime factor p, say, then $p \geq X^{1/u}$, that is, $1 - u(\log p)/\log X \leq 0$. It follows that

$$\sum_{\substack{X^{1/v} \leq p < X^{1/u} \\ p \in \mathcal{P},\, p | a'}} \left(1 - \frac{u \log p}{\log X}\right) \geq \sum_{\substack{p \geq X^{1/v} \\ p \in \mathcal{P},\, p | a'}}^{*} \left(1 - \frac{u \log p}{\log X}\right) = \Omega(a') - \frac{u \log |a'|}{\log X},$$

11.3 Arithmetic interpretation

where Σ^* denotes summation with appropriate multiplicity. Thus

$$W(\mathcal{A}', \mathcal{P}, v, u, \lambda) \leq \sum_{\substack{a' \in \mathcal{A}' \\ (a', P(X^{1/v}))=1}} \left(\lambda - \left\{\Omega(a') - \frac{u \log |a'|}{\log X}\right\}\right)$$

$$\leq \sum_{\substack{a' \in \mathcal{A}' \\ (a', P(X^{1/v}))=1}} (\lambda - \Omega(a') + \tau\mu_0 u)$$

by (11.4). Let r be a natural number such that $r+1 > \tau\mu_0 u$ and choose

(11.15) $$\lambda := r + 1 - \tau\mu_0 u.$$

Now

$$W(\mathcal{A}', \mathcal{P}, v, u, \lambda) \leq \sum_{\substack{a' \in \mathcal{A}' \\ (a', P(X^{1/v}))=1}} \{r + 1 - \Omega(a')\}.$$

If $\Omega(a') \geq r+1$, then the weight for a' in the last sum is negative. Thus

$$W(\mathcal{A}', \mathcal{P}, v, u, \lambda) \leq \sum_{\substack{a' \in \mathcal{A}', \, \Omega(a') \leq r \\ (a', P(X^{1/v}))=1}} \{r + 1 - \Omega(a')\} \leq \sum_{\substack{a' \in \mathcal{A}', \, \Omega(a') \leq r \\ (a', P(X^{1/v}))=1}} \{r + 1\}.$$

With P_r denoting (as usual) a number having at most r prime factors, it follows that

$$|\{P_r \in \mathcal{A}\}| \geq |\{P_r \in \mathcal{A}' : p(P_r) \geq X^{1/v}\}|$$

$$= \sum_{\substack{a' \in \mathcal{A}', \, \Omega(a') \leq r \\ (a', P(X^{1/v}))=1}} 1 \geq \frac{1}{r+1} W(\mathcal{A}', \mathcal{P}, v, u, \lambda)$$

$$\geq \frac{1}{r+1} W(\mathcal{A}, \mathcal{P}, v, u, \lambda) - C_4 XV(X^{1/v})/\log X.$$

Hence, by (11.14),

$$|\{P_r \in \mathcal{A}\}| \geq \frac{XV(X^{1/v})}{r+1} \times$$

$$\left\{\lambda f_\kappa(\tau v) - \kappa \int_u^v F_\kappa\left(v\left(\tau - \frac{1}{s}\right)\right)\left(1 - \frac{u}{s}\right)\frac{ds}{s} - C_5 \delta(X)\right\}$$

$$\gg XV(X^{1/v})$$

for large values of X, provided that

$$\lambda > \frac{\kappa}{f_\kappa(\tau v)} \int_u^v F_\kappa\left(v\left(\tau - \frac{1}{s}\right)\right)\left(1 - \frac{u}{s}\right)\frac{ds}{s}.$$

Thus we have proved

Theorem 11.1. *Given a sieve problem of dimension $\kappa \geq 1$ concerning a finite integer sequence $\mathcal{A}$ and a set $\mathcal{P}$ of primes, suppose that $\mathcal{A}$ and $\mathcal{P}$ possess the properties $\Omega(\kappa)$, $\mathbf{Q_0}$, $\mathbf{R_0}$, $\mathbf{M_0}$ and also are such that no prime from $\mathcal{P}^c$ divides an element of $\mathcal{A}$. Let r be a natural number satisfying*
$$r > N(u, v; \kappa, \mu_0, \tau),$$
where

(11.16) $\quad N(\kappa, \mu_0, \tau; u, v)$
$$:= \tau\mu_0 u - 1 + \frac{\kappa}{f_\kappa(\tau v)} \int_u^v F_\kappa\left(v\left(\tau - \frac{1}{s}\right)\right)\left(1 - \frac{u}{s}\right)\frac{ds}{s},$$

with u and v satisfying $\tau v > \beta_\kappa$ and $1/\tau < u < v$. Then
$$|\{P_r : P_r \in \mathcal{A}\}| \gg XV(X^{1/v}).$$

In particular, $\mathcal{A}$ contains many integers having at most r prime divisors.

When $\kappa = 1$, Theorem 11.1 takes an especially simple and elegant form (Theorem 9.3 of [Ric69]), with many applications to diverse linear sieve problems:

Corollary 11.2. *Suppose that $\mathcal{A}$ and $\mathcal{P}$ are as in Theorem 11.1 with $\kappa = 1$. Let*
$$\Lambda_r := r + 1 - \log\{4/(1 + 3^{-r})\}/\log 3, \quad r \in \mathbb{N},$$

and let δ be a real number satisfying $0 < \delta < 2/3$. Also, let r be the least positive integer satisfying

(11.17) $\qquad\qquad\qquad \mu_0 \leq \Lambda_r - \delta.$

Then $\mathcal{A}$ contains many P_rs, and the smallest prime factor of each such P_r is at least as large as $X^{\tau/4}$.

Proof. We choose
$$v = 4/\tau \quad \text{and} \quad u = (1 + 3^{-r})/\tau.$$

Then, by Theorem 6.1,
$$f_1(\tau v) = f_1(4) = (1/2)e^\gamma \log 3 = 0.97835\ldots,$$

11.3 Arithmetic interpretation

which is already quite close to the limiting value 1. On the other hand, the argument of F_1 in the integral occurring in (11.16) is

$$v(\tau - 1/s) = 4 - 4/(\tau s) \leq 3,$$

and so

$$F_1(v(\tau - 1/s)) = \tau e^\gamma s/(2\tau s - 2),$$

again by Theorem 6.1.

We have

$$N(u, v; \kappa, \mu_0, \tau)$$

$$= \mu_0(1 + 3^{-r}) - 1 + \frac{1}{\log 3} \int_{(1+3^{-r})/\tau}^{4/\tau} \frac{\tau s}{\tau s - 1} \left(1 - \frac{1 + 3^{-r}}{\tau s}\right) \frac{ds}{s}$$

$$\leq (\Lambda_r - \delta)(1 + 3^{-r}) - 1 + \frac{1}{\log 3} \int_{1+3^{-r}}^{4} \frac{1}{s - 1} \left(1 - \frac{1 + 3^{-r}}{s}\right) ds$$

$$= (\Lambda_r - \delta)(1 + 3^{-r}) - 1 + \frac{1}{\log 3} \int_{1+3^{-r}}^{4} \left(\frac{1 + 3^{-r}}{s} - \frac{3^{-r}}{s - 1}\right) ds$$

$$= (1 + 3^{-r})\left(\Lambda_r - \delta + \frac{\log\{4/(1 + 3^{-r})\}}{\log 3}\right) - 1 - (r + 1)3^{-r}$$

$$= (1 + 3^{-r})(r + 1 - \delta) - 1 - (r + 1)3^{-r}$$

$$= r - \delta(1 + 3^{-r}) < r,$$

but we cannot claim more, since $\delta(1 + 3^{-r}) < 8/9 < 1$ for $r \geq 1$. □

Note that

$$\Lambda_r = r - \frac{1}{\log 3} \log \frac{4/3}{1 + 3^{-r}};$$

thus $\Lambda_1 = 1$, and for $r \geq 2$,

$$r - 0.26186 < \Lambda_r < r - 0.16595,$$

and thus r is the least integer exceeding $\Lambda_r - \delta$. We see from (11.17) that the magnitude of μ_0 alone leads to a host of results for sieve problems satisfying the conditions of Corollary 11.2. For example, the reader will easily check that *if h is an irreducible polynomial of degree g (≥ 2) with integer coefficients and without any fixed prime factors, then $h(n)$ is infinitely often a P_{g+1}; and that if h also satisfies $\rho(p) < p - 1$ when both conditions $p \nmid h(0)$ and $p \leq g + 1$ hold, then $h(p)$ is infinitely often a P_{2g+1}*. For the details of proofs of these and many related results, see [Ric69] or Chapter 9 of [HR74].

Comparing (11.16) with Example 10.3, with $H(n) = L(n)$, we should have $\mu_0 = g$, $\tau = 1$; the choice $u = v > \beta_g$ would then lead to the condition $r > g\beta_g - 1$, the same as the result of the example. Here we can do better by choosing u smaller than β_g.

For the best applications of Theorem 11.1, we have to find choices of the parameters $u < \beta_\kappa/\tau$ and $v > \beta_\kappa/\tau$ for which $N(u, v; \kappa, \mu_0, \tau)$ is as small as possible. There is no straightforward way to accomplish this, although in Section A1.10 we outline a computational method for finding good approximations to the optimal u and v.

11.4 A simple estimate

The following "bare hands" estimation of $N(u, v; \kappa, \mu_0, \tau)$ for $\kappa > 1$ yields results which when checked against those given for $\kappa = 1$, above, are surprisingly good, and which indicate where improvements may be possible with more information and more delicate procedures. Moreover, for large κ, procedures that are delicate are also less reliable; for example, in studying a 50-dimensional problem, it is likely that a ball-park value of r is all that we can hope for. Here the approximate method given below needs only a value, or estimate, of the sieving limit β_{50} (cf. (11.25)), and that we can find: we have shown in [DH97b] that β_κ is smaller than the corresponding Ankeny–Onishi sieve lower bound, and that quantity is asymptotic to $c\kappa$, with $c = 2.445\ldots$.

Let

(11.18) $$I := \kappa \int_u^v F_\kappa\left(v\left(\tau - \frac{1}{s}\right)\right)\left(1 - \frac{u}{s}\right)\frac{ds}{s},$$

and make the substitution $t - 1 = v(\tau - 1/s)$, so that $s = v/(\tau v + 1 - t)$. Then

$$I = \kappa \int_{\tau v + 1 - v/u}^{\tau v} F_\kappa(t-1)\frac{u}{v}\left(\frac{v}{u} - 1 - \tau v + t\right)\frac{dt}{\tau v + 1 - t}$$

$$= \frac{u}{v}\kappa \int_\xi^{\tau v} F_\kappa(t-1)\frac{t - \xi}{\tau v + 1 - t}dt$$

on writing

(11.19) $\xi = \tau v + 1 - v/u = (\tau u - 1)v/u + 1 > 1$,

with the last inequality coming from (11.8). Also note that $\xi < \tau v$ since

11.4 A simple estimate

$v > u$. Integrating I by parts, we obtain

$$\begin{aligned}
I &= \kappa \frac{u}{v} \Big\{ -F_\kappa(t-1)(t-\xi)\log(\tau v + 1 - t)\Big|_\xi^{\tau v} \\
&\quad + \int_\xi^{\tau v} ((t-\xi)F_\kappa(t-1))' \log(\tau v + 1 - t) dt \Big\} \\
&= \frac{u}{v}\kappa \int_\xi^{\tau v} \{F_\kappa(t-1) + (t-\xi)F'_\kappa(t-1)\}\log(\tau v + 1 - t) dt \\
&\leq \frac{u}{v}\kappa \int_\xi^{\tau v} F_\kappa(t-1)\log(\tau v + 1 - t) dt,
\end{aligned}$$

since $F'_\kappa(t-1) < 0$ for all $t > 1$. If ξ is not too close to 1, one may hope that not too much has been lost here, because $(t-\xi)\log(\tau v + 1 - t)$ vanishes at both limits of integration, and $F'_\kappa(t-1)$ is very small when t exceeds β_κ (see Figure A1.1(b)).

Prompted by this last remark, we shall require

(11.20) $$\xi \geq \beta_\kappa.$$

Recall from (6.4) that $\kappa t^{\kappa-1} F_\kappa(t-1) = (t^\kappa f_\kappa(t))'$ when $t \geq \beta_\kappa$. Thus

$$\begin{aligned}
I &\leq \frac{u}{v} \int_\xi^{\tau v} (t^\kappa f_\kappa(t))' t^{1-\kappa} \log(\tau v + 1 - t) dt \\
&= \frac{u}{v}\Big\{ tf_\kappa(t) \log(\tau v + 1 - t)\Big|_\xi^{\tau v} \\
&\quad - \int_\xi^{\tau v} t^\kappa f_\kappa(t) \big((1-\kappa)t^{-\kappa}\log(\tau v + 1 - t) - t^{1-\kappa}(\tau v + 1 - t)^{-1}\big) dt \Big\} \\
&= \frac{u}{v} \int_\xi^{\tau v} f_\kappa(t)\Big((\kappa-1)\log(\tau v + 1 - t) + \frac{t}{\tau v + 1 - t}\Big) dt - \frac{u}{v}\xi f_\kappa(\xi)\log\frac{v}{u} \\
&\leq \frac{u}{v} f_\kappa(\tau v) \int_\xi^{\tau v} \Big((\kappa-1)\log(\tau v + 1 - t) + \frac{t}{\tau v + 1 - t}\Big) dt - \frac{u}{v}\xi f_\kappa(\xi)\log\frac{v}{u},
\end{aligned}$$

since $f_\kappa(t)$ increases strictly with t when $t \geq \beta_\kappa$. The integral on the right is elementary and, after multiplication by u/v, equals

$$\Big(\kappa + \frac{u}{v}\xi\Big)\log\frac{v}{u} - \kappa\Big(1 - \frac{u}{v}\Big).$$

Hence, by (11.18),

(11.21)
$$\frac{\kappa}{f_\kappa(\tau v)} \int_u^v F_\kappa(v(\tau - \frac{1}{s}))(1 - \frac{u}{s})\frac{ds}{s}$$
$$< (\kappa + \frac{u}{v}\xi)\log\frac{v}{u} - \kappa(1 - \frac{u}{v}) - \xi\frac{f_\kappa(\xi)}{f_\kappa(\tau v)}\frac{u}{v}\log\frac{v}{u}$$
$$= \begin{cases} (\kappa + \frac{u}{v}\beta_\kappa)\log\frac{v}{u} - \kappa(1 - \frac{u}{v}), & \xi = \beta_\kappa, \\ (\kappa + \frac{u}{v}\xi\{1 - f_\kappa(\xi)/f_\kappa(\tau v)\})\log\frac{v}{u} - \kappa(1 - \frac{u}{v}), & \xi \geq \beta_\kappa. \end{cases}$$

The temptation is to choose $\xi = \beta_\kappa = v/u$, and we do not resist it. On substituting in (11.21), we obtain the upper bound
$$(\kappa + 1)\log\beta_\kappa - \kappa(1 - 1/\beta_\kappa),$$
and the result of Theorem 11.1 becomes

(11.22) $r > \tau\mu_0(v/\beta_\kappa) - 1 + (\kappa + 1)\log\beta_\kappa - \kappa(1 - 1/\beta_\kappa).$

By (11.19), with our choices, $\tau v = 2\beta_\kappa - 1$ and hence $(\tau v)/\beta_\kappa = 2 - 1/\beta_\kappa$. The inequality for r now reads
$$r > \mu_0(2 - 1/\beta_\kappa) - 1 + (\kappa + 1)\log\beta_\kappa - \kappa(1 - 1/\beta_\kappa)$$
$$= \mu_0 - 1 + (\mu_0 - \kappa)(1 - 1/\beta_\kappa) + (\kappa + 1)\log\beta_\kappa.$$

Let us test the quality of this inequality in the case of $L(n)$, the product of g linear forms, when $g = 2$ and 3. In either case $\tau = 1$, $\kappa = g$ and condition $\boldsymbol{M_0}$ in (11.4) allows us to take $\mu_0 = g$; also we shall need the values of β_g from Table 17.1. We obtain when $g = 2$,
$$r > 1 + 3\log 4.2665 = 5.352\ldots,$$
so that $r = 6$ is a valid choice; and when $g = 3$,
$$r > 2 + 4\log 6.641 = 9.573\ldots,$$
so that $r = 10$ is admissible. For the product of g linear forms we have
$$r > (g + 1)\log\beta_g + g - 1.$$
Since, as $g \to \infty$, we have $\beta_g \leq \nu_g$ ([DH97b], Theorem 2) and $\nu_g \sim cg$ with $c \approx 2.445$ [AO65], we may conclude that
$$r = [(g + 1)\log g + 2g]$$
is an admissible choice for large g.

The simple inequality (11.22) yields fairly satisfactory results in a wide range of applications (see [HR74], Chapter 10.3, for consequences of an inequality inferior to (11.22) only in having a very slightly larger number in place of β_κ); but the second inequality in (11.21), where $\xi > \beta_\kappa$, is sharper and sometimes gives better results than (11.22). It asserts that

$$r > \tau\mu_0 u - 1 - \kappa\left(1 - \frac{u}{v}\right) + \left\{\kappa + \frac{u}{v}\xi\left(1 - \frac{f_\kappa(\xi)}{f_\kappa(\tau v)}\right)\right\}\log\frac{v}{u},$$

but we transform it to a more readily applicable form by introducing a new parameter ζ via the relation $1 < v/u = \xi/\zeta$. Since $\xi = \tau v + 1 - v/u$, we have

$$\tau v = \zeta + \xi/\zeta - 1, \quad \tau u = \zeta + 1 - \zeta/\xi,$$

whence

(11.23) $$r > (1+\zeta)\mu_0 - 1 - \kappa - (\mu_0 - \kappa)\frac{\zeta}{\xi}$$
$$+ \left\{\kappa + \xi\left(1 - \frac{f_\kappa(\xi)}{f_\kappa(\xi + \xi/\zeta - 1)}\right)\right\}\log\left(\frac{\xi}{\zeta}\right).$$

For applications we need a value $\xi > \beta_\kappa$ such that $f_\kappa(\xi)$ is not much smaller than 1, while $f_\kappa(\xi + \zeta/\xi - 1)$ is essentially equal to 1.

11.5 Products of irreducible polynomials

Suppose $\mathcal{A} = \{H(n): 1 \leq n \leq x\}$, where H is a polynomial with integer coefficients of degree G that is the product of g irreducible factors, and $\mathcal{P}$ is the set of all primes. Here $\kappa = g$ and $\tau = 1$, and since $H(n) = O_H(n^G)$, we may take μ_0 to be any constant larger than G if x is sufficiently large. Taking $\zeta = g/G$, (11.23) becomes in this class of problems
(11.24)
$$r > G - 1 - \frac{G-g}{\xi}\frac{g}{G} + \left\{g + \frac{g}{G}\left(1 - \frac{f_g(\xi)}{f_g(\xi + G\xi/g - 1)}\right)\right\}\log\left(\frac{G}{g}\xi\right).$$

If all the polynomial factors of H have equal degree h, so that $G = gh$, (11.24) becomes

(11.25) $$r > gh - 1 - \frac{h-1}{h}\frac{g}{\xi} + \left\{g + \frac{1}{h}\left(1 - \frac{f_g(\xi)}{f_g((h+1)\xi - 1)}\right)\right\}\log(h\xi),$$

and if all the factors are linear, so that $h = 1$, (11.25) becomes

(11.26) $$r > g - 1 + \left\{g + 1 - \frac{f_g(\xi)}{f_g(2\xi - 1)}\right\}\log\xi.$$

Example 11.3. With (11.26) (and $\xi > \beta$) we are back to the original Example 1.2, and when $g = 2$ or 3 we can do a little better. When $g = 2$ we use $f_2(7) = 0.9797\ldots$, so that $f_2(13) \approx 1$. Then (11.26) gives $r > 4.93\ldots$, so that we can take $r = 5$. When $g = 3$ we use $f_3(10) = 0.9804\ldots$, so that $f_3(19)$ is indeed very nearly 1, $r > 8.95\ldots$ and we can take $r = 9$.

When we have recourse to numerical integration, we find for $g = 2$ that we narrowly miss $r = 4$ and so cannot do better than $r = 5$; but when $g = 3$ we are able in this way to reach $r = 8$ and indeed we obtain the admissible values for r shown in Table 11.1.

Table 11.1. r values for small g

g :	3	4	5	6	7	8	9	10
r :	8	12	16	20	25	29	34	39

Inequality (11.25) generally yields sharper estimates when we take $\xi > \beta_\kappa$, but here is an instance when it does no better: let H be the product of two irreducible polynomials each of degree 5, so $g = \kappa = 2$ and $h = 5$. Then (11.25) with $\xi = 6$ and $f_2(6) = 0.8844\ldots$ leads to an admissible value $r = 16$, while (11.25) with $\xi = \beta_2 = 4.26645\ldots$ leads to the same r value. Numerical integration gives even $r = 14$. Indeed, when H is the product of g irreducible polynomials each of the same degree k, numerical integration yields the admissible values of r given in Table 11.2.

Table 11.2. r values for small g and h

$g \backslash h$	2	3	4	5	6	7
2	7	9	11	14	16	18
3	12	16	19	22	25	28
4	17	22	27	31	35	40
5	23	29	35	40	46	51
6	29	36	43	50	57	63
7	35	44	52	60	68	75

Also, when $g = 2$ and the degree G of H ranges from 3 through 10, numerical integration leads to the admissible values for r given in Table 11.3:

11.6 Polynomials at prime arguments

Table 11.3. r values for small G

G :	3	4	5	6	7	8	9	10
r :		6	7	8	9	10	11	12	14

11.6 Polynomials at prime arguments

Turning to the existence of almost-primes in sequences generated by polynomials with *prime* arguments, H is again the product of $g > 1$ irreducible nonconstant polynomials; but now we require not only that $\rho(p) < p$ hold for all primes p, but also that

$$\rho(p) < p - 1 \text{ if } p \mid H(0).$$

We have $\mathcal{A} = \{H(p) : p \leq x\}$, $\mathcal{P}$ is again the set of all primes, $X = \operatorname{li} x$, and

$$\omega(p) = \frac{p}{p-1} \rho_1(p),$$

where $\rho_1(p)$ gives the number of solutions of $H(m) \equiv 0 \bmod d$ with $(m, d) = 1$, i.e.,

$$\rho_1(p) = \begin{cases} \rho(p), & p \nmid H(0), \\ \rho(p) - 1, & p \mid H(0). \end{cases}$$

Condition $\boldsymbol{R_o}$ holds with $\tau = 1/2$ by Bombieri's Theorem, and $\boldsymbol{M_o}$ with $\tau = 1/2$ and $\mu_0 > 2G$, supposing x to be large. Apply Theorem 11.1, with x large, $u = U/\tau = 2U$, $v = V/\tau = 2V$, to obtain

$$|\{p : p \leq x, H(p) = P_r\}| \gg \frac{x}{(\log x)^{g+1}}$$

for any natural number r satisfying

$$r > 2GU - 1 + \frac{g}{f_g(V)} \int_1^{V/U} H_g(V-s) \left(1 - \frac{Us}{V}\right) \frac{ds}{s},$$

with $1 < U < V$ and $\beta_g < V$. With the aid of numerical integration we illustrate the result with two sets of data, giving admissible values of r in each case:

Example 11.4. Let all the g irreducible factors of H have the same degree k, so that $G = gk$. For $g = 2, 3, 4$ and $2 \leq k \leq 7$, Table 11.4 gives admissible values of r:

Table 11.4. r values for small g and k and prime arguments

$g \backslash k$	2	3	4	5	6	7
2	11	16	20	24	28	32
3	19	25	32	38	44	50
4	27	35	44	52	61	69

Finally, consider the cases $g = 2$, $3 \leq G \leq 8$. Here admissible values of r are given in Table 11.5.

Table 11.5. r values for small G and prime arguments

G:	3	4	5	6	7	8
r:	9	11	14	16	18	20

11.7 Other weights

There are other weighting procedures in the literature, and there is scope for further development in this area. We mention here two variations:

(i) A refinement of the method of this chapter.

There may well be negative terms present in our W-expressions that diminish their effectiveness. There is a method described in [HR85] that avoids this problem for $\kappa = 1$, i.e., for linear sieve problems. It would be desirable to extend the procedure to larger values of κ.

(ii) The Selberg method and its refinements.

In [Sel91], (45, Lectures on Sieves, §23), Selberg gave what he described as an early approach to the twin prime problem: it is in effect a weighted form of his λ^2 method. The aim is to find λ_ds that make the ratio

$$\frac{Q_2}{Q_1} := \frac{\sum_{x<n\leq 2x}\{\tau(n)+\tau(n+2)\}\left(\sum_{d|n(n+2)}\lambda_d\right)^2}{\sum_{x<n\leq 2x}\left(\sum_{d|n(n+2)}\lambda_d\right)^2}$$

as small as possible, but this is apparently intractable. Selberg did, however, find λ_ds that minimize just Q_2 and for which $Q_2/Q_1 < 14 + \epsilon$ for any fixed $\epsilon > 0$. This implies the existence of infinitely many n such that $\tau(n)+\tau(n+2) \leq 14$, i.e., of infinitely many pairs n, $n+2$ of which one is a P_2 and the other a P_3.

Heath–Brown [H-B97] generalized this approach to apply to the prime g-tuple problem by considering

$$Q = \sum_{n \leq x} \{1 - \ell \sum_{i=1}^{g} \tau(a_i n + b_i)\} \Big(\sum_{d|L(n)} \lambda_d\Big)^2 =: Q_1 - \ell Q_2,$$

where ℓ is a constant and $L(\cdot)$ is as in Example 1.2. He too focused on achieving a good upper bound only for his Q_2 and chose

$$\lambda_d = \mu(d) \Big(\frac{\log(y/d)}{\log y}\Big)^{g+1}, \quad d \leq y,$$
$$= 0, \quad d > y.$$

Although the λ_ds here have a simple structure, this is close to Selberg's optimal choice (when $g = 2$).

Ho and Tsang [HoTs06] have recently refined Heath–Brown's approach by introducing convex combinations, taking

$$\lambda_d = \mu(d) \Big\{\delta \Big(\frac{\log(y/d)}{\log y}\Big)^{g+1} + (1-\delta)\Big(\frac{\log(y/d)}{\log y}\Big)^{g+2}\Big\}, \quad d \leq y,$$

with a free parameter $\delta \in [0, 1]$; or even

$$\lambda_d = \mu(d) \sum_{\nu=g+1}^{G} \delta_\nu \Big(\frac{\log(y/d)}{\log y}\Big)^\nu, \quad d \leq y,$$

with $\lambda_d = 0$ when $d > y$. In this way they have obtained results as good as or slightly better than those of Table 11.1 by giving explicit formulas for the numbers δ_ν.

A variant of the Selberg approach features also in the recent work of Goldston et al. [GPY] on small gaps between consecutive primes and on a whole range of other important problems in that area.

All these methods deploy deeper analytic devices than those present in this monograph.

11.8 Notes on Chapter 11

Selberg's weighted sieve dates from about 1950.

Iwaniec, by using Linnik's dispersion method, has shown even that $n^2 + 1 = P_2$ infinitely often [Iwa78].

There are many other applications given in Chapter 9 of [HR74] that could serve as exercises, and there are useful references and historical comments in the Notes of that chapter.

The tables of this chapter were created using Theorem 11.1, software developed by Wheeler and Bradley to perform numerical integration, and a hand search for suitable values of u and v. These tables were first published in [DH97a]; they have subsequently been confirmed by an independent calculation using the methods described in Section A1.10.

Theorem 11.1 was used also in [DH97a] to show that there are infinitely many primitive Pythagorean triples (pairwise relatively prime integer solutions of $x^2 + y^2 - z^2 = 0$) for which xyz has at most 17 prime factors. This theme has been broadly generalized by Liu and Sarnak [LS] to indefinite quadratic forms in three variables. Also, Marasingha [Mrs06] has shown that certain products of two binary irreducible quadratic forms have at most five prime factors for infinitely many values of the arguments.

As we noted at the end of Chapter 8, if (11.3) holds with $\tau = 0.5453$, then, by Corollary 11.2, we would have $p + 2 = P_2$ infinitely often, even in a quantitative form. Further, as Goldston et al. [GPY] have shown (see the survey of Soundararajan [Snd07]), a proof of the Bombieri–Vinogradov theorem with exponent $1 - \epsilon$ in place of $1/2$— the so-called Elliott–Halberstam conjecture—would have some spectacular consequences. As another example, Table 11.6 shows how larger admissible values of τ would improve upon $r = 16$ for $g = 2$, $k = 3$ in Table 11.4. In the table, $N_{\min}(\kappa, \mu_0, \tau)$ denotes the minimal value of $N(u, v; \kappa, \mu_0, \tau)$ when u and v are free to vary subject to the conditions imposed by Theorem 11.1.

Table 11.6. *Values of τ_r for which $r \approx N_{\min}(g, gk/\tau_r, \tau_r)$*

r	τ_r
15	0.503682
14	0.548199
13	0.601098
12	0.664916
11	0.743295
10	0.841628
9	0.968205

Part II
Proof of the Main Analytic Theorem

12
Dramatis personae and preliminaries

12.1 P and Q and their adjoints

Our sieve theory has two components: a combinatorial part, presented in Part I, and an analytical part, which we take up here. The central result still to be established is our *Main Analytic Theorem*, Theorem 6.1, which asserts the solvability of a certain coupled system of difference differential equations satisfying given boundary conditions. We show that the system has a solution pair F_κ and f_κ for each $\kappa \geq 1$ with $2\kappa \in \mathbb{N}$. These solutions provide the upper and lower bound functions used to estimate $S(\mathcal{A}, \mathcal{P}, z)$ in Chapters 7 and 9. Also, we establish properties of these functions and intervening auxiliary functions.

Theorem 6.1 was established laboriously in stages over several years in a series of rather technical papers, [DHR88]–[DHR96], written with H.-E. Richert. Our chief obstacle was to prove this result for *each real value of* $\kappa > 1$. To our surprise, it was the small values of κ, particularly those lying a little below 2, that gave by far the most trouble. The aim of Part II is to prove the theorem for integer and half integer values of $\kappa \geq 1$, the ones most interesting for applications and for which many of the technical problems arising with other κ values are avoided.

In this chapter we introduce the several functions that will be used in our argument. We begin by restating Theorem 6.1 in terms of the functions

(12.1) $\qquad P(u) := F(u) + f(u), \quad Q(u) := F(u) - f(u).$

The boundary value problem translates into

(12.2) $\qquad P(u) = Q(u) = 1/\sigma(u), \quad 0 < u \leq \beta,$

$$\text{(12.3)} \qquad u^\kappa P(u) = \frac{u^\kappa}{\sigma(u)} + \kappa \int_\beta^u \frac{t^{\kappa-1}}{\sigma(t-1)} dt, \quad \beta < u \leq \alpha,$$

$$\text{(12.4)} \qquad u^\kappa Q(u) = \frac{u^\kappa}{\sigma(u)} - \kappa \int_\beta^u \frac{t^{\kappa-1}}{\sigma(t-1)} dt, \quad \beta < u \leq \alpha,$$

$$\text{(12.5)} \qquad (u^\kappa P(u))' = \kappa u^{\kappa-1} P(u-1), \quad u > \alpha,$$

$$\text{(12.6)} \qquad (u^\kappa Q(u))' = -\kappa u^{\kappa-1} Q(u-1), \quad u > \alpha,$$

and

$$\text{(12.7)} \qquad P(u) = 2 + O(e^{-u}), \quad Q(u) = O(e^{-u}).$$

The σ function is what is known, and what is sought are α, β, P, and Q; all of these, of course, depend on κ. To determine P and Q (or equivalently $F = (P+Q)/2$ and $f = (P-Q)/2$), we proceed step-by-step, solving (12.5) and (12.6); but it is evident that we cannot even begin this procedure until α_κ and β_κ have been found. Our chief remaining task will be to determine the parameters α_κ and β_κ, which is actually an assignment of some independent interest. We take this up next in Chapter 13.

The monotonicities of F and f asserted in the Theorem will be seen to follow from the inequality

$$\text{(12.8)} \qquad Q(u) > 0, \quad u > 0,$$

which will be established in Chapter 18.

We turn our attention to solving the system of difference differential equations (12.5) and (12.6). To do this we follow the method of Iwaniec [Iwa80] and introduce "adjoint" functions. These will be combined with P and Q via the so-called Iwaniec "inner products," which are described near the end of this section.

For $\kappa > 0$, the adjoint functions are determined by the equations

$$\text{(12.9)} \qquad (up(u))' = \kappa p(u) - \kappa p(u+1), \quad u > 0,$$

and

$$\text{(12.10)} \qquad (uq(u))' = \kappa q(u) + \kappa q(u+1), \quad u > 0.$$

Whereas we cannot solve (12.5) or (12.6) in closed form, we do have explicit solutions $p = p_\kappa(u)$ and $q = q_\kappa(u)$ for (12.9) and (12.10). We

12.1 P and Q and their adjoints

remark that these solutions are unique up to a multiplicative constant if we assume that each has the behavior of a rational function at infinity ([BrD97], [DHR93b]), and we shall choose for each a suitable normalization.

The Laplace transform representation

$$(12.11) \qquad p(u) = p_\kappa(u) := \int_0^\infty e^{-ut - \kappa \, \text{Ein}(t)} \, dt, \quad u > 0,$$

satisfies (12.9), as we soon show; here

$$(12.12) \qquad \text{Ein}(t) := \int_0^t (1 - e^{-s}) \frac{ds}{s} = \sum_{n=1}^\infty (-1)^{n-1} \frac{t^n}{n! \, n}, \quad t \in \mathbb{C},$$

an entire function of t. For t bounded away from the origin, we have the useful formula (see [AS94], Ch. 5, footnote 3; [Olv97], (3.05))

$$(12.13) \qquad \text{Ein}(t) = \log t + \gamma + \int_t^\infty e^{-s} \frac{ds}{s}, \quad |\arg t| < \pi$$

(with $\log t$ denoting the principal value for complex t).

To see that the integral (12.11) satisfies (12.9), first integrate it by parts, next multiply by u, and then differentiate with respect to u. It is clear from the integral representation that $p(u)$ is positive and strictly decreasing in u for $u > 0$. Also, since $0 < \text{Ein}(t) < t$ for all $t > 0$,

$$(12.14) \qquad 1/(u + \kappa) < p(u) < 1/u, \quad u > 0.$$

We deduce that $up(u) \to 1$ as $u \to \infty$, and hence, on integrating (12.9) from u to ∞, that

$$(12.15) \qquad up(u) + \kappa \int_{u-1}^u p(t+1) \, dt = 1, \quad u > 0.$$

Next, one can check (just as was done for p) that

$$(12.16) \qquad q(u) = q_\kappa(u) := \frac{\Gamma(2\kappa)}{2\pi i} \int_C z^{-2\kappa} e^{uz + \kappa \, \text{Ein}(-z)} \, dz, \quad u > 0,$$

satisfies (12.10); here C is the path from $-\infty$ back to $-\infty$ which envelopes the negative real axis in the positive sense. For 2κ a positive integer, $z^{-2\kappa}$ has a pole at the origin and we can close the path C and apply Cauchy's Theorem. We find that $q(u)$, as given by (12.16), is a monic polynomial of degree $2\kappa - 1$.

When (12.10) is integrated from u to $u+1$, we obtain

(12.17) $$(u+1)q(u+1) - uq(u) = \kappa \int_u^{u+2} q(t)\,dt.$$

Several other properties of the functions p and q that we need are given in Chapter 15.

We verify by differentiation that each of the so-called Iwaniec "inner products"

$$\langle P, p \rangle_{-\kappa} = \langle P, p \rangle_{-\kappa}(u) := up(u)P(u) + \kappa \int_{u-1}^u P(t)p(t+1)\,dt$$

and

$$\langle Q, q \rangle_\kappa = \langle Q, q \rangle_\kappa(u) := uq(u)Q(u) - \kappa \int_{u-1}^u Q(t)q(t+1)\,dt$$

is constant for $u \geq \alpha$. The conditions on P, Q, p, and q as $u \to \infty$ show these two constants to be 2 and 0 respectively, so that we have

(12.18) $$up(u)P(u) + \kappa \int_{u-1}^u P(t)p(t+1)\,dt = 2, \quad u \geq \alpha,$$

and

(12.19) $$uq(u)Q(u) - \kappa \int_{u-1}^u Q(t)q(t+1)\,dt = 0, \quad u \geq \alpha.$$

These two equations play a crucial role in our quest for a pair of numbers α, β and functions P, Q that satisfy (12.2)–(12.8). For further discussion of this inner product see the Notes to this chapter and Section A1.1.

12.2 Rapidly vanishing functions

As a warm-up, we show, in the converse direction, that *if* there exist functions P and Q for which (12.18) and (12.19) hold, then the estimates of (12.7) follow directly. A stronger form of this assertion is a consequence of the following extension of a method used by de Bruijn and by Hua.

Lemma 12.1. *Suppose that ϕ is a piecewise continuous function on an interval $[\delta, \infty)$, with $\delta > 0$, and that $\phi \in L^1$ locally and satisfies the inequality*

(12.20) $$|\phi(u)| \leq \frac{\kappa}{u} \int_{u-1}^u |\phi(x)|\,dx, \quad u \geq \delta + 1.$$

Then

(12.21)
$$\phi(u) \ll \kappa^u/\Gamma(u+1).$$

Proof. ϕ is locally bounded, as we see from (12.20) and the fact that $\phi \in L^1$ locally. Next, ϕ is bounded on $[\delta+1, \infty)$, for otherwise there would exist a sequence $u_n \to \infty$ on which $|\phi(u_n)| > |\phi(u)|$ for all $u < u_n$. But for $u_n > \kappa$, we then would have

$$|\phi(u_n)| \le \frac{\kappa}{u_n} \int_{u_n-1}^{u_n} |\phi(x)|\, dx < |\phi(u_n)|,$$

which is impossible.

Now set $s(u) := \sup\{|\phi(x)| : x \ge u\}$. This function is finite and decreasing on $[\delta+1, \infty)$ and satisfies (12.20) there. Indeed, we have

$$|\phi(u)| \le \frac{\kappa}{u} \int_{u-1}^{u} s(x)\, dx \le \frac{\kappa}{u} s(u-1),$$

and, since the right side of this inequality is decreasing in u, we deduce that

$$s(u) \le \frac{\kappa}{u} s(u-1), \quad u \ge \delta+2,$$

whence

$$s(u) \le \frac{\kappa}{u} s(u-1) \le \ldots \le \frac{\kappa^L}{u(u-1)\ldots(u-L+1)} s(u-L),$$

for any positive integer $L \le u - \delta - 2$. We conclude that

$$s(u) \ll \kappa^u/\Gamma(u+1), \quad u \ge \delta+2,$$

and thus (12.21) holds. □

Remark 12.2. Further analysis, based on the inequality

$$s(u) \le c\frac{\kappa}{u} \int_{u-1}^{u} \frac{\kappa^x}{\Gamma(x+1)}\, dx,$$

yields an estimate $s(u) = o(\kappa^u/\Gamma(u+1))$.

If we combine (12.18) with formula (12.15), we obtain

$$up(u)(P(u)-2) = -\kappa \int_{u-1}^{u} (P(t)-2)p(t+1)\, dt$$

and hence, since p is positive and decreasing,

$$|P(u)-2| \le \frac{\kappa}{u} \int_{u-1}^{u} |P(t)-2|\, dt.$$

Now the first estimate of (12.7) follows from the lemma.

As for Q, we note that
$$\max_{u-1\le t\le u} q(t+1)/q(u) \to 1, \quad u\to\infty,$$
since q is a polynomial. It follows from (12.19) that, for any $\lambda > \kappa$
$$|Q(u)| \le \frac{\lambda}{u} \int_{u-1}^{u} |Q(t)|\, dt$$
holds for all sufficiently large values of u, and hence the second estimate of (12.7) also follows from the lemma.

For later use, we note that (12.5) can be rewritten in the form
$$(12.22) \qquad uP'(u) = -\kappa(P(u) - P(u-1)) = -\kappa \int_{u-1}^{u} P'(t)\, dt, \quad u > \alpha,$$
so that
$$|P'(u)| \le \frac{\kappa}{u} \int_{u-1}^{u} |P'(t)|\, dt, \quad u > \alpha;$$
and, by a third application of the lemma, we deduce that
$$(12.23) \qquad P'(u) \ll \kappa^u / \Gamma(u+1) \ll \exp(-u).$$

12.3 The Π and Ξ functions

Relations (12.18) and (12.19) taken at $u = \alpha_\kappa$ determine pairs of equations which α_κ and β_κ need to satisfy if Theorem 6.1 is to be true. Solving these equations (for $\alpha_\kappa, \beta_\kappa$) will be our chief task. It will be more convenient to work with a related pair of expressions that we now introduce and discuss. Set

$$(12.24) \qquad \Pi_\kappa(u,v) := \frac{up(u)}{\sigma(u)} + \kappa \int_{v-1}^{u} \frac{p(t+1)}{\sigma(t)} dt$$
$$= \frac{(v-1)p(v-1)}{\sigma(v-1)} + \kappa \int_{v-1}^{u} \frac{p(t)\sigma(t-2)}{\sigma^2(t)} dt$$

and

$$(12.25) \qquad \Xi_\kappa(u,v) := \frac{uq(u)}{\sigma(u)} - \kappa \int_{v-1}^{u} \frac{q(t+1)}{\sigma(t)} dt$$
$$= \frac{(v-1)q(v-1)}{\sigma(v-1)} + \kappa \int_{v-1}^{u} \frac{q(t)\sigma(t-2)}{\sigma^2(t)} dt,$$

each defined for $u > 0$, $v > 1$; in each case, the second representation derives from the first by integration by parts. Indeed, in the case of (12.24), if we rewrite (12.9) as

$$(12.26) \qquad (u^{1-\kappa} p(u))' = -\kappa u^{-\kappa} p(u+1),$$

then

$$\kappa \int_{v-1}^{u} \frac{p(t+1)}{\sigma(t)} dt = -\int_{v-1}^{u} \frac{t^\kappa}{\sigma(t)} (t^{1-\kappa} p(t))' dt$$

$$= -\frac{up(u)}{\sigma(u)} + \frac{(v-1)p(v-1)}{\sigma(v-1)} + \int_{v-1}^{u} \left(\frac{t^\kappa}{\sigma(t)}\right)' t^{1-\kappa} p(t) \, dt,$$

and the second result in (12.24) follows from (6.8), rewritten in the form

$$(12.27) \qquad \left(\frac{t^\kappa}{\sigma(t)}\right)' = \kappa \frac{t^{\kappa-1} \sigma(t-2)}{\sigma^2(t)}.$$

A similar argument, based on the last equation and (12.10) in the form

$$(12.28) \qquad (u^{1-\kappa} q(u))' = \kappa u^{-\kappa} q(u+1),$$

leads from the first representation of $\Xi_\kappa(u, v)$ in (12.25) to the second.

The following instances of Π and Ξ will be very useful in later chapters:

$$\Pi(u) = \Pi_\kappa(u) := \Pi_\kappa(u, u), \qquad \Xi(u) = \Xi_\kappa(u) := \Xi_\kappa(u, u) \quad (u > 1),$$
$$\widetilde{\Pi}(u) = \widetilde{\Pi}_\kappa(u) := \Pi_\kappa(u, u-1), \qquad \widetilde{\Xi}(u) = \widetilde{\Xi}_\kappa(u) := \Xi_\kappa(u, u-1) \quad (u > 2).$$

12.4 Notes on Chapter 12

For $\kappa = 1$, the functions P and Q assume an especially simple form:

$$P_1(u) = 2e^\gamma \omega(u), \qquad Q_1(u) = 2e^\gamma \rho(u-1)/u, \qquad u > 1,$$

where ω is Buchstab's function (no relation to any of our previously introduced omegas!), defined by

$$u\omega(u) = 1, \ 1 \le u \le 2, \qquad (u\omega(u))' = w(u-1), \quad u > 2,$$

and ρ is Dickman's function (see [Ten01], Section III.5), defined by

$$\rho(u) = 1, \ 0 \le u \le 1, \qquad u\rho'(u) = -\rho(u-1), \quad u > 1.$$

(This ρ is not related to ρ_κ, the largest zero of $q_\kappa(\cdot)$!)

The Iwaniec inner product was introduced in [Iwa80] and arises in the following way. Consider the difference differential equation

$$uG'(u) = -aG(u) - bG(u-1), \quad u > u_0,$$

with a retarded argument and parameters a, b. (The functions P_κ and Q_κ discussed at the beginning of this chapter are both of this type as is j_κ, but p_κ and q_κ defined in (12.9) and (12.10) are *not* of this type.) Associated with G is an "adjoint function" satisfying the differential equation

$$(ug(u))' = ag(u) + bg(u+1), \quad u > u_0,$$

with an advanced argument (p_κ and q_κ are functions of this type) so that

$$\langle G, g \rangle_b := uG(u)g(u) - b\int_{u-1}^{u} G(t)g(t+1)\,dt = \text{Constant}, \quad u > u_0.$$

Note that the sign between the terms in $\langle G, g \rangle_b$ depends upon the sign of b. The G functions can be wild (ours are not too bad), but g is analytic in a half plane and $g(u)/u^{a+b-1}$ has a limit as $u \to \infty$.

The functions $\Pi_\kappa(u,v)$, $\Xi_\kappa(u,v)$ were introduced in [ILR80]. The functions $\Pi_\kappa(u)$ and $\Xi_\kappa(u)$ played a key role in [DHR90a] in showing that $\alpha_\kappa > \beta_\kappa$ holds for all $\kappa > 1$, a result that we do not need here.

13
Strategy and a necessary condition

13.1 Two different sieve situations

Originally, we had attempted to prove Theorem 6.1 using hints provided by the theorem of [ILR80] (see Lemma 13.1 below), the Ph.D. thesis of Rawsthorne [Raw80], and the article [DHR88], by distinguishing between the two cases $\beta_\kappa \leq \alpha_\kappa \leq \beta_\kappa + 1$ and $\beta_\kappa + 1 < \alpha_\kappa$. (This distinction appeared naturally in the generalized system of Rosser–Iwaniec inequalities—cf. the remarks on page 76. Also, we shall see how these cases affect the differential equations occurring in Lemma 13.1.)

What we needed to do was to identify the ranges of values of κ corresponding to each of these cases (the possibility $\alpha_\kappa < \beta_\kappa$ having been ruled out in [DHR90a]); but we had no success with this approach, even though we "knew" on the basis of numerical evidence that $1 < \kappa \leq \kappa_0 \,(= 1.8344\ldots)$ belonged to the first case and $\kappa > \kappa_0$ to the second. It took two separate perceptions to put us on the right path: first, that an analysis of the families of Π and Ξ functions would be central to any successful approach and, second, that classification of the values of κ would come from the relative magnitude, not of α_κ and $\beta_\kappa + 1$, but, instead, of the zeros of $\widetilde{\Pi}_\kappa - 2$ and $\widetilde{\Xi}_\kappa$ (with equality holding at $\kappa = \kappa_0$).

To better understand the main issue in the proof of Theorem 6.1, we consider the role of the numbers α_κ and β_κ that occur there. For *any* two numbers α and β satisfying $2 \leq \beta \leq \alpha$, we can determine *some* functions F and f by starting with formulas (6.1) and (6.2) of the theorem and driving the functions forward by successively integrating (6.3) and (6.4). For arbitrary numbers α and β there is no reason for functions that were so constructed to satisfy the boundary conditions given in (6.5). However, we shall show, for given κ, that there is a pair α_κ, β_κ (which is probably unique) for which these boundary conditions hold. (See Figure A1.1 in Section A1.2 of the Appendix, where a choice of "arbitrary values" and the "correct values" are illustrated.) Thus, the main step in determining functions F_κ and f_κ that satisfy the conditions of the theorem is to determine suitable values for α_κ and β_κ. A description of the computation of α_κ and β_κ is given in Section A1.9.

13.2 A necessary condition

As a start, we show that numbers α_κ and β_κ that occur in Theorem 6.1 must satisfy certain equations involving Π_κ and Ξ_κ. Later, we shall solve these equations to find α_κ and β_κ. To simplify notation, we generally omit the subscript κ.

Lemma 13.1 ([ILR80]). *Suppose that α, β is a pair of numbers for which Theorem 6.1 is valid. If $\beta \leq \alpha \leq \beta + 1$, then*

$$(13.1) \qquad \Pi(\alpha, \beta) = 2 \text{ and } \Xi(\alpha, \beta) = 0.$$

If $\beta + 1 \leq \alpha$, then

$$(13.2) \qquad \begin{cases} \widetilde{\Pi}(\alpha) + (\alpha - 1)p(\alpha - 1)f(\alpha - 1) = 2, \\ \widetilde{\Xi}(\alpha) - (\alpha - 1)q(\alpha - 1)f(\alpha - 1) = 0. \end{cases}$$

Here $f = f_\kappa$ is the lower bound sieve function, specifically

$$(13.3) \qquad (\alpha - 1)^\kappa f(\alpha - 1) = \kappa \int_\beta^{\alpha - 1} \frac{t^{\kappa - 1}}{\sigma(t - 1)} dt.$$

(Note that (13.1) and (13.2) coincide when $\alpha = \beta + 1$.)

Proof. (i) By definition, $f(t) = 0$ for $t \leq \beta$ and $F(t) = 1/\sigma(t)$ for $t \leq \alpha$, so if $\alpha = \beta$, then $P(t) = 1/\sigma(t)$ for $t \leq \alpha$. Now $\Pi(\alpha, \beta) = 2$ follows immediately from (12.18) and $\Xi(\alpha, \beta) = 0$ from (12.19). This case occurs for $\kappa = 1$, where $\alpha_1 = \beta_1 = 2$, cf. Chapter 17. We henceforth suppose that $\beta < \alpha$.

(ii) Suppose next that $\beta < \alpha \leq \beta + 1$. In view of (12.2), we may write (12.18) at $u = \alpha$ as

$$(13.4) \qquad \alpha p(\alpha) P(\alpha) + \kappa \int_{\alpha - 1}^\beta \frac{p(t+1)}{\sigma(t)} dt + \kappa \int_\beta^\alpha P(t)p(t+1) dt = 2,$$

13.2 A necessary condition

and, if we substitute from (12.3), we obtain $I_1 + I_2 + I_3 = 2$, where

$$I_1 := \alpha p(\alpha)\left\{\frac{1}{\sigma(\alpha)} + \kappa\alpha^{-\kappa}\int_\beta^\alpha \frac{t^{\kappa-1}}{\sigma(t-1)}dt\right\}$$

$$I_2 := \kappa\int_{\alpha-1}^\beta \frac{p(t+1)}{\sigma(t)}dt$$

$$I_3 := \int_\beta^\alpha \left\{\frac{\kappa}{\sigma(t)} + \kappa^2 t^{-\kappa}\int_\beta^t \frac{s^{\kappa-1}}{\sigma(s-1)}ds\right\}p(t+1)dt.$$

By (12.26), p satisfies

$$\kappa t^{-\kappa}p(t+1) = -(t^{1-\kappa}p(t))',$$

so that the double integral in I_3 equals

$$-\kappa\int_\beta^\alpha (t^{1-\kappa}p(t))'\int_\beta^t \frac{s^{\kappa-1}}{\sigma(s-1)}ds\,dt$$

$$= -\kappa\alpha^{1-\kappa}p(\alpha)\int_\beta^\alpha \frac{s^{\kappa-1}}{\sigma(s-1)}ds + \kappa\int_\beta^\alpha \frac{p(t)}{\sigma(t-1)}dt.$$

Substituting in $I_1 + I_2 + I_3$ and recalling (12.24), we find that

$$\Pi(\alpha,\beta) := \frac{\alpha p(\alpha)}{\sigma(\alpha)} + \kappa\int_{\beta-1}^\alpha \frac{p(t+1)}{\sigma(t)}dt = 2.$$

Using a similar procedure, with (12.2), (12.4), and (12.28), we see by (12.25) that (12.19) with $u = \alpha$ may be rewritten as $\Xi(\alpha,\beta) = 0$.

(iii) Finally, suppose that $\beta + 1 < \alpha$. This time, let us deal in detail with (12.19) at $u = \alpha$, namely

$$\alpha q(\alpha)Q(\alpha) - \kappa\int_{\alpha-1}^\alpha Q(t)q(t+1)dt = 0.$$

Since $\alpha - 1 > \beta$, only (12.4) is relevant. Substituting from there yields

$$(13.5)\quad \alpha q(\alpha)\left\{\frac{1}{\sigma(\alpha)} - \kappa\alpha^{-\kappa}\int_\beta^\alpha \frac{t^{\kappa-1}}{\sigma(t-1)}dt\right\}$$

$$-\kappa\int_{\alpha-1}^\alpha \left\{\frac{1}{\sigma(t)} - \kappa t^{-\kappa}\int_\beta^t \frac{s^{\kappa-1}}{\sigma(s-1)}ds\right\}q(t+1)dt = 0.$$

We integrate by parts the double integral using (12.28):

$$\kappa \int_{\alpha-1}^{\alpha} \{t^{1-\kappa}q(t)\}' \int_{\beta}^{t} \frac{s^{\kappa-1}}{\sigma(s-1)} \, ds \, dt = \kappa \alpha^{1-\kappa} q(\alpha) \int_{\beta}^{\alpha} \frac{s^{\kappa-1}}{\sigma(s-1)} \, ds$$

$$- \kappa(\alpha-1)^{1-\kappa} q(\alpha-1) \int_{\beta}^{\alpha-1} \frac{s^{\kappa-1}}{\sigma(s-1)} \, ds - \kappa \int_{\alpha-1}^{\alpha} \frac{q(t)}{\sigma(t-1)} \, dt;$$

the last two terms on the right side can be expressed as

$$-(\alpha-1)q(\alpha-1)f(\alpha-1) - \kappa \int_{\alpha-2}^{\alpha-1} \frac{q(t+1)}{\sigma(t)} \, dt$$

using (13.3) and a change of variable respectively. Now substituting back into (13.5) gives the $\widetilde{\Xi}$ formula in (13.2):

$$\frac{\alpha q(\alpha)}{\sigma(\alpha)} - \kappa \int_{\alpha-2}^{\alpha} \frac{q(t+1)}{\sigma(t)} \, dt - (\alpha-1)q(\alpha-1)f(\alpha-1) = 0.$$

The formula for $\widetilde{\Pi}$ is established in an analogous way. □

Returning to our strategy, the main task to establish Theorem 6.1 is to determine suitable numbers α and β. This we shall do using the pairs of equations (13.1) and (13.2) of the preceding lemma. How do we determine which of the preceding pairs of simultaneous differential equations is the appropriate one to use for a given integer or half integer value of κ? Let $z_{\widetilde{\Pi}}(\kappa)$ denote the root of $\widetilde{\Pi}_\kappa(u) - 2 = 0$ and $z_{\widetilde{\Xi}}(\kappa)$ denote the root of $\widetilde{\Xi}_\kappa(u) = 0$; the answer will be given in terms of the relative size of $z_{\widetilde{\Pi}}(\kappa)$ and $z_{\widetilde{\Xi}}(\kappa)$.

After finding α and β, we next determine F and f and then show that these functions have the expected properties.

13.3 A program for determining F and f

Our steps and where they occur below are as follows:

(i) First show that

    (a) $z_{\widetilde{\Xi}}(\kappa) < z_{\widetilde{\Pi}}(\kappa)$ for $\kappa = 1$ or $\kappa = 1.5$ (Proposition 16.2),
    (b) $z_{\widetilde{\Pi}}(\kappa) < z_{\widetilde{\Xi}}(\kappa)$ for $\kappa = 2, 2.5, 3, \ldots$ (Proposition 16.3).

(ii) Next show that

    (a) when $z_{\widetilde{\Xi}}(\kappa) < z_{\widetilde{\Pi}}(\kappa)$, the equations (13.1) have a common solution pair α, β with $\beta \leq \alpha < \beta + 1$ (Theorem 17.1),
    (b) when $z_{\widetilde{\Xi}}(\kappa) \geq z_{\widetilde{\Pi}}(\kappa)$, the equations (13.2) have a common solution pair α, β with $\alpha \geq \beta + 1$ (Theorem 17.2).

13.3 A program for determining $\boldsymbol{F}$ and $\boldsymbol{f}$

(iii) Finally (in Chapter 18), we show, in each of cases (a) and (b), that the functions F_κ and f_κ defined in an initial range by (6.1) and (6.2) and continued forward by the equations (6.3) and (6.4) satisfy the required monotonicity and limiting conditions of the theorem.

To carry out this program, we must of course use various properties of σ_κ, p_κ, and q_κ. The next two chapters are devoted to studying these functions.

14
Estimates of $\sigma_\kappa(u) = j_\kappa(u/2)$

14.1 Lower bounds on σ

We shall need reasonably accurate and computationally effective lower bounds on $\sigma_\kappa(u)$ for $u \geq 2\kappa$ in order to compare the zeros of $\widetilde{\Pi}_\kappa(u) - 2$ and $\widetilde{\Xi}_\kappa(u)$ in Chapter 16. Here we shall establish several properties of σ. We begin by stating a useful lower bound for σ; we shall revisit this topic in Proposition 14.18.

Theorem 14.1. *Let $\kappa \geq 1$. For $u \geq 2\kappa$, we have*

(14.1) $$\sigma_\kappa(u) > 1 - \frac{1}{2}\left(\frac{u}{2\kappa}\right)^{2\kappa} \exp(2\kappa - u)$$

(14.2) $$> 1 - \frac{2\kappa}{u},$$

and for every positive integer n

(14.3) $\sigma_\kappa(2n + 2\kappa)$
$$> 1 - \frac{1}{2}\left(1 - \frac{\kappa}{2\kappa + 2}\right)\left(1 + \frac{\kappa}{2n + 2 + \kappa}\right)\Gamma\!\left(\frac{\kappa}{2}\right)\frac{(\kappa/2)^{n+1}}{\Gamma(n+1+\kappa/2)}$$
$$> 1 - \frac{1}{2}\Gamma\!\left(\frac{\kappa}{2}\right)\frac{(\kappa/2)^{n+1}}{\Gamma(n+1+\kappa/2)}.$$

We defer our proof of Theorem 14.1 to the end of Section 14.4.

Many sieve formulas have nicer expressions when written in terms of σ_κ rather than j_κ (introduced in Chapter 5). But it is more convenient to study the function itself in the form $j(u) = j_\kappa(u) := \sigma_\kappa(2u)$. We have, by (5.29) and (5.30),

(14.4) $$j(u) = \begin{cases} 0, & u \leq 0, \\ e^{-\gamma\kappa} u^\kappa / \Gamma(\kappa + 1), & 0 < u \leq 1, \end{cases}$$

and j is continued forward as the continuous solution of

(14.5) $$uj'(u) = \kappa j(u) - \kappa j(u-1) = \kappa \int_{u-1}^{u} j'(t)dt, \quad u > 1;$$

in fact, (14.5) holds for all $u \geq 0$.

We begin studying j with some observations about the continuity of its derivatives. If $u > 0$ and $\kappa \geq 1$, then $j'(u)$ is continuous for $u > 0$ by the preceding equation and the continuity of j, and, more generally, by differentiating (14.5), we see that $j_\kappa^{(n)}(u)$ is continuous for $u \geq 0$ for all positive integers $n < \kappa$. If κ is a positive integer, then $j_\kappa^{(\kappa)}(u)$ has a jump discontinuity at $u = 0$, and $j_\kappa^{(\kappa+n)}(u)$ has jump discontinuities at $u = 1, \ldots, n$. If $\kappa > 1$ is not an integer, then $j^{([\kappa]+n)}$ has infinite jump discontinuities from the right at $u = 0, 1, \ldots, n-1$ for each positive integer n. In each of the preceding cases, the function is continuous at all other values of $u > 0$.

We show next for each $\kappa \geq 1$ that $j_\kappa(u)$ is a positive strictly increasing function for $u > 0$. By (14.4), $j'(u) > 0$ when $0 < u \leq 1$, and by (14.5), it remains positive for some distance to the right side of 1. Suppose there were a point $u_0 > 1$ with $j'(u_0) \leq 0$. By the continuity of j', we may assume that u_0 is the first such point, i.e., that $j'(u_0) = 0$ and $j'(t) > 0$ for $0 < t < u_0$. Upon evaluating the integral form of (14.5) at $u = u_0$, we obtain a contradiction, since the left side is 0 and the right side is the integral of a positive function. Hence

(14.6) $$j'(u) > 0, \quad u > 0;$$

and we deduce immediately that

(14.7) $$j(u) > 0, \quad u > 0.$$

The integral form of (14.5) and Lemma 12.1 imply that

(14.8) $$0 < j'(u) \ll \kappa^u / \Gamma(u+1) \ll e^{-u},$$

so that

$$0 < j(u_2) - j(u_1) = \int_{u_1}^{u_2} j'(t)dt \ll e^{-u_1}, \quad 0 < u_1 < u_2,$$

and therefore

$$\lim_{u \to \infty} j(u) \text{ exists.}$$

We shall evaluate this limit in two ways. The first method is based on the asymptotic behavior at 0 of $\widehat{j}(s)$, the Laplace transform of j.

14.1 Lower bounds on σ

Lemma 14.2. *Let $\kappa > 0$ and set*

$$\widehat{j}(s) := \int_0^\infty e^{-su} j(u) du, \quad \Re s > 0.$$

Then

(14.9) $$\widehat{j}(s) = \frac{1}{s} \exp\{-\kappa \operatorname{Ein}(s)\}$$

and

(14.10) $$\widehat{j'}(s) = \exp\{-\kappa \operatorname{Ein}(s)\},$$

where Ein *is defined in* (12.12).

Proof. Since j is a bounded, continuous function on $[0, \infty)$, its Laplace transform exists for $\Re s > 0$. By integration by parts,

$$\widehat{j}(s) = \frac{1}{s} \int_0^\infty e^{-su} j'(u) du, \quad \Re s > 0.$$

Thus $\widehat{j'}(s) = s\widehat{j}(s)$. Also, by (14.5),

$$(s\widehat{j}(s))' = -\int_0^\infty e^{-us} u j'(u) du = -\kappa \int_0^\infty e^{-us}\{j(u) - j(u-1)\} du$$

$$= -\kappa \widehat{j}(s) + \kappa \int_0^\infty e^{-(u+1)s} j(u) du = -\kappa(1 - e^{-s})\widehat{j}(s),$$

so that

$$\frac{\widehat{j'}(s)}{\widehat{j}(s)} = \frac{1}{s}(-1 - \kappa + \kappa e^{-s}).$$

The right side of the last formula has a primitive

$$\log b - (1+\kappa)\log s - \kappa \int_s^\infty e^{-t} \frac{dt}{t}$$

with b a constant, and thus, for $\Re s > 0$,

(14.11) $$\widehat{j}(s) = b s^{-1-\kappa} \exp\left(-\kappa \int_s^\infty e^{-t} \frac{dt}{t}\right) = \frac{b e^{\gamma\kappa}}{s} \exp\{-\kappa \operatorname{Ein}(s)\}$$

by (12.13).

On the other hand, by (14.4) with $B_\kappa := e^{\kappa\gamma} \Gamma(\kappa+1)$, we have

$$\widehat{j}(s) = B_\kappa^{-1} \int_0^1 e^{-su} u^\kappa du + \int_1^\infty e^{-su} j(u) du$$

$$= B_\kappa^{-1} s^{-\kappa-1} \int_0^s e^{-t} t^\kappa dt + \int_1^\infty e^{-su} j(u) du,$$

and letting $s \to \infty$, the last integral tends to 0 exponentially fast, so that
$$\widehat{j}(s) \sim e^{-\gamma\kappa} s^{-\kappa-1}, \quad s \to \infty.$$

By (14.11), we obtain
$$\widehat{j}(s) \sim bs^{-\kappa-1}, \quad s \to \infty,$$

since $\int_s^\infty e^{-t} dt/t \to 0$ as $s \to \infty$. Comparing the two expressions for $\widehat{j}$, we conclude that $b = e^{-\gamma\kappa}$. Hence, by (14.11) again, (14.9) and therefore (14.10) hold. □

Now we evaluate the limit of $j(u)$ at infinity. This asymptotic formula "justifies" the normalization constant $e^{-\kappa\gamma}/\Gamma(\kappa+1) =: B_\kappa^{-1}$ chosen for j_κ in (14.4).

Lemma 14.3. *For each $\kappa \geq 1$,*

(14.12)
$$\lim_{u \to \infty} j_\kappa(u) = 1,$$

and $j_\kappa(u) < 1$ for each positive number u.

Proof. We observed earlier in the chapter that $j(u)$ has a limit at infinity. If we let ℓ denote this limit, we have on the one hand
$$\widehat{j}(s) = \frac{1}{s} \int_0^\infty e^{-t} j(t/s) dt \sim \frac{\ell}{s} \text{ as } s \to 0+,$$

and on the other hand
$$\widehat{j}(s) = (1/s) \exp\{-\kappa \operatorname{Ein}(s)\} \sim 1/s \text{ as } s \to 0+,$$

since $\lim_{s \to 0} \operatorname{Ein}(s) = 0$ by (12.12). Hence $\ell = 1$.

Also, by (14.6), $j \uparrow$; thus $j(u) < 1$ for each positive number u. □

The last lemma and the estimate (14.8) for j' together imply that $1 - j(u)$ vanishes at infinity at a faster-than-exponential rate. Indeed,
$$1 - j(u) = \int_u^\infty j'(t) dt \ll \int_u^\infty \frac{\kappa^t \, dt}{\Gamma(t+1)} \ll \frac{\kappa^u}{\Gamma(u-1)} \int_u^\infty \frac{dt}{t(t-1)} \ll \frac{\kappa^u}{\Gamma(u)}.$$

We shall revisit this inequality below, establishing it in several explicit forms of various degrees of precision. Also, we shall obtain another proof of Lemma 14.3 from the identity (14.26) below.

14.2 Differential relations

The higher derivatives of $j(u)$ satisfy difference differential equations similar to (14.5). Upon differentiating (14.5), and then once again, we obtain

(14.13) $$uj''(u) = (\kappa - 1)j'(u) - \kappa j'(u-1)$$

and

(14.14) $$uj'''(u) = (\kappa - 2)j''(u) - \kappa j''(u-1).$$

In (14.5) itself, if we integrate by parts on the right, we obtain

$$uj'(u) = \kappa(t - \kappa + 1)j'(t)\Big|_{u-1}^{u} - \kappa \int_{u-1}^{u} (t - \kappa + 1)j''(t)dt$$

or

$$(u - \kappa)\{(\kappa - 1)j'(u) - \kappa j'(u-1)\} = \kappa \int_{u-1}^{u} (t - \kappa + 1)j''(t)dt;$$

hence, by (14.13), valid for $\kappa \geq 1$ and $u > 0$,

(14.15) $$u(u - \kappa)j''(u) = \kappa \int_{u-1}^{u} (t - \kappa + 1)j''(t)dt.$$

Lemma 14.4. *Suppose $\kappa > 1$. Then there exists a unique number, call it u_κ, between $\kappa - 1$ and κ, such that $j''(u) > 0$ for $0 < u < u_\kappa$ and $j''(u) < 0$ for all $u > u_\kappa$. For $\kappa = 1$, we have $j''(u) = 0$ for all $u < u_1 = \kappa = 1$ and $j''(u) < 0$ for all $u > 1$.*

Proof. For $\kappa = 1$ we have by (14.13) that $uj''(u) = -j'(u-1)$, an expression that is 0 for $u < 1$ and is negative (by (14.6)) for $u > 1$.

Now suppose $\kappa > 1$. On taking $u = \kappa$ in (14.15), we find that

$$\int_{\kappa-1}^{\kappa} (t - \kappa + 1)j''(t)dt = 0.$$

Since $t - \kappa + 1 > 0$ on $(\kappa - 1, \kappa)$, it follows that $j''(t)$ changes sign in this interval. By (14.4), $j''(u) > 0$ on $(0,1]$, and it follows from (14.13) and the continuity of j' that j'' is continuous on $[0, \infty)$. Thus there exists some number u_κ, the smallest value of $u > 1$ at which $j''(u) = 0$. By (14.14) at $u = u_\kappa$,

$$u_\kappa j'''(u_\kappa) = -\kappa j''(u_\kappa - 1) < 0$$

since $j''(u) > 0$ for $0 < u < u_\kappa$; whence u_κ is a simple zero of j''.

Suppose if possible that j'' has other zeros beyond u_κ, and let v be the least of these. We claim that

$$v < u_\kappa + 1;$$

for if, on the contrary, $v \geq u_\kappa + 1$, then $j''(v) = 0$ and $j''(u) < 0$ when $u_\kappa < u < v$. But then, by (14.13) at $u = v$,

$$0 = vj''(v) = (\kappa - 1)j'(v) - \kappa j'(v - 1),$$

so that

$$0 < j'(v) = \kappa\{j'(v) - j'(v-1)\} = \kappa j''(w)$$

for some w strictly between $v - 1 (\geq u_\kappa)$ and v, a contradiction.

Next suppose that $u_\kappa < v < u_\kappa + 1$. We know that $j''(u)$ is non-decreasing at $u = v$, so that $j'''(v) \geq 0$; yet, by (14.14),

$$vj'''(v) = -\kappa j''(v-1) < 0$$

since $v - 1 < u_\kappa$: also an impossibility. Hence v does not exist, and j'' has just the one zero u_κ, which is simple and lies in $(\kappa - 1, \kappa)$. □

Lemma 14.5. *Let $\kappa > 1$. Then*

$$\left(\frac{j'}{j}(u)\right)' < 0, \quad u > 0.$$

Proof. We may assume $u > 1$, since $\{(j'/j)(u)\}' = -\kappa/u^2$ for $0 < u \leq 1$. By (14.5) and (14.13),

$$(14.16) \quad u\left(\frac{j'}{j}(u)\right)'$$

$$= \frac{u}{j(u)^2}\{j(u)j''(u) - j'(u)^2\}$$

$$= j(u)^{-2}\{-j(u)j'(u) - \kappa j(u)j'(u-1) + \kappa j'(u)j(u-1)\}$$

$$= -\frac{j'}{j}(u) + \kappa \frac{j(u-1)}{j(u)}\left\{\frac{j'}{j}(u) - \frac{j'}{j}(u-1)\right\}$$

$$= -\frac{j'}{j}(u) + \kappa \frac{j(u-1)}{j(u)}\int_{u-1}^{u}\left(\frac{j'}{j}(t)\right)' dt,$$

and the integral expression on the right is valid because $j'(t)/j(t)$ has a continuous derivative for $0 < t < \infty$.

If the lemma were false, there would exist a least number, call it v, such that $v > 1$, $(j'/j)'(v) = 0$ and $(j'/j)'(u) < 0$ on $0 < u < v$. But if

14.2 Differential relations

we take $u = v$ in (14.16), we get 0 on the left and a negative quantity on the right, which is impossible. Thus the logarithmic derivative of j is strictly decreasing. □

Corollary 14.6. *If $\kappa > 1$ and $u > 0$, then $(1/j(u))'' > 0$.*

Proof. We have

$$\left(\frac{1}{j(u)}\right)'' = -\left(\frac{1}{j(u)} \cdot \frac{j'}{j}(u)\right)' = -\frac{1}{j(u)}\left(\frac{j'}{j}(u)\right)' + \frac{1}{j(u)}\left(\frac{j'}{j}(u)\right)^2.$$

The last term is clearly positive, and by the preceding lemma, the first term is as well. □

Corollary 14.7. *Let $\kappa > 1$ and $u > 1$. Then $j(u-1)/j(u)$ is increasing.*

Proof. It suffices to show that $\log\{j(u-1)/j(u)\}$ is increasing. The derivative of this function is

$$\frac{j'}{j}(u-1) - \frac{j'}{j}(u),$$

which is positive by Lemma 14.5. □

The last corollary and the identity

$$u\frac{j'}{j}(u) = \kappa - \kappa\frac{j(u-1)}{j(u)}$$

together imply that

$$(14.17) \qquad \left(u\frac{j'}{j}(u)\right)' < 0 \quad \text{if } u > 1 \text{ and } \kappa > 1.$$

A similar monotonicity relation holds for j''/j'. The following result will be the basis of our later analysis of j', $1-j$, and their quotient.

Lemma 14.8. *Let $2\kappa \in \mathbb{N}$, $\kappa \geq 3/2$. Then*

$$\left(\frac{j''}{j'}(u)\right)' < 0, \quad u > 0 \quad (u \neq 1 \text{ for } \kappa = 3/2, 2).$$

Proof. On the interval $(0,1)$ we have

$$\left(\frac{j''}{j'}(u)\right)' = \left(\frac{\kappa-1}{u}\right)' = -\frac{\kappa-1}{u^2} < 0.$$

For $\kappa > 2$, each of the functions j, j', j'', and j''' is continuous on $(0, \infty)$, so $(j''/j')' < 0$ holds on an open interval that contains the point 1.

For $\kappa = 3/2, 2$, each of j, j', and j'' is continuous on $(0, \infty)$, but

$(j''/j')'$ does not exist at $u = 1$. A small calculation shows that $j''' < 0$ immediately to the right of $u = 1$, while $j'(u) > 0$ for all $u > 0$. Thus

$$\left(\frac{j''}{j'}(u)\right)' = \frac{j'(u)j'''(u) - j''(u)^2}{j'(u)^2} < 0$$

on some interval $1 < u < 1 + \epsilon$.

We proceed to give two formulas involving derivatives of j, which will be combined to yield our main identity. We start with the basic differential equation (14.5) for $j'(u)$, divide it by $\kappa j'(u-1)$, and make a small manipulation:

(14.18)
$$\frac{u}{\kappa} \frac{j'(u)}{j'(u-1)} = \int_0^1 \frac{j'(u-1+t)}{j'(u-1)} dt = \int_0^1 \exp\left(\int_{u-1}^{t+u-1} \frac{j''}{j'}(s)\, ds\right) dt.$$

Next, we have by (14.13)

$$uj''(u) = (\kappa - 1)j'(u) - \kappa j'(u-1).$$

Dividing by $j'(u)$ and doing a little algebra yields

$$0 < \frac{u}{\kappa} \frac{j'(u)}{j'(u-1)} = 1 \Big/ \left(\frac{\kappa - 1}{u} - \frac{j''}{j'}(u)\right), \quad u > 1.$$

If we equate the last expression with (14.18), we get

$$1 \Big/ \left(\frac{\kappa - 1}{u} - \frac{j''}{j'}(u)\right) = \int_0^1 \exp\left(\int_{u-1}^{t+u-1} \frac{j''}{j'}(s)\, ds\right) dt.$$

We differentiate each side of this equation with respect to u to obtain a formula for $(j''/j')'(u)$.

The derivative of the left side is

$$\left\{\frac{\kappa - 1}{u^2} + \left(\frac{j''}{j'}(u)\right)'\right\} \Big/ D^2, \quad \text{where} \quad D = \frac{\kappa - 1}{u} - \frac{j''}{j'}(u) > 0,$$

and the derivative of the right side is

$$\int_0^1 \exp\left(\int_{u-1}^{t+u-1} \frac{j''}{j'}(s)\, ds\right) \left(\frac{j''}{j'}(u-1+t) - \frac{j''}{j'}(u-1)\right) dt.$$

Equating the last two expressions and solving for $(j''/j')'$, we obtain

(14.19) $$\left(\frac{j''}{j'}(u)\right)' = -\left(\frac{\kappa - 1}{u^2}\right) + D^2 \int_0^1 \exp\left(\int_{u-1}^{t+u-1} \frac{j''}{j'}(s)\, ds\right) \times$$
$$\left(\frac{j''}{j'}(u-1+t) - \frac{j''}{j'}(u-1)\right) dt.$$

If the lemma were false, there would be a smallest point $u > 1$ such that $(j''/j')'(t) < 0$ for $0 < t < u$ but $(j''/j')'(u) = 0$. Since $u - 1 < u - 1 + t < u$ for $0 < t < 1$, the j''/j' difference on the second line of (14.19) is negative, and so the entire right side of (14.19) is negative. Thus there can exist no such point u. □

14.3 The adjoint function of j

We introduce an adjoint function as an aid in studying j (cf. the occurrence of p and q in (12.9) and (12.10)). This is a function $r(u) = r_\kappa(u)$ defined for $\kappa > 0$ by

(14.20) $\qquad (ur(u))' = \kappa r(u+1) - \kappa r(u), \quad u > 0,$

with the normalization

$$\lim_{u \to \infty} ur(u) = 1.$$

Formally, $r_\kappa(u) = p_{-\kappa}(u)$, but p_κ has been defined only for positive κ.

A normalized solution of (14.20) is provided by the Laplace transform

(14.21) $\qquad r_\kappa(u) = \int_0^\infty \exp\{-ut + \kappa \operatorname{Ein}(t)\} dt$

(cf. (12.11)), where Ein was defined in (12.12). To see that this integral satisfies (14.20), first integrate it by parts, next multiply by u, and then differentiate with respect to u. The behavior of $r(u)$ as $u \to \infty$ is no harder to derive: by an earlier remark based on (12.12), $0 \le \operatorname{Ein}(t) \le t$ when $t \ge 0$, whence

$$\int_0^\infty \exp(-ut) dt < r(u) < \int_0^\infty \exp(-ut + \kappa t) dt,$$

and it follows at once that

$$u^{-1} < r_\kappa(u) \ (u > 0) \quad \text{and} \quad r_\kappa(u) < (u - \kappa)^{-1} \ (u > \kappa).$$

The integral representation of $r(u)$ shows that $(-1)^\nu r^{(\nu)}(u) > 0$ for $u > 0$ and $\nu = 0, 1, 2, \ldots$, and, in particular, that $r'(u) < 0$ and $r''(u) > 0$.

Since $r(u)$ is decreasing in u, (14.20) implies that $\{ur(u)\}' < 0$, and hence that $ur(u) > (u+1)r(u+1)$ or

(14.22) $\qquad r(u+1)/r(u) < u/(u+1), \quad u > 0.$

Moreover, again by (14.20),

(14.23) $\qquad \{(u - \kappa)r(u)\}' = \kappa\{r(u+1) - r(u) - r'(u)\} > 0$

by Taylor's theorem, since $r'' > 0$ on $(0, \infty)$, whence
$$r(u+1)/r(u) > (u-\kappa)/(u-\kappa+1), \quad u \geq \kappa.$$

The Iwaniec "inner product" (introduced in Chapter 12)
$$\langle j, r \rangle_\kappa := uj(u)r(u) - \kappa \int_{u-1}^{u} j(t)r(t+1)\,dt, \quad u > 0,$$
is constant, as one can verify by differentiating it and using the defining equations of r and j. To evaluate this constant let $u \to 0+$; we have by (14.21) and (12.13)
$$r(u) = \int_0^1 O(1)\,dt + \int_1^\infty \exp\{-ut + \kappa \log t + \gamma\kappa + o(1)\}\,dt$$
$$\sim e^{\gamma\kappa} \int_0^\infty \exp(-ut)\,t^\kappa\,dt = e^{\gamma\kappa}\Gamma(\kappa+1)u^{-\kappa-1}, \quad u \to 0+.$$
Hence, by (14.4), $uj(u)r(u) \to 1$ as $u \to 0$ and so

(14.24) $\qquad uj(u)r(u) - \kappa \int_{u-1}^{u} j(t)r(t+1)\,dt = 1, \quad u > 0.$

In the same vein, $\langle 1, r \rangle_\kappa(u)$ is constant by (14.20), and, since $ur(u) \to 1$ as $u \to \infty$, we see that

(14.25) $\qquad \langle 1, r \rangle_\kappa = ur(u) - \kappa \int_{u-1}^{u} r(t+1)\,dt = 1, \quad u > 0.$

14.4 Inequalities for $1 - j$

We showed above that $1 - j(u)$ vanishes rapidly at infinity. Here we give explicit estimates that will be used to prove Theorem 14.1. We begin by establishing a convexity relation involving j and r.

Lemma 14.9. *Suppose $\kappa \geq 1$ and $u \geq \kappa$. Then*
$$J(t) := J_\kappa(t) := \{1 - j(t)\}r(t+1)$$
is convex in t on the interval $u - 1 \leq t \leq u$.

Proof. We have by (14.21),
$$J''(t) = -r(t+1)j''(t) + 2(-j'(t))r'(t+1) + (1-j(t))r''(t+1)$$
$$= \int_0^\infty \{-j''(t) + 2(-j'(t))(-u) + (1-j(t))u^2\}$$
$$\times \exp\{-(t+1)u + \kappa\,\mathrm{Ein}(u)\}\,du.$$

14.4 Inequalities for $1 - j$

By (14.13), the expression in curly brackets is equal to

$$-\frac{1}{t}((\kappa - 1)j'(t) - \kappa j'(t-1)) + 2uj'(t) + (1 - j(t))u^2$$
$$= \left(2u - \frac{\kappa - 1}{t}\right)j'(t) + \frac{\kappa}{t}j'(t-1) + (1 - j(t))u^2;$$

The second and third terms here are positive, and the coefficient of $j'(t)$ is at least

$$2u - \frac{\kappa - 1}{u - 1} \geq 2u - 1 > 0$$

since $u \geq \kappa$. Hence $J'' > 0$. □

If we subtract equations (14.24) and (14.25), we obtain the formula

$$(14.26) \quad ur(u)\{1 - j(u)\} = \kappa \int_{u-1}^{u} \{1 - j(t)\}r(t+1)\,dt, \quad u > 0.$$

This relation is the springboard to reach the estimate (14.3). We briefly reprove two earlier results here, to show the strength of this formula.

First, $j(u) < 1$ for all $u \geq 0$: recall that

$$1 - j(u) \geq 1 - e^{-\gamma\kappa}/\Gamma(\kappa + 1) > 0$$

in the initial interval, $j(u)$ is continuous for $u \geq 0$, and also $r(u) > 0$ for $u > 0$; hence there cannot be a first point $u^* \geq 1$ at which $1 - j(u^*) \leq 0$.

Next, $1 - j(u)$ converges to 0 rapidly at infinity: since $r(u)$ decreases, we have from (14.26)

$$u\{1 - j(u)\} < \kappa \int_{u-1}^{u} \{1 - j(t)\}\,dt, \quad u > 0.$$

It now follows from Lemma 12.1 that $1 - j(u) \ll \kappa^u/\Gamma(u+1)$.

Returning to our main project, we change (14.26) to remove the integral, producing a form that is more convenient for explicit inequalities. A trapezoidal estimate, using the preceding convexity lemma, yields

$$ur(u)\{1 - j(u)\} < \frac{\kappa}{2}\{(1 - j(u))r(u+1) + (1 - j(u-1))r(u)\}, \quad u \geq \kappa.$$

This inequality may be rewritten in two different ways:

(i) In iterative form

$$(1 - j(u))\{2ur(u) - \kappa r(u+1)\} < \kappa\{1 - j(u-1)\}r(u), \quad u \geq \kappa.$$

The last inequality and (14.22) imply that

$$1 - j(u) < \frac{\kappa}{2u - \kappa u/(u+1)} \{1 - j(u-1)\}$$

or

(14.27) $$1 - j(u) < \frac{u+1}{u} \frac{\kappa/2}{u+1-\kappa/2} \{1 - j(u-1)\},$$

an inequality that we shall later use iteratively for $u \geq \kappa + 1$.

(ii) As a differential inequality

$$2ur(u)\{1 - j(u)\} < \kappa\{1 - j(u)\}\{r(u+1) + r(u)\} + \kappa\{j(u) - j(u-1)\}r(u)$$

for $u \geq \kappa$. Another application of (14.22) yields

$$2u\{1 - j(u)\} < \kappa\{1 - j(u)\}\{u/(u+1) + 1\} + \kappa\{j(u) - j(u-1)\}.$$

If we replace the last term, using (14.5), we obtain

$$\{2u - \kappa - \kappa u/(u+1)\}\{1 - j(u)\} < uj'(u),$$

i.e.,

(14.28) $$\{1 - j(u)\}' + \left(2 - \frac{\kappa}{u} - \frac{\kappa}{u+1}\right)\{1 - j(u)\} < 0, \quad u \geq \kappa.$$

Multiplying by the integrating factor $\exp\{2u - \kappa \log u - \kappa \log(u+1)\}$ and integrating from κ to u, we find that

$$\{1 - j(u)\}e^{2u}\{u(u+1)\}^{-\kappa} \leq \{1 - j(\kappa)\}e^{2\kappa}\{\kappa(\kappa+1)\}^{-\kappa}, \quad u \geq \kappa,$$

i.e.,

(14.29) $$1 - j(u) \leq \left(\frac{u(u+1)}{\kappa(\kappa+1)}\right)^{\kappa} e^{2\kappa - 2u} \{1 - j(\kappa)\}, \quad u \geq \kappa.$$

Here is another property of $j_\kappa(u)$ that is complementary to Theorem 14.11 and follows readily from the last inequality.

Corollary 14.10. [GrR88] *If $c > 1$ is a constant, then $j_\kappa(c\kappa) \to 1$ from below as $\kappa \to \infty$.*

Proof. Let $\delta = c - 1 > 0$, a constant, and $u = (1+\delta)\kappa$. Since we have $(u+1)/(\kappa+1) < u/\kappa$ and $1 - j(u) < 1$ for $u > 0$,

$$0 < 1 - j(\{1+\delta\}\kappa) < \{(1+\delta)\kappa/\kappa\}^{2\kappa} e^{2\kappa - 2(1+\delta)\kappa} = \left(\frac{1+\delta}{e^\delta}\right)^{2\kappa} \to 0$$

as $\kappa \to \infty$. □

14.4 Inequalities for $1 - j$

Now we give a striking uniform lower bound for j_κ, which we shall combine with the preceding work to obtain explicit upper bounds for $1 - j(u)$.

Theorem 14.11. [GrR88] *If $\kappa > 1$ then $j_\kappa(\kappa) > \frac{1}{2}$.*

Proof. From (14.9) it follows that

$$(1 - j_\kappa)\hat{\,}(s) = \int_0^\infty e^{-su}(1 - j_\kappa(u))\,du = \frac{1 - \exp(-\kappa\,\mathrm{Ein}(s))}{s},$$

valid for $\Re s \geq 0$, since $1 - j(u)$ vanishes rapidly at infinity. By Fourier inversion (Laplace inversion on the imaginary axis), we have for any $u > 0$

$$1 - j_\kappa(u) = \lim_{T\to\infty} \frac{1}{2\pi}\int_{-T}^T e^{iuy}\{1 - \exp(-\kappa\,\mathrm{Ein}(iy))\}\frac{dy}{iy}.$$

Since j is real valued, we have at $u = \kappa$

$$1 - j_\kappa(\kappa) = \Re\left\{\frac{1}{2\pi i}\int_{-\infty}^\infty e^{i\kappa y}(1 - \exp\{-\kappa\,\mathrm{Ein}(iy)\})\frac{dy}{y}\right\}$$

$$= \frac{1}{\pi}\int_0^\infty \sin\kappa y\,\frac{dy}{y} - \Re\left\{\frac{1}{2\pi i}\int_{-\infty}^\infty e^{-\kappa(\mathrm{Ein}(iy)-iy)}\frac{dy}{y}\right\}.$$

The first expression on the right is known to be equal to $1/2$. In the second expression,

$$\mathrm{Ein}(iy) - iy = \int_0^y \frac{1 - \cos t}{t}\,dt + i\int_0^y \frac{\sin t - t}{t}\,dt$$

$$= C(y) + iS(y),$$

say, where $C(y)$ is an even function of y and $S(y)$ an odd function. Hence

$$(14.30) \qquad j_\kappa(\kappa) - \frac{1}{2} = \frac{1}{\pi}\int_0^\infty e^{-\kappa C(y)}\sin(-\kappa S(y))\frac{dy}{y}.$$

We complete the proof by showing that the integral on the right is positive.

Since

$$\sin(-\kappa S(y)) = \left(\kappa\frac{\sin y - y}{y}\right)^{-1}\frac{d}{dy}\cos(-\kappa S(y)),$$

the integral equals, after integrating by parts,

$$\frac{1}{\kappa}\left\{\frac{e^{-\kappa C(y)}}{y - \sin y}(1 - \cos\{\kappa S(y)\})\right\}\bigg|_0^\infty - \frac{1}{\kappa}\int_0^\infty (1 - \cos\{\kappa S(y)\})\left(\frac{e^{-\kappa C(y)}}{y - \sin y}\right)'dy.$$

The integrated term vanishes at infinity, since $C(y) \sim \log y$ as $y \to \infty$, and it vanishes also at 0, since

$$1 - \cos(\kappa S(y)) \sim \frac{1}{2!}(\kappa S(y))^2 \sim \frac{\kappa^2}{648} y^6 \quad \text{as } y \to 0$$

whereas

$$y - \sin y \sim \frac{1}{6} y^3 \quad \text{as } y \to 0.$$

As for the integral, we observe that each of $e^{-\kappa C(y)}$ and $(y - \sin y)^{-1}$ is positive and decreasing as y increases, so that

$$-\frac{d}{dy}\left(\frac{e^{-\kappa C(y)}}{y - \sin y}\right) > 0.$$

Since $1 - \cos(\kappa S(y)) > 0$, this completes the proof that the integral on the right side of (14.30) is positive. □

We can now provide our promised proof of Theorem 14.1:

Proof of Theorem 14.1. The O-bounds we have given for $1 - j(u)$ do not provide numerical estimates. Here we shall combine the preceding theorem with the inequalities (14.29) and (14.27) to obtain the explicit inequalities given in Theorem 14.1.

Starting with (14.29), we recall the inequality $(u+1)/(\kappa+1) \le u/\kappa$, valid for $u \ge \kappa$, and the inequality $1 - j_\kappa(\kappa) < 1/2$ from Theorem 14.11. We find that

$$1 - j_\kappa(u) < \frac{1}{2}\left(\frac{u}{\kappa}\right)^{2\kappa} \exp(2\kappa - 2u), \quad u \ge \kappa.$$

Changing back to the σ function, we obtain (14.1). Surprisingly, this inequality shows only an exponential rate of decay of $1 - j(u)$ as $u \to \infty$, but its simplicity is appealing. We revisit this matter at the end of the chapter.

To see that (14.2) holds, set $x := u/(2\kappa) \ge 1$ and note that

$$\frac{x}{e^x - 1} \le \frac{x}{1 + (x-1) + (x-1)^2/2} = \frac{x}{x + (x-1)^2/2} = \frac{2x}{1 + x^2} \le 1.$$

Thus

$$\left(\frac{x}{e^x - 1}\right)^{2\kappa} \le \frac{x}{e^x - 1} \le \frac{2x}{1 + x^2} < \frac{2}{x},$$

and so, for $u \ge 2\kappa$,

$$1 - \sigma_\kappa(u) < \frac{1}{2}\left(\frac{u}{2\kappa}\right)^{2\kappa} \exp\left\{2\kappa\left(1 - \frac{u}{2\kappa}\right)\right\} < \frac{2\kappa}{u}.$$

Finally, we establish (14.3), which, since $j(u) < 1$ for all real u, shows the decay of $1 - j$ or $1 - \sigma$ at ∞ to be faster than exponential. We substitute (14.27) into itself $n - 1$ times, starting with $u = \kappa + 1$ and use Theorem 14.11 to estimate $1 - j_\kappa(\kappa)$. We find

$$1 - j_\kappa(\kappa + n) < \frac{1}{2}\frac{\kappa + n + 1}{\kappa + 1} \frac{(\kappa/2)^n}{(\kappa/2 + 2)(\kappa/2 + 3)\cdots(\kappa/2 + n + 1)},$$

which is, after simplification, equivalent to the first inequality of (14.3). The second inequality follows immediately from the first one and

$$\left(1 - \frac{\kappa}{2\kappa + 2}\right)\left(1 + \frac{\kappa}{2n + 2 + \kappa}\right) < 1,$$

which is valid for all positive values of κ and n. □

14.5 Relations between σ' and $1 - \sigma$

We develop here better upper and lower estimates of $(1 - \sigma)/\sigma'$. This work was motivated by the unfavorable comparison of (14.1) with (14.3): the latter shows that the function $1 - \sigma(u)$ converges to 0 as $u \to \infty$ about as fast as $1/\Gamma(u)$, while the first formula implies only exponential decay. These inequalities follow from the differential inequality (14.29) and the recurrence (14.27) respectively. Here we develop an improved differential relation to show the faster decay. As usual, we carry out the argument for j and convert to σ at the end.

As a starter, we show the quotient of j' and $1 - j$ to be monotonic.

Lemma 14.12. *Suppose that $\kappa \geq 3/2$. Then*

$$\psi(u) := j'(u)/(1 - j(u)) \uparrow, \quad u > 0.$$

Proof. We have

$$\frac{1}{\psi(u)} = \frac{1 - j(u)}{j'(u)} = \int_0^\infty \frac{j'(t + u)}{j'(u)} dt.$$

Now

$$\left\{\frac{1}{\psi(u)}\right\}' = \int_0^\infty \frac{j'(t + u)}{j'(u)} \left\{\frac{j''}{j'}(t + u) - \frac{j''}{j'}(u)\right\} dt < 0,$$

since $j''/j' \downarrow$ on $(0, \infty)$ by Lemma 14.8. □

The key step in our program to analyze j' and $1 - j$ is to establish the following two-sided bounds on $\exp(-j''/j')$. This, in turn, will lead to inequalities for j''/j' itself and then to new inequalities for $1 - j$.

Lemma 14.13. *Suppose $\kappa \geq 3/2$ and $u > 1$. Then*

$$(14.31) \qquad \frac{\exp\left(-(j''/j')(u-1)\right)-1}{-(j''/j')(u-1)} < \frac{u}{\kappa} < \frac{\exp\left(-(j''/j')(u)\right)-1}{-(j''/j')(u)}.$$

(If either of $(j''/j')(u-1)$, $(j''/j')(u)$ is 0, the corresponding quotient above is interpreted as 1.)

Proof. Starting from the basic equation (14.5) for j', we have

$$\frac{u}{\kappa} = \int_{u-1}^{u} \frac{j'(t)}{j'(u)}\,dt = \int_{u-1}^{u} \exp\left\{\log\frac{j'(t)}{j'(u)}\right\}dt = \int_{u-1}^{u} \exp\left\{\int_{t}^{u} -\frac{j''}{j'}(s)\,ds\right\}dt.$$

By Lemma 14.8, $-j''/j' \uparrow$, and the result follows at once. □

14.6 The ξ function

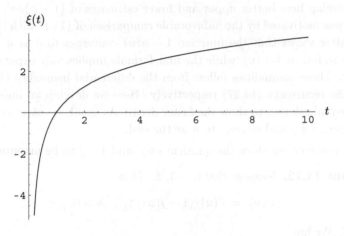

Fig. 14.1. The function $\xi(t)$

Each side of (14.31) involves a composition of j''/j' with the function

$$X : x \mapsto (e^x - 1)/x.$$

Since we are interested in estimating j''/j' itself, we should consider the inverse of the function X. For $t > 0$, define $\xi(t)$ to be the solution h of the equation $X(h) = t$ (see Figure 14.1). This is a near-logarithmic function that is familiar, e.g., in Dickman's theory. It is easy to see that

14.6 The ξ function

Table 14.1. *Values of* ξ, $2\log t$, ℓ_1, *and* ℓ_2

t	$\xi(t)$	$2\log t$	$\ell_1(t)$	$\ell_2(t)$
1.1	0.18769	0.19062	0.18635	0.20360
3.0	1.90381	2.19722	1.83989	1.93946
10.0	3.61495	4.60517	3.49729	3.62783
30.0	5.02155	6.80239	4.88307	5.02679
100.0	6.47460	9.21034	6.32886	6.48100
300.0	7.75219	11.4075	7.60645	7.76360
1000.0	9.11813	13.8155	8.97560	9.13662

X is positive and strictly increasing. Thus ξ is well defined on $(0, \infty)$ and increasing there. Also, its power series

$$X(x) = \sum_{n=0}^{\infty} \frac{x^n}{(n+1)!}$$

shows that $X(\cdot)$ is convex on $(0, \infty)$, and so ξ is concave on $(1, \infty)$. There is a geometric interpretation of $\xi(t)$ as the unique number h for which the line joining the points $(0, 1)$ and (h, e^h) has slope t.

The last lemma and the monotonicity of ξ yield inequalities for $-j''/j'$:

Corollary 14.14. *Suppose* $\kappa \geq 3/2$ *and* $u > 1$. *Then*

$$\xi\left(\frac{u}{\kappa}\right) < -\frac{j''}{j'}(u) < \xi\left(\frac{u+1}{\kappa}\right).$$

It is not hard to approximate ξ. We have, for all $t > 1$,

(14.32) $\quad \xi(t) < 2\log t,$

(14.33) $\quad \xi(t) > \ell_1(t) := \log t + \log(1 + \log t),$

(14.34) $\quad \xi(t) < \ell_2(t) := \log t + \log(1 + 1.2 \log t).$

(These relationships are illustrated in Table 14.1 and Figure 14.2.)

To see (14.32), for example, it suffices by monotonicity to show that

$$\frac{\exp(2\log t) - 1}{2\log t} > t$$

holds for all $t > 1$, an exercise in calculus. Similar reasoning establishes (14.33) and also the asymptotic formula

$$\xi(t) = \log t + \log\log t + O\left(\frac{\log\log t}{\log t}\right), \quad t \to \infty.$$

A more precise expansion of this function is given in [HlTn93].

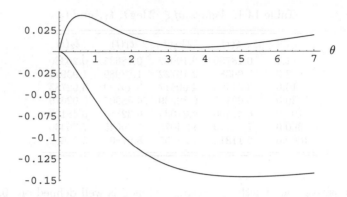

Fig. 14.2. The differences $\ell_2(t) - \xi(t)$ (upper curve) and $\ell_1(t) - \xi(t)$ (lower curve), where $t = e^\theta$.

The proof of (14.34) is more elaborate, and we give the details here.

Lemma 14.15. $\xi(t) < \ell_2(t)$ for all $t > 1$.

Proof. Using the monotonicity of ξ again, it suffices to show that
$$\frac{\exp\{\log t + \log(1 + 1.2\log t)\} - 1}{\log t + \log(1 + 1.2\log t)} > t, \quad t > 1.$$
After simplification, the last inequality is equivalent to
$$(14.35) \quad g(t) := t + 0.2t\log t - 1 - t\log(1 + 1.2\log t) > 0, \quad t > 1.$$
We have $g(1) = 0$ and
$$g'(t) = 1.2 + 0.2\log t - \log(1 + 1.2\log t) - \frac{1.2}{1 + 1.2\log t};$$
we shall show that $g' > 0$ on $(1, \infty)$. It is convenient to express $g'(t)$ in terms of the variable $v := 1 + 1.2\log t$. For $t \geq 1$, we have $v \geq 1$ and
$$g'(t) = h(v) := \frac{6}{5} - \frac{1}{6} + \frac{v}{6} - \log v - \frac{6}{5v}.$$
Clearly $g'(1) = h(1) = 0$; we shall show that $h > 0$ on $(1, \infty)$.

We have
$$6v^2 h'(v) = v^2 - 6v + 36/5,$$
a quadratic expression that has zeros at $v = 3 \pm 3/\sqrt{5}$. Thus $h'(v) > 0$ except on the interval $3 - 3/\sqrt{5} \leq v \leq 3 + 3/\sqrt{5}$. It follows that $h(v) > 0$

for $1 < v \leq 3 - 3/\sqrt{5}$, and the minimal value of $h(v)$ on $(3 - 3/\sqrt{5}, \infty)$ occurs at the point $v = 3 + 3/\sqrt{5}$. Since

$$h(3 + 3/\sqrt{5}) = \frac{31}{30} + \frac{1}{\sqrt{5}} - \log(3 + 3/\sqrt{5}) = 0.01229\ldots > 0,$$

$h(v) = g'(t) > 0$ on $(1, \infty)$. Thus (14.35) holds, and this in turn establishes (14.34). □

Another estimate that we shall need is one for ξ'/ξ. Implicit differentiation of the defining relation for ξ gives $\xi'(t) = \xi(t)/(1 + t\xi(t) - t)$; note that the last denominator is positive for $t > 1$, since $\xi(t)$ and $\xi'(t)$ are each positive. By (14.33) and a small calculation, we obtain

(14.36) $\quad \dfrac{\xi'}{\xi}(t) < \dfrac{1}{1 + t\{\log t + \log(1 + \log t)\} - t} < \dfrac{1}{t \log t}, \quad t > 1.$

Next, we establish two-sided inequalities for $\psi(u) := j'(u)/(1 - j(u))$.

Lemma 14.16. *Assume that $u > \kappa \geq 3/2$. Then*

$$\xi\left(\frac{u}{\kappa}\right) < \psi(u) < \xi\left(\frac{u+1}{\kappa}\right) + \frac{2}{\kappa} \frac{\xi'}{\xi}\left(\frac{u+1}{\kappa}\right).$$

Proof. Starting as in the proof of Lemma 14.12, we have

$$\frac{1}{\psi(u)} = \int_0^\infty \frac{j'(t+u)}{j'(u)} \, dt = \int_0^\infty \exp\left(\int_u^{u+t} \frac{j''}{j'}(s) \, ds\right) dt.$$

Since $j''/j' \downarrow$, by Lemma 14.8, it follows that

$$t \frac{j''}{j'}(u+t) < \int_u^{u+t} \frac{j''}{j'}(s) \, ds < t \frac{j''}{j'}(u), \quad t, u > 0.$$

Then, estimating j''/j' by Corollary 14.14, we get

$$-t\xi\left(\frac{u+t+1}{\kappa}\right) < \int_u^{u+t} \frac{j''}{j'}(s) \, ds < -t\xi\left(\frac{u}{\kappa}\right), \quad t > 0, \, u > 1.$$

Now,

$$\frac{1}{\psi(u)} < \int_0^\infty \exp\left(-\xi\left(\frac{u}{\kappa}\right)t\right) dt = \frac{1}{\xi(u/\kappa)},$$

which gives the first inequality of the lemma. (Note that the last integral is convergent, because $\xi(u/\kappa) > 0$ for $u > \kappa$.)

For the other inequality, recall that ξ is concave on $(0, \infty)$, so by Taylor's formula,

$$\xi\left(\frac{u+t+1}{\kappa}\right) < \xi\left(\frac{u+1}{\kappa}\right) + \frac{t}{\kappa}\xi'\left(\frac{u+1}{\kappa}\right) =: A + Bt,$$

where

$$A := \xi\left(\frac{u+1}{\kappa}\right), \quad B := \frac{1}{\kappa}\xi'\left(\frac{u+1}{\kappa}\right).$$

Thus

$$\frac{1}{\psi(u)} > \int_0^\infty \exp\left(-\xi\left(\frac{u+1+t}{\kappa}\right)t\right)dt > \int_0^\infty \exp\left(-At - Bt^2\right)dt.$$

Changing the variable, we find that

(14.37) $$\qquad 1/\psi(u) > L(A/\sqrt{B})/\sqrt{B},$$

where

(14.38) $$\qquad L(s) := \int_0^\infty e^{-st} e^{-t^2} dt$$

is the Laplace transform of the Gaussian function. This integral can be evaluated numerically in terms of the error function

$$L(s) = e^{s^2/4} \int_{s/2}^\infty e^{-t^2} dt = \frac{\sqrt{\pi}}{2} e^{s^2/2} \{1 - \mathrm{Erf}(s/2)\}.$$

A simple lower bound for $L(s)$ for positive values of s is

(14.39) $$\qquad L(s) > s/(s^2 + 2), \quad s > 0.$$

To show this inequality, we first establish a differential equation for L. Differentiating (14.38) and then integrating by parts, we obtain

$$L'(s) = -\int_0^\infty e^{-st} te^{-t^2} dt = -\frac{1}{2} + \frac{s}{2}L(s).$$

If we differentiate the last expression and then substitute for L', we get

$$L''(s) = \frac{L(s)}{2} + \frac{s}{2}L'(s) = L(s)\left\{\frac{1}{2} + \frac{s^2}{4}\right\} - \frac{s}{4}.$$

This is positive, since

$$L''(s) = \int_0^\infty e^{-st} t^2 e^{-t^2} dt > 0.$$

Thus (14.39) holds. Incidentally, a small calculation shows that $L''(s) \ll 1/s^3$ as $s \to \infty$, so the two sides of (14.39) differ by a term of order s^{-5}.

Finally, if we combine (14.37) and (14.39), we obtain the claimed upper bound for ψ. □

The following corollary plays an important role in the proof of Proposition 17.3; in particular in establishing 17.21.

Corollary 14.17. *Suppose $u/2 > \kappa \geq 3/2$. Then*

$$\frac{\sigma'(u)}{1-\sigma(u)} < \begin{cases} 1.12, & 2\kappa < u \leq 4.6\kappa - 2, \\ 0.64 + 0.57 \log\{(u+2)/(2\kappa)\}, & u > 4.6\kappa - 2. \end{cases}$$

Proof. Recall that $\sigma(u) = j(u/2)$. Thus

$$\frac{\sigma'(u)}{1-\sigma(u)} = \frac{(1/2)j'(u/2)}{1-j(u/2)} = \frac{1}{2}\psi(u/2).$$

Applying the upper estimate for ψ from the preceding lemma and then estimating ξ and ξ'/ξ by (14.34) and (14.36), we find that

(14.40) $$\frac{\sigma'(u)}{1-\sigma(u)} \leq \frac{1}{2}v + \frac{1}{2}\log\{1+1.2v\} + \frac{1}{(3/2)ve^v},$$

where

$$v = \log\frac{u/2+1}{\kappa} = \log\frac{u+2}{2\kappa} > 0.$$

Suppose first that u is in the initial range. We showed in Lemma 14.12 that $\psi(t) \uparrow$ for all $t > 0$. It follows from (14.40) that

$$\frac{\sigma'(u)}{1-\sigma(u)} \leq \frac{\sigma'(4.6\kappa-2)}{1-\sigma(4.6\kappa-2)} < 1.111, \quad 2\kappa < u \leq 4.6\kappa - 2.$$

For the later range, we find a simple upper bound for the last two terms of (14.40). Let

$$h(v) := 0.07v + 0.64 - \left\{\frac{1}{2}\log(1+1.2v) + \frac{2}{3ve^v}\right\}, \quad v \geq \log 2.3.$$

We have $h(\log 2.3) > 0.00385 > 0$. Also,

$$h'(v) = .07 - \frac{.6}{1+1.2v} + \frac{2}{3ve^v} + \frac{2}{3v^2e^v},$$

from which we see that $h'(v) > .5 > 0$ for v near $\log 2.3$ and $h'(v) > 0$ for $v \geq 265/42 \approx 6.30952$. A Mathematica calculation shows that h has roots $v_1 \approx 1.60724$ and $v_2 \approx 6.28625$, so h is increasing on $[\log 2.3, v_1)$, decreasing on (v_1, v_2), and increasing again on (v_2, ∞). It follows that h has one interior minimum, at v_2, and direct evaluation shows that

$h(v_2) > 0.00725 > 0$. Thus $h > 0$ throughout its range, so the right side of (14.40) is less than $0.5v + 0.07v + 0.64$ for $u > 4.6\kappa - 2$. □

14.7 An improved upper bound for $1 - j$

We conclude this chapter by giving a sharper, though more complicated, upper bound for $1 - j_\kappa$ than that of Theorem 14.1 (which is expressed in terms of σ_κ).

Proposition 14.18. *For $u > \kappa \geq 3/2$ we have*

$$1 - j_\kappa(u) < \frac{1}{2} \exp\left\{ -\kappa \int_1^{u/\kappa} \{\log v + \log(1 + \log v)\}\, dv \right\}$$

$$= \frac{1}{2} \exp\left\{ -u \log\left(\frac{u}{\kappa} \log \frac{eu}{\kappa}\right) + u - \kappa + \frac{\kappa}{e}(\mathrm{li}(eu/\kappa) - \mathrm{li}\, e) \right\}.$$

Proof. By the preceding estimates,

$$\frac{-j'(t)}{1 - j(t)} < -\xi(t/\kappa) < -\log(t/\kappa) - \log(1 + \log(t/\kappa)), \qquad t > \kappa.$$

Integrating and changing the variable, we obtain

$$\log \frac{1 - j(u)}{1 - j(\kappa)} = \int_\kappa^u \frac{-j'(t)\, dt}{1 - j(t)} < -\kappa \int_1^{u/\kappa} \{\log v + \log(1 + \log v)\}\, dv$$

$$= -\kappa \left\{ \frac{u}{\kappa} \log \frac{u}{\kappa} - \frac{u}{\kappa} + 1 + \frac{u}{\kappa} \log\left(1 + \log \frac{u}{\kappa}\right) - \int_1^{u/\kappa} \frac{dv}{1 + \log v} \right\}.$$

The last integral equals

$$\int_1^{u/\kappa} \frac{dv}{\log ev} = \frac{1}{e} \int_e^{eu/\kappa} \frac{dw}{\log w} = \frac{1}{e}\{\mathrm{li}(eu/\kappa) - \mathrm{li}(e)\}.$$

Putting together the calculations and estimating $1 - j(\kappa)$ by the result of Theorem 14.11, we get the claimed inequality. □

14.8 Notes on Chapter 14

The function $e^\gamma j_1'(u) = \rho(u)$, the Dickman function, appears in estimates of the number of positive integers having only small prime factors (see [Ten01], Section III.5).

Estimates of the type given in Theorem 14.1 are useful also in the

14.8 Notes on Chapter 14

Ankeny–Onishi sieve method [AO65] and are shown there (rather more simply) for $u \geq 2\kappa + 2$.

For all $\kappa \geq 3/2$, it is known that u_κ, the zero of j''_κ, lies in the interval $(\kappa - 1/2, \kappa)$ [GrR88] and even in $(\kappa - 1/2, \kappa - 1/4)$ (unpublished notes). Moreover, we have (again from [GrR88]),

$$\max_u j'_\kappa(u) = j'_\kappa(u_\kappa) \sim \frac{1}{\sqrt{\pi\kappa}}, \quad \kappa \to \infty,$$

and (from [Whe88])

(14.41) $$j_\kappa(\kappa) = \frac{1}{2} + \frac{1}{9\sqrt{\pi\kappa}} + O(\kappa^{-3/2}), \quad \kappa \geq 1.$$

The proof of Theorem 14.11 given here first appears in [Whe88].

Our first attempt at proving Lemma 14.8 was to mimic the proof of Lemma 14.5. Indeed, we have for $u > 0$

$$u\left(\frac{j''}{j'}(u)\right)' = -\frac{j''}{j'}(u) + \kappa \frac{j'(u-1)}{j'(u)} \int_{u-1}^{u} \left(\frac{j''}{j'}(t)\right)' dt.$$

This identity yields the desired monotonicity for $u \leq u_\kappa$, but it is not clear how to use this formula for larger u, where $-j''/j' > 0$. Following a suggestion of A. J. Hildebrand, we have based our argument on one he used in proving Lemma 1 of [Hil86], which treats an analogous problem for the Dickman function.

15
The p_κ and q_κ functions

15.1 The p functions

The p and q functions, which were introduced in Chapter 12, will play an important role in approximating the zeros of $\widetilde{\Pi}_\kappa(u) - 2$ and $\widetilde{\Xi}_\kappa(u)$. Here we present further properties, and some quite sharp estimates, of these functions.

We start with the p_κ functions, which were defined for $\kappa > 0$ by

$$p(u) = p_\kappa(u) := \int_0^\infty e^{-ut - \kappa \operatorname{Ein}(t)} \, dt, \quad u > 0.$$

Ein is given by

$$\operatorname{Ein}(t) := \int_0^t (1 - e^{-s}) \frac{ds}{s} = \sum_{n=1}^\infty (-1)^{n-1} \frac{t^n}{n!\, n}, \quad t \in \mathbb{C}.$$

These functions satisfy the family of difference differential equations

$$(up(u))' = \kappa p(u) - \kappa p(u+1), \quad u > 0,$$

and the normalization $p(u) \sim 1/u$ as $u \to \infty$. Note that p is clearly decreasing by (12.11), and hence, by the last equation, $up(u)$ increases with u. Equations equivalent to (12.9) are

(15.1) $$\{u^{1-\kappa} p(u)\}' = -\kappa u^{-\kappa} p(u+1)$$

and

(15.2) $$\{u^{1-2\kappa} p(u)\}' = -\kappa u^{-2\kappa} \{p(u) + p(u+1)\}.$$

We saw earlier that

$$1/(u+\kappa) < p(u) < 1/u, \quad u > 0.$$

We improve this in

Lemma 15.1. *For all $u > 0$, $\kappa > 0$, we have*

$$\left\{u + \kappa(u+\kappa)\log\left(1 + \frac{1}{(u+\kappa)}\right)\right\}^{-1} < p_\kappa(u) < \left\{u + \kappa u \log\left(1 + \frac{1}{u}\right)\right\}^{-1}.$$

Proof. First the upper bound. As noted above, $up(u)$ increases with u. Now it follows from (12.15) that

$$1 = up(u) + \kappa \int_u^{u+1} tp(t)\frac{dt}{t} > up(u)\left\{1 + \kappa \int_u^{u+1} \frac{dt}{t}\right\}$$
$$= up(u)\left\{1 + \kappa \log\left(1 + \frac{1}{u}\right)\right\}.$$

Before proving the lower bound, we show that $(u+\kappa)p_\kappa(u)$ is decreasing as a function of u for each fixed $\kappa > 0$. Indeed, we have first by differentiating (12.11) two times, that $-p'$ is decreasing. Then, using (12.9),

$$\{(u+\kappa)p(u)\}' = \kappa \int_u^{u+1} \{p'(u) - p'(t)\}dt < 0.$$

Now we start with (12.15), use the inequality

$$\kappa \int_u^{u+1} p(t)(t+\kappa)\frac{dt}{t+\kappa} < \kappa p(u)(u+\kappa)\int_u^{u+1} \frac{dt}{t+\kappa},$$

and proceed as we did for the upper bound. □

We can deduce from the last lemma a simpler upper bound for p_κ that is sharper than (12.14).

Corollary 15.2. *For $\kappa > 0$ and all $u \geq \kappa$, we have*

$$p_\kappa(u) < 1/(u + \kappa - 1/2).$$

Proof. Combine the bound $1/p_\kappa(u) > u + \kappa u \log(1+1/u)$ with the simple inequality $\log(1+x) > x - x^2/2$, which is valid for all positive x, to obtain

(15.3) $$\frac{1}{p_\kappa(u)} > u + \kappa - \frac{\kappa}{2u} \geq u + \kappa - 1/2.$$ □

As an indication of the quality of these bounds, we have, in rounded figures, $p_2(5) = 0.14567$, while by the lemma, $0.14557 < p_2(5) < 0.14656$. From the simpler estimates (12.14), we obtain $0.1428 < p_2(5) < 0.2000$. The corollary gives the upper bound 0.15385. The condition $u \geq \kappa$

was arbitrary (it is chosen for an application in Chapter 17); a further restriction would yield a somewhat sharper estimate.

We shall need one specific value of p:

(15.4) $$p_1(1) = e^{-\gamma}.$$

Indeed, by (12.12) and (12.13) (the definition of Ein and its asymptotic formula), we have

$$p_1(1) := \int_0^\infty e^{-x-\text{Ein}(x)}\, dx = \int_0^\infty d(x\, e^{-\text{Ein}(x)}) = e^{-\gamma}.$$

For further information and still sharper bounds for p_κ, see [DHR90b, DHR93b].

15.2 The q functions

Recall that the family of q_κ functions is defined for $\kappa > 0$ by the contour integral

$$q(u) = q_\kappa(u) := \frac{\Gamma(2\kappa)}{2\pi i} \int_C z^{-2\kappa} e^{uz+\kappa\,\text{Ein}(-z)}\, dz, \quad u > 0,$$

where C is the path that goes from $-\infty$ back to $-\infty$ traveling around the negative axis once in the positive sense. As we noted earlier, q satisfies the difference differential equation (12.10). Equivalent forms of this equation that we shall employ in the sequel are

(15.5) $$uq'(u) = (\kappa - 1)q(u) + \kappa q(u+1),$$
(15.6) $$\{u^{1-\kappa} q(u)\}' = \kappa u^{-\kappa} q(u+1),$$
(15.7) $$\{u^{1-2\kappa} q(u)\}' = \kappa u^{-2\kappa}\{q(u+1) - q(u)\}.$$

Higher derivatives of q satisfy

(15.8) $$uq_\kappa^{(\nu)}(u) = (\kappa - \nu)q_\kappa^{(\nu-1)}(u) + \kappa q_\kappa^{(\nu-1)}(u+1), \quad \nu \geq 1.$$

When $2\kappa \in \mathbb{N}$ the contour integral for q_κ can be evaluated by Cauchy's residue theorem, and we find that q_κ is a real monic polynomial of degree $2\kappa - 1$. It is given by the formula

(15.9) $$q_\kappa(u) = \sum_{n=0}^{2\kappa-1} (-1)^n \binom{2\kappa - 1}{n} c_n(\kappa) u^{2\kappa-1-n}$$

Table 15.1. *The first few q functions*

$$q_1(u) = u - 1$$
$$q_{3/2}(u) = u^2 - 3u + 3/2$$
$$q_2(u) = u^3 - 6u^2 + 9u - 8/3$$
$$q_{5/2}(u) = u^4 - 10u^3 + 30u^2 - 85u/3 + 55/12$$
$$q_3(u) = u^5 - 15u^4 + 75u^3 - 145u^2 + 90u - 18/5$$
$$q_{7/2}(u) = u^6 - 21u^5 + \tfrac{315}{2}u^4 - \tfrac{1540}{3}u^3 + \tfrac{2765}{4}u^2 - \tfrac{5229}{20}u - \tfrac{11333}{360}.$$

with

$$c_n(\kappa) = \left(e^{\kappa\,\mathrm{Ein}(x)}\right)^{(n)}\Big|_{x=0}.$$

In particular, we have

$$c_0(\kappa) = 1, \quad c_1(\kappa) = \kappa, \quad c_2(\kappa) = (\kappa - 1/2)\kappa,$$

and in general $c_n(\kappa)$ is given by the recurrence relation

$$c_{n+1} = \kappa \sum_{\nu=0}^{n} \binom{n}{\nu} c_{n-\nu} \frac{(-1)^\nu}{\nu + 1}, \quad n \ge 0.$$

This structure increases in complexity with the growth of κ, and we limit ourselves to recording in Table 15.1 the first few q functions.

15.3 Zeros of the q functions

We have

(15.10) $$q_\kappa(u) \sim u^{2\kappa - 1}, \quad u \to \infty,$$

and thus $q_\kappa(u)$ is ultimately positive for each $\kappa \ge 1$. These functions do have positive zeros, however. Since $q_1(u) = u - 1$, we see that $q_1(1) = 0$ and $q_1(u)$ is positive and increasing for $u > 1$. For $\kappa \ge 1.5$, relation (12.10) at $u = 0$ yields

$$(\kappa - 1)q_\kappa(0) + \kappa q_\kappa(1) = 0,$$

from which we infer that $q_\kappa(u)$ has at least one positive zero. Let ρ_κ denote the largest positive zero of $q_\kappa(u)$. Using q formulas and computer algebra, we find the values of ρ_κ given in Table 15.2. In subsequent sections, we shall obtain lower and upper bounds on ρ_κ.

Table 15.2. Values of ρ_κ

κ	ρ_κ	κ	ρ_κ
1	1.0000	1.5	2.3660
2	3.8340	2.5	5.3581
3	6.9191	3.5	8.5064
4	10.1137	4.5	11.7368
5	13.3727	5.5	15.0194
6	16.6751	6.5	18.3386
7	20.0089	7.5	21.6852

15.4 Monotonicity and convexity relations

It is immediate from the definition of ρ_κ that, for each $\kappa \geq 1$, we have $q_\kappa(u) > 0$ on $\rho_\kappa < u < \infty$ and thus, by (15.5), $q'_\kappa(u) > 0$ for $u \geq \rho_\kappa$, i.e., q_κ is strictly increasing on $[\rho_\kappa, \infty)$. Now consider second derivatives. By inspection, $q''_1(u) = 0$ and $q''_{3/2}(u) = 2$ for all u. For $\kappa \geq 2$ we have

$$(15.11) \qquad uq''_\kappa(u) = (\kappa - 2)q'_\kappa(u) + \kappa q'_\kappa(u+1) > 0, \quad u \geq \rho,$$

i.e., $q_\kappa(u)$ is strictly convex for $u \geq \rho$ and $\kappa \geq 2$.

Also, for all $\kappa \geq 2$,

$$(15.12) \qquad q'''_\kappa(u) > 0, \quad u \geq \rho_\kappa.$$

Indeed, by inspection, $q'''_2(u) = 6 > 0$ and $q'''_{2.5}(u) = 24u - 60 > 0$ for $u \geq \rho_{2.5} = 5.358\ldots$. For $\kappa \geq 3$, by (15.11) we have

$$uq'''_\kappa(u) = (\kappa - 3)q''_\kappa(u) + \kappa q''_\kappa(u+1) > 0, \quad u \geq \rho.$$

We apply (15.12) to obtain a useful lower bound for q''/q'. Let $\kappa \geq 2$ and $u \geq \rho_\kappa$. Then

$$(15.13) \qquad q''_\kappa(u)/q'_\kappa(u) > (2\kappa - 2)/(u - \kappa).$$

(We are assuming here that $\rho_\kappa > \kappa$ holds for all $\kappa \geq 2$, which will be shown shortly by an argument independent of (15.13).) We have by (15.8) and Taylor's formula

$$uq''(u) = (\kappa - 2)q'(u) + \kappa q'(u+1)$$
$$= (\kappa - 2)q'(u) + \kappa\{q'(u) + q''(u) + q'''(u+\vartheta)/2\}$$

for some $\vartheta \in (0, 1)$. As we have just seen, $q''' > 0$ on $[\rho, \infty)$, so

$$(u - \kappa)q''(u) > (2\kappa - 2)q'(u),$$

which yields the claimed result.

Next we establish a convexity relation for the function occurring on the right side of (15.6).

Lemma 15.3. *Suppose $\kappa \geq 2$ and $2\kappa \in \mathbb{N}$. Then $u^{-\kappa}q(u+1)$ is convex on $\{u: u > \rho - 1\}$.*

Proof. By direct calculation,
$$\{u^{-2}q_2(u+1)\}'' = 8u^{-4} > 0, \quad u > 0,$$
and, for $u \geq 1$,
$$4u^{9/2}\{u^{-5/2}q_{5/2}(u+1)\}'' = 3u^4 + 6u^3 + 18u^2 + 85u - 385/4 > 0.$$

Henceforth we suppose that $\kappa \geq 3$. We know already that $q(t)$, $q'(t)$ and $q''(t)$ are positive for $t > \rho$, and from
$$tq'''(t) = (\kappa - 3)q''(t) + \kappa q''(t+1)$$
it follows that the same is true of $q'''(t)$.

By (12.10) and (15.11), we have
$$u^{\kappa+2}\{u^{-\kappa}q(u+1)\}'' = \kappa^2 q(u+3) - 3\kappa q(u+2) - 3\kappa q'(u+2)$$
$$+ 2q(u+1) + 4q'(u+1) + q''(u+1);$$
since the last three terms on the right are positive, it is enough to show that
$$\kappa q(u+3) > 3q(u+2) + 3q'(u+2).$$
But by Taylor's theorem, using the fact that $q''(t) > 0$ for $t > \rho$, we have
$$\kappa q(u+3) \geq 3q(u+3) > 3q(u+2) + 3q'(u+2),$$
and thus the claimed convexity holds. □

To end this section, we show that $\log q$ is concave on (ρ, ∞), a result which has several useful consequences.

Lemma 15.4. *Let $\kappa \geq 1$. Then $\{\log q_\kappa\}'' < 0$ on (ρ, ∞).*

Proof. We begin with the remark that $q_\kappa^{(\nu)}(u) \sim \{u^{2\kappa-1}\}^{(\nu)}$ holds for $\nu = 1, 2, \ldots$ as $u \to \infty$. This follows from the asymptotic relation (15.10) and the differential equation (15.5).

Next we show that $\{\log q(u)\}'' = \{q'/q\}'(u) < 0$ for $u \to \infty$. Indeed,

$$\{\log q(u)\}'' = \{q(u)q''(u) - q'(u)^2\}q(u)^{-2}$$
$$\sim \{u^{2\kappa-1} \cdot (2\kappa-1)(2\kappa-2)u^{2\kappa-3} - (2\kappa-1)^2 u^{4\kappa-4}\}u^{-4\kappa+2}$$
$$= -(2\kappa-1)u^{-2}, \quad u \to \infty.$$

If the lemma were false, there would exist a maximal number $w > \rho$ with $\{q'/q\}'(w) \geq 0$. Then we would have $q(w)q''(w) - q'(w)^2 \geq 0$, i.e., by (15.8),

$$0 \leq q(w)\{(\kappa-2)q'(w) + \kappa q'(w+1)\} - q'(w)\{(\kappa-1)q(w) + \kappa q(w+1)\}$$
$$= -q(w)q'(w) + \kappa q(w)q(w+1)\left\{\frac{q'}{q}(w+1) - \frac{q'}{q}(w)\right\},$$

or

$$0 < \frac{q'(w)}{\kappa q(w+1)} \leq \frac{q'}{q}(w+1) - \frac{q'}{q}(w).$$

It would follow by the mean value theorem that $\{q'/q\}'(w+\theta) > 0$ for some $\theta > 0$, which is impossible. □

Corollary 15.5. *When $\rho < u < \infty$, each of*

$$\frac{q(u+1)}{q(u)} = \exp \int_u^{u+1} \frac{q'}{q}(t)\, dt,$$

$$u \frac{q'}{q}(u) = \kappa - 1 + \kappa \frac{q(u+1)}{q(u)}$$

is decreasing in u.

15.5 Some lower bounds for ρ_κ

By (12.17) at $u = \rho$, we have

$$(\rho+1)q(\rho+1) = \kappa \int_\rho^{\rho+2} q(t)dt;$$

also, the integrand is convex for $\kappa \geq 2$, as we noted above. Thus, by a familiar integral estimate,

$$(\rho+1)q(\rho+1) \geq 2\kappa q(\rho+1), \quad \kappa \geq 2,$$

that is,

(15.14) $\qquad \rho_\kappa \geq 2\kappa - 1, \quad \kappa \geq 1,\ 2\kappa \in \mathbb{N}.$

(This result holds by inspection for $\kappa = 1, 3/2$.)

Lemma 15.6. *Suppose $\kappa \geq 2$ and $2\kappa \in \mathbb{N}$. Let $R = 2.84305\ldots$ be the positive solution of the exponential equation $e^{1/R} = R/2$. Then*

$$\rho_\kappa > R\kappa - 1.87.$$

In particular, $\rho_\kappa > 3.8$ for $\kappa \geq 2$, and also

(15.15) $$\rho_\kappa > 2.7\kappa - 1, \quad \kappa \geq 3.5.$$

Proof. We integrate (15.6) from $u = \rho$ to $u = \rho + 2$ to obtain

$$(\rho+2)^{1-\kappa} q(\rho+2) = \kappa \int_\rho^{\rho+2} u^{-\kappa} q(u+1) du > 2\kappa(\rho+1)^{-\kappa} q(\rho+2)$$

by the convexity of the integrand, established in Lemma 15.3. Hence

$$\rho + 2 > 2\kappa \left(\frac{\rho+2}{\rho+1}\right)^\kappa = 2\kappa \exp\left\{\kappa \log \frac{1+(2\rho+3)^{-1}}{1-(2\rho+3)^{-1}}\right\}$$

$$> 2\kappa \exp\left(\frac{\kappa}{\rho+3/2}\right).$$

Writing $x_0 = \kappa/(\rho + 3/2)$, the preceding inequality takes the form

(15.16) $$x_0 e^{x_0} < \frac{1}{2} + \frac{x_0}{4\kappa}.$$

If $x_0 \leq 1/R$, then we have $\rho \geq R\kappa - 1.5$, a better lower bound than is claimed in the statement of the lemma. We may suppose therefore that

(15.17) $$x_0 > 1/R.$$

Let $\xi(x) = xe^x$, so that $\xi(1/R) = \frac{1}{2}$. We have $\xi'(x) = (1+x)e^x$ and $\xi''(x) = (2+x)e^x$, so that both $\xi'(x)$ and $\xi''(x)$ are positive for $x > 0$ and $\xi'(x)$ is increasing in $x > 0$. Hence, by (15.16) and (15.17),

$$(x_0 - 1/R)\xi'(1/R) < \xi(x_0) - \xi(1/R) < \frac{x_0}{4\kappa},$$

so that

$$x_0 \left(\xi'(1/R) - \frac{1}{4\kappa}\right) < \frac{1}{R}\xi'(1/R).$$

Since

$$\xi'(1/R) = \left(1 + \frac{1}{R}\right)e^{1/R} = \frac{1}{2}(R+1),$$

we obtain

$$x_0 \left(\frac{1}{2}(R+1) - \frac{1}{4\kappa}\right) < \frac{R+1}{2R},$$

Table 15.3. $\rho_\kappa + 1$ compared with 2.7κ

κ	$\rho_\kappa + 1$	2.7κ
3.5	9.5064	9.45
4.0	11.1137	10.80
4.5	12.7368	12.15
5.0	14.3727	13.50
5.5	16.0194	14.85
6.0	17.6751	16.20

or
$$\rho + 3/2 = \kappa/x_0 > R\kappa - R/(2R+2).$$
Hence
$$\rho > R\kappa - 2 + \frac{1}{2(R+1)} = R\kappa - 1.86989\ldots.$$

To establish (15.15), we made explicit calculations in the initial range $3.5 \leq \kappa < 6.5$, with the results shown in Table 15.3. For all $\kappa \geq 6.5$, we apply the inequalities
$$\rho_\kappa > 2.843\kappa - 1.87 > 2.7\kappa - 1;$$
the first of these was just established for all $\kappa \geq 2$, and the second one holds for all $\kappa \geq 6.5$. □

The first estimate of the lemma is fairly accurate for smaller values of κ: it gives $\rho_2 > 3.816$, $\rho_{2.5} > 5.237$, and $\rho_3 > 6.659$, as compared respectively with $\rho_2 = 3.833\ldots$, $\rho_{2.5} = 5.358\ldots$, and $\rho_3 = 6.919\ldots$. We need nothing better for our purposes, but sharper estimates of this type are known (see [Gru88]).

15.6 An upper bound for ρ_κ

It is known that (see [Iwa80], [Tsa89])
$$\lim_{\kappa \to \infty} \rho_\kappa/\kappa = 3.591121\ldots.$$
Here we establish an upper estimate of this type.

Proposition 15.7. *Let* $\kappa \geq 1$. *Then*
$$\rho_\kappa < r_c\kappa - 2r_c/(1+r_c),$$

where $r_c = 3.591121\ldots$ is defined by $r_c \log(r_c/e) = 1$. In particular, we have $\rho_\kappa < 3.6\kappa - 1.56$ for all $\kappa \geq 1$.

Proof. By Corollary 15.5, $(q'/q)(u)$ decreases on (ρ, ∞) from ∞ at $\rho+0$ to 0 as $u \to \infty$. Thus there exists a number $u_0 > \rho$ with $(q'/q)(u_0) = \log r_c$. Now

$$u(q'/q)(u) = \kappa - 1 + \kappa q(u+1)/q(u)$$

$$= \kappa - 1 + \kappa \exp\left\{\int_u^{u+1} \frac{q'}{q}(t)\,dt\right\} < \kappa - 1 + \kappa \exp\frac{q'}{q}(u),$$

using again the fact that q'/q decreases (beyond ρ). Therefore, at $u = u_0$, we have $u_0 \log r_c < \kappa - 1 + \kappa r_c$ or, since $\log r_c = (1 + r_c)/r_c$,

$$u_0 < \frac{r_c\{\kappa(1+r_c) - 1\}}{1 + r_c} = r_c\kappa - \frac{r_c}{1 + r_c}.$$

Next, we apply one step of Newton's method to approximate the root ρ of the equation $q(u) = 0$, starting at u_0. Since q is convex on (ρ, ∞), by (15.11), the first Newton step satisfies

$$u_1 = u_0 - q(u_0)/q'(u_0) > \rho,$$

and thus $r_c\kappa - 2r_c/(1 + r_c) > \rho$. □

Further information and sharper differential estimates for q_κ are given in [DHR90b], [DHR93b], and [DH99].

15.7 The integrands of $\widetilde{\Pi}$ and $\widetilde{\Xi}$

Here we establish the positivity, monotonicity, and convexity of the integrands of $\widetilde{\Pi}_\kappa$ and of $\widetilde{\Xi}_\kappa$. In Sections 16.2 and 16.3 we use these properties to give quite accurate one-sided bounds on $\widetilde{\Pi}_\kappa$ and $\widetilde{\Xi}_\kappa$ while using a modest number of function evaluations.

Lemma 15.8. *Let $\kappa > 1$. Then $p(t+1)/\sigma(t)$ is positive, decreasing, and convex on $0 < t < \infty$.*

Proof. We have

$$(-1)^\nu p^{(\nu)}(u) > 0, \quad u > 0, \quad \nu = 0, 1, 2, \ldots,$$

by (12.11) and that

$$(-1)^\nu (1/\sigma(t))^{(\nu)} > 0, \quad t > 0,$$

for $\nu = 0$, by (14.7); for $\nu = 1$, by (14.6); and for $\nu = 2$, by Lemma 14.6.

15.7 The integrands of $\widetilde{\Pi}$ and $\widetilde{\Xi}$

It is immediate that $p(t+1)/\sigma(t)$ is positive and decreasing. To see that the function is convex, write

$$\left(\frac{p(t+1)}{\sigma(t)}\right)'' = p''(t+1)\frac{1}{\sigma(t)} + 2p'(t+1)\left(\frac{1}{\sigma(t)}\right)' + p(t+1)\left(\frac{1}{\sigma(t)}\right)'',$$

and note that each product on the right is positive. □

Since the integrand in

$$\widetilde{\Pi}_\kappa(u) := \frac{up(u)}{\sigma(u)} + \kappa \int_{u-2}^{u} \frac{p(t+1)}{\sigma(t)} dt$$

is convex, the preceding lemma ensures that a trapezoidal (resp. midpoint) approximation provides an upper (resp. lower) bound on the last integral.

We now show that similar reasoning applies to the function

$$\widetilde{\Xi}_\kappa(u) := \frac{uq(u)}{\sigma(u)} - \kappa \int_{u-2}^{u} \frac{q(t+1)}{\sigma(t)} dt.$$

Lemma 15.9. *Let $\kappa \geq 2$ be an integer or half integer. For $t > \rho_\kappa - 1$ the quotient $q_\kappa(t+1)/\sigma_\kappa(t)$ is positive, increasing, and convex.*

Proof. The positivity is immediate since $\sigma(t) > 0$ for all positive t and $q(t+1) > 0$ for $t+1 > \rho$.

To establish monotonicity, we show that, on this ray,

$$\frac{d}{dt} \log \frac{q(t+1)}{\sigma(t)} = \frac{q'}{q}(t+1) - \frac{\sigma'}{\sigma}(t) > 0.$$

We have

$$(t+1)\frac{q'}{q}(t+1) = \kappa - 1 + \kappa \frac{q(t+2)}{q(t+1)} > 2\kappa - 1$$

by the differential equation (12.10) and the consequent result that $q(u)$ is increasing for $u > \rho$. Also,

(15.18) $$t\frac{\sigma'}{\sigma}(t) = \kappa - \kappa\frac{\sigma(t-2)}{\sigma(t)} \leq \kappa$$

by the differential equation (6.8') and (14.7), the positivity of $\sigma(u) > 0$ for $u > 0$. By Lemma 15.6, $\rho_\kappa > 3.8$, and thus $t > 2.8$ here. Now

$$\frac{q'}{q}(t+1) - \frac{\sigma'}{\sigma}(t) > \frac{2\kappa - 1}{t+1} - \frac{\kappa}{t} = \frac{\kappa t - t - \kappa}{t(t+1)} = \frac{(\kappa - 1)(t-1) - 1}{t(t+1)} > 0,$$

since $(\kappa-1)(t-1) > 1.8 > 1$, which establishes the claimed monotonicity.

For convexity, we show that

$$\left\{\frac{q(t+1)}{\sigma(t)}\right\}''$$

$$= \left\{\frac{q(t+1)}{t^\kappa} \cdot \frac{t^\kappa}{\sigma(t)}\right\}''$$

$$= \left\{\frac{q(t+1)}{t^\kappa}\right\}'' \frac{t^\kappa}{\sigma(t)} + 2\left\{\frac{q(t+1)}{t^\kappa}\right\}' \left\{\frac{t^\kappa}{\sigma(t)}\right\}' + \frac{q(t+1)}{t^\kappa}\left\{\frac{t^\kappa}{\sigma(t)}\right\}''$$

$$=: A + B + C > 0.$$

By Lemma 15.3, $A > 0$ for all $\kappa \geq 2$ and $u > \rho_\kappa - 1$.

In B, first note that, by the differential equation (15.18) for σ_κ and a small calculation, we have

$$\left\{\frac{t^\kappa}{\sigma(t)}\right\}' = \kappa\, t^{\kappa-1}\frac{Y(t)}{\sigma(t)}, \quad Y(t) := \frac{\sigma(t-2)}{\sigma(t)}.$$

Next,

$$\{t^{-\kappa} q(t+1)\}' = -\kappa t^{-\kappa-1} q(t+1) + \frac{t^{-\kappa}}{t+1}(t+1)\, q'(t+1)$$

$$= \frac{t^{-\kappa-1}}{t+1}\{t\kappa q(t+2) + [t(\kappa - 1) - \kappa(t+1)]\, q(t+1)\}$$

$$> \frac{t^{-\kappa-1}}{t+1}\{(\kappa-1)(t-1) - 1\}\, q(t+1)$$

$$\geq \frac{t^{-\kappa-1}}{t+1}\frac{3t-5}{2}\, q(t+1) \quad \text{for } \kappa \geq 5/2.$$

Thus

$$B \geq \frac{\kappa t^{-2}}{t+1}\frac{Y(t)}{\sigma(t)}(3t-5)q(t+1), \quad \kappa \geq 5/2.$$

On the other hand, for $\kappa = 2$,

$$t^{-2} q_2(t+1) = t^{-2}(t^3 - 3t^2 + 4/3) = t - 3 + (4/3)t^{-2},$$

and so

$$B = 2\left\{1 - \frac{8}{3t^3}\right\} \cdot 2t\frac{Y(t)}{\sigma(t)} = \frac{4Y(t)}{t^2\sigma(t)}\left\{t^3 - \frac{8}{3}\right\}, \quad \kappa = 2.$$

In C we have

$$\left\{\frac{t^\kappa}{\sigma(t)}\right\}'' = \left\{\kappa t^{\kappa-1}\frac{Y(t)}{\sigma(t)}\right\}'$$

$$= \frac{\kappa t^{\kappa-2}}{\sigma(t)}\left\{(\kappa-1)Y(t) + tY'(t) - t\frac{\sigma'}{\sigma}(t)Y(t)\right\}$$

$$= \frac{\kappa t^{\kappa-2}}{\sigma(t)}Y(t)\left\{\kappa - 1 + t\frac{Y'}{Y}(t) - \kappa + \kappa Y(t)\right\}$$

$$> -\frac{\kappa t^{\kappa-2}}{\sigma(t)}Y(t),$$

since $Y(t)$ and $Y'(t) > 0$, the latter by Corollary 14.7 (restated for σ). Thus

$$C > -\frac{\kappa t^{-2}}{\sigma(t)}Y(t)q(t+1).$$

Now we combine B and C. For $\kappa \geq 5/2$,

$$B + C > \frac{1}{t+1}\frac{\kappa t^{-2}}{\sigma(t)}Y(t)q(t+1)\{3t - 5 - (t+1)\} > 0,$$

since $2t - 6 > 0$ for $t > \rho_{5/2} - 1 > 4.358$. For $\kappa = 2$,

$$B + C > \frac{4Y(t)}{t^2\sigma(t)}\left\{t^3 - \frac{8}{3}\right\} - \frac{2Y(t)}{t^2\sigma(t)}q(t+1)$$

$$= \frac{2Y(t)}{t^2\sigma(t)}\left\{2t^3 - 16/3 - \left(t^3 - 3t^2 + 4/3\right)\right\} > 0,$$

since $t^3 + 3t^2 - 20/3 > 0$ for $t > \rho_2 - 1 = 2.833\ldots$.

In both cases, $A + B + C > 0$ and hence $q(t+1)/\sigma(t)$ is convex for all $\kappa \geq 2$ and $t > \rho_\kappa - 1$. □

16
The zeros of $\widetilde{\Pi} - 2$ and $\widetilde{\Xi}$

16.1 Properties of the Π and Ξ functions

In this chapter we carry out Step 1 of the program described at the end of Chapter 13. Let $\widetilde{\Pi}_\kappa$ and $\widetilde{\Xi}_\kappa$ be the functions introduced in Chapter 12. We shall show that there exist unique zeros $z_{\widetilde{\Pi}}(\kappa)$ and $z_{\widetilde{\Xi}}(\kappa)$ of $\widetilde{\Pi}_\kappa - 2$ and $\widetilde{\Xi}_\kappa$ respectively, and that

(i) $z_{\widetilde{\Xi}}(\kappa) < z_{\widetilde{\Pi}}(\kappa)$ for $\kappa = 1$ or $\kappa = 1.5$ (Proposition 16.2),
(ii) $z_{\widetilde{\Pi}}(\kappa) < z_{\widetilde{\Xi}}(\kappa)$ for $\kappa = 2, 2.5, 3, \ldots$ (Proposition 16.3).

We shall study at the same time the related functions $\Pi(u) := \Pi(u, u)$ and $\Xi(u) := \Xi(u, u)$, which also play a role in our theory.

We begin by establishing the monotonicity of several functions, which will be used to show the solvability of equations and make estimates. From (12.24) we see, for $u > 0$, $v > 1$, that

$$\frac{\partial}{\partial u}\Pi(u, v) = \kappa \frac{p(u)\sigma(u-2)}{\sigma^2(u)} \quad \text{and} \quad \frac{\partial}{\partial v}\Pi(u, v) = -\kappa \frac{p(v)}{\sigma(v-1)}.$$

For $t > 0$, we have $p(t) > 0$ by (12.11) and $\sigma(t) > 0$ by (14.7). Thus

(16.1) $\quad \dfrac{\partial}{\partial u}\Pi(u, v) > 0 \;\; (u > 2) \quad \text{and} \quad \dfrac{\partial}{\partial v}\Pi(u, v) < 0 \;\; (v > 1)$.

Similarly, we deduce from (12.25), for $u > 0$, $v > 1$, that

$$\frac{\partial}{\partial u}\Xi(u, v) = \kappa \frac{q(u)\sigma(u-2)}{\sigma^2(u)} \quad \text{and} \quad \frac{\partial}{\partial v}\Xi(u, v) = \kappa \frac{q(v)}{\sigma(v-1)}.$$

In the last chapter we found that each polynomial $q_\kappa(u) = q(u)$ possesses positive zeros; that the largest of these, which we call $\rho_\kappa = \rho$, exceeds 2 for all values of $\kappa \geq 1.5$; and that $q(u) \to \infty$ as $u \to \infty$. Thus $q(u) > 0$ when $u > \rho$. Hence we have, for $\kappa \geq 1.5$,

(16.2) $\quad \dfrac{\partial}{\partial u}\Xi(u,v) > 0 \quad (u > \rho) \quad \text{and} \quad \dfrac{\partial}{\partial v}\Xi(u,v) > 0 \quad (v > \rho).$

For $\kappa = 1$, $q_1(u) = u - 1$ and $\rho_1 = 1$; the preceding differential inequalities hold in this case for $u > 2$ and $v > 1$ respectively.

First, recall from Chapter 12 the definitions

(16.3) $\quad \Pi(u) := \Pi_\kappa(u) := \Pi_\kappa(u,u) = \dfrac{up(u)}{\sigma(u)} + \kappa \int_{u-1}^{u} \dfrac{p(t+1)}{\sigma(t)} dt$

and

(16.4) $\quad \Xi(u) := \Xi_\kappa(u) := \Xi_\kappa(u,u) = \dfrac{uq(u)}{\sigma(u)} - \kappa \int_{u-1}^{u} \dfrac{q(t+1)}{\sigma(t)} dt$

for $u > 1$. We have the formulas

$$\Pi'(u) = \dfrac{\partial}{\partial u}\Pi(u,v)\Big|_{v=u} + \dfrac{\partial}{\partial v}\Pi(u,v)\Big|_{v=u}$$
$$= \kappa p(u)\left\{\dfrac{\sigma(u-2)}{\sigma^2(u)} - \dfrac{1}{\sigma(u-1)}\right\} < 0,$$

since $p(u) > 0$ and $\sigma(u)$ exceeds each of $\sigma(u-1)$ and $\sigma(u-2)$ by (14.7) and (14.6) (restated for σ). Thus

$\Pi(u)$ is strictly decreasing in $u > 1$.

Similarly, by (16.3),

$$\Xi'(u) = \dfrac{\partial}{\partial u}\Xi(u,v)\Big|_{v=u} + \dfrac{\partial}{\partial v}\Xi(u,v)\Big|_{v=u}$$
$$= \kappa q(u)\left\{\dfrac{\sigma(u-2)}{\sigma^2(u)} + \dfrac{1}{\sigma(u-1)}\right\} > 0, \quad u > \rho,$$

since $q(u) > 0$ when $u > \rho$ and $\rho = \rho_\kappa > 1$ for $\kappa > 1$. Hence

$\Xi(u)$ is strictly increasing in $u > \rho$.

Next, recalling the definitions

$\widetilde{\Pi}(u) = \widetilde{\Pi}_\kappa(u) := \Pi_\kappa(u, u-1), \quad \widetilde{\Xi}(u) = \widetilde{\Xi}_\kappa(u) := \Xi_\kappa(u, u-1), \quad u > 2,$

we have, from above,

(16.5) $\quad \widetilde{\Pi}'(u) = \kappa \dfrac{p(u)}{\sigma(u-2)}\left\{\left(\dfrac{\sigma(u-2)}{\sigma(u)}\right)^2 - \dfrac{p(u-1)}{p(u)}\right\} < 0, \quad u > 2;$

here we have used the information that, for $u > 0$ (and fixed $\kappa > 1$), $\sigma(u)$ increases by (14.6) and $p(u)$ decreases by (12.11). Also, we have

$$(16.6) \quad \widetilde{\Xi}'(u) = \kappa \frac{q(u)}{\sigma(u-2)} \left\{ \left(\frac{\sigma(u-2)}{\sigma(u)} \right)^2 + \frac{q(u-1)}{q(u)} \right\} > 0, \quad u > \rho + 1.$$

Thus $\widetilde{\Pi}(u)$ is strictly decreasing in $u > 2$ and $\widetilde{\Xi}(u)$ is strictly increasing in $u > \rho + 1$.

16.2 Solution of some Π and Ξ equations

At several points in the sequel we shall need a statement that functions defined by a class of integrals involving the σ function are unbounded. It is convenient to express this simple result once as a lemma.

Lemma 16.1. *Let $\kappa \geq 1$, $B > 0$, and $\phi(x)$ a positive valued continuous function on $[0, B]$. Then*

$$I(x) := \int_x^B \frac{\phi(t)}{\sigma_\kappa(t)} \, dt \to \infty \quad \text{as} \quad x \to 0+.$$

Proof. For $0 < t \leq b := \min(2, B)$, the integrand of I is at least as large as C/t^κ for some positive constant C (depending on ϕ and κ). Thus

$$I(x) \geq \int_x^b C t^{-\kappa} \, dt,$$

which diverges to infinity as $x \to 0+$. □

We see from the formula (16.3) for $\Pi(u)$ that this function is continuous. Also, we have shown that it is strictly decreasing in $u > 1$. Moreover, since $p(t+1)$ is bounded away from 0 near $t = 1$ (by its integral representation (12.11)), the last lemma yields

$$\lim_{u \to 1+0} \Pi(u) = +\infty.$$

Moreover, $\Pi(u) \to 1$ as $u \to \infty$, since $up(u) \sim 1$ (by (12.14)) and $\sigma(u) \to 1$ as $u \to \infty$ (by Lemma 14.3). Hence *the equation $\Pi(u) = 2$ possesses a unique root, call it $z_\Pi(\kappa)$, which exceeds 1*.

We have shown that $\Xi(u)$ is strictly increasing for $u > \rho$. By inspection of formula (16.4), we see that this function is continuous, $\Xi(\rho) < 0$, and

$$\Xi(u) \sim u^{2\kappa} \to +\infty, \quad u \to \infty.$$

Thus *the equation* $\Xi(u) = 0$ *possesses a root, call it* $z_\Xi(\kappa)$, *exceeding* ρ_κ, *that is unique on* (ρ_κ, ∞).

Next, (12.24), the positivity of p on $(0, \infty)$, and the last lemma yield

$$\widetilde{\Pi}(u) = \frac{up(u)}{\sigma(u)} + \kappa \int_{u-2}^{u} \frac{p(t+1)}{\sigma(t)}\, dt \to \infty, \quad u \to 2+0.$$

Just as for $\Pi(u)$, we have $\widetilde{\Pi}(u) \to 1$ as $u \to \infty$. Also, $\widetilde{\Pi}$ is continuous and decreasing on $(2, \infty)$. It follows that, for each $\kappa \geq 1$, *the equation* $\widetilde{\Pi}(u) - 2 = 0$ *possesses a unique root exceeding* 2, *to be denoted by* $z_{\widetilde{\Pi}}(\kappa)$.

Similarly, by the formulas in (12.25),

$$\widetilde{\Xi}(u) > \frac{(u-2)q(u-2)}{\sigma(u-2)} \geq 0$$

for $u \geq \rho + 2$, whereas

$$\widetilde{\Xi}(\rho + 1) = \frac{(\rho+1)q(\rho+1)}{\sigma(\rho+1)} - \kappa \int_\rho^{\rho+2} \frac{q(t)}{\sigma(t-1)}\, dt$$

$$< \frac{1}{\sigma(\rho+1)}\left\{(\rho+1)q(\rho+1) - \kappa \int_\rho^{\rho+2} q(t)\, dt\right\} = 0,$$

since the expression in brackets vanishes by (12.17), upon using the definition of ρ. Hence, for each $\kappa \geq 1$, *the equation* $\widetilde{\Xi}_\kappa(u) = 0$ *possesses a unique root exceeding* $\rho_\kappa + 1$, *to be denoted by* $z_{\widetilde{\Xi}}(\kappa)$. Indeed, by the preceding argument, we have

(16.7) $$\rho_\kappa + 1 < z_{\widetilde{\Xi}}(\kappa) < \rho_\kappa + 2.$$

Now we come to the main results of this chapter.

Proposition 16.2. *Let* $z_{\widetilde{\Xi}}(\kappa)$, $z_{\widetilde{\Pi}}(\kappa)$ *be as above. Then, for* $\kappa = 1$ *and for* $\kappa = 1.5$,

$$z_{\widetilde{\Xi}}(\kappa) < z_{\widetilde{\Pi}}(\kappa).$$

Proof. For $\kappa = 1$, we can avoid computation. Recall that $\rho_1 = 1$, and hence $z_{\widetilde{\Xi}}(1) < 3$ by (16.7). We shall show that $3 < z_{\widetilde{\Pi}}(1)$. Since $\widetilde{\Pi}(u) \downarrow$ for $u > 2$ (by (16.5)), it suffices to show that $\widetilde{\Pi}(3) - 2 > 0$. Applying the second formula for $\widetilde{\Pi}$ in (12.24) at $u = 3$, we get

$$\widetilde{\Pi}_1(3) = \frac{p_1(1)}{\sigma_1(1)} + \int_1^3 \frac{p_1(t)\sigma_1(t-2)}{\sigma^2(t)}\, dt > \frac{p_1(1)}{\sigma_1(1)},$$

since the integral is positive. From the definition of σ in its initial region, $\sigma_1(1) = e^{-\gamma}/2$. Also, by (15.4), $p_1(1) = e^{-\gamma}$. It follows that $\widetilde{\Pi}_1(3) - 2 > 0$ and hence $z_{\widetilde{\Xi}}(1) < 3 < z_{\widetilde{\Pi}}(1)$, as claimed. (In fact, $z_{\widetilde{\Xi}}(1) = 2.89803\ldots$ and $z_{\widetilde{\Pi}}(1) = 3.07424\ldots$.)

For $\kappa = 1.5$, recall that the integrand $p(t+1)/\sigma(t)$ in the formula

$$\widetilde{\Pi}_{1.5}(u) := \frac{up_{1.5}(u)}{\sigma_{1.5}(u)} + 1.5 \int_{u-2}^{u} \frac{p_{1.5}(t+1)}{\sigma_{1.5}(t)}\,dt$$

is convex (Lemma 15.8). Thus we can obtain a lower bound on the integral via a midpoint approximation (with two intervals) using three computer evaluations of $p_{1.5}$ and $\sigma_{1.5}$. We find that $\widetilde{\Pi}_{1.5}(4.1) > 2.01 > 2$. It follows that $z_{\widetilde{\Pi}}(1.5) > 4.1$.

We make a similar calculation for

$$\widetilde{\Xi}_{1.5}(u) := \frac{uq_{1.5}(u)}{\sigma_{1.5}(u)} - 1.5 \int_{u-2}^{u} \frac{q_{1.5}(t+1)}{\sigma_{1.5}(t)}\,dt,$$

using the fact that the integrand $q(t+1)/\sigma(t)$ is a convex function of t for $t > \rho - 1$ (Lemma 15.9). This time we apply a trapezoidal approximation, obtaining $\widetilde{\Xi}_{1.5}(4.1) > 0.68 > 0$, so that $z_{\widetilde{\Xi}}(1.5) < 4.1$. Hence

$$z_{\widetilde{\Xi}}(1.5) < z_{\widetilde{\Pi}}(1.5).$$

(In fact, $z_{\widetilde{\Xi}}(1.5) = 4.04009\ldots$ and $z_{\widetilde{\Pi}}(1.5) = 4.15694\ldots$.) □

For all other κ under consideration we have

Proposition 16.3. *Let $z_{\widetilde{\Xi}}(\kappa)$, $z_{\widetilde{\Pi}}(\kappa)$ be as above. Then, for $\kappa = 2, 2.5, 3, \ldots,$*

$$z_{\widetilde{\Pi}}(\kappa) < z_{\widetilde{\Xi}}(\kappa).$$

Proof. Our proof runs through the end of Section 16.3 and includes the statements and proofs of two lemmas. The argument is based on a simple set of inequalities. Recall that $\widetilde{\Pi}(u) \downarrow$ for $u > 2$; if $\widetilde{\Pi}(u^*) - 2 < 0$ for some $u^* > 2$, then $z_{\widetilde{\Pi}} < u^*$. Also, $\widetilde{\Xi}(u) \uparrow$ for $u > \rho+1$, and if $\widetilde{\Xi}(u^*) < 0$ for some $u^* > \rho+1$, then $z_{\widetilde{\Xi}} > u^*$.

To give upper bounds on

(16.8) $$\widetilde{\Pi}(u) := \frac{up(u)}{\sigma(u)} + \kappa \int_{u-2}^{u} \frac{p(t+1)}{\sigma(t)}\,dt, \quad u > 2,$$

we again use the fact that $p(t+1)/\sigma(t)$ is convex. This time we bound the integral from above by the trapezoidal method. We start with four intervals:

$$(16.9) \quad \widetilde{\Pi}(u) < \frac{up(u)}{\sigma(u)} + \frac{\kappa}{4}\left\{\frac{p(u+1)}{\sigma(u)} + \frac{2p(u+\frac{1}{2})}{\sigma(u-\frac{1}{2})}\right.$$
$$\left. + \frac{2p(u)}{\sigma(u-1)} + \frac{2p(u-\frac{1}{2})}{\sigma(u-\frac{3}{2})} + \frac{p(u-1)}{\sigma(u-2)}\right\}.$$

Using Lemma 15.1, we approximate $p_\kappa(u)$ from above by

$$(16.10) \quad \left\{u + u\kappa \log \frac{u+1}{u}\right\}^{-1} =: p_\kappa^*(u) =: p^*(u).$$

This bound is valid for all $\kappa > 0$ and $u > 0$.

We turn now to the case $\kappa = 2$. Here (16.9), (16.10), and computer evaluations of $\sigma_2(5.28 - n/2)$ for $n = 0, 1, 2, 3, 4$ yield

$$\widetilde{\Pi}_2(5.28) < 1.9960 < 2,$$

whence $z_{\widetilde{\Pi}}(2) < 5.28$. (Computation: $z_{\widetilde{\Pi}}(2) = 5.2405\ldots$.)

For $\widetilde{\Xi}$, we have

$$(16.11) \quad \widetilde{\Xi}_\kappa(u) := \widetilde{\Xi}(u) := \frac{uq(u)}{\sigma(u)} - \kappa \int_{u-2}^{u} \frac{q(t+1)}{\sigma(t)}\,dt, \quad u > 2.$$

By Lemma 15.9, the last integrand is a convex function of t for $t > \rho - 1$. Thus a midpoint approximation yields a lower bound on the integral (for $u > \rho + 1$) and in turn an upper bound on $\widetilde{\Xi}(u)$. For $\kappa = 2$, we have $q_2(u) = u^3 - 6u^2 + 9u - 8/3$, a polynomial whose largest real zero occurs at $\rho_2 = 3.83398\ldots$. We find that $\widetilde{\Xi}_2(5.28) < -0.02 < 0$. (Computation: $z_{\widetilde{\Xi}}(2) = 5.3192\ldots$.) It follows that $z_{\widetilde{\Xi}}(2) > 5.28 > z_{\widetilde{\Pi}}(2)$. This establishes the proposition for $\kappa = 2$.

For all integer or half integer values of $\kappa \geq 2.5$, we shall show that

$$(16.12) \quad z_{\widetilde{\Pi}}(\kappa) < \rho_\kappa + 1,$$

and since $\rho_\kappa + 1 < z_{\widetilde{\Xi}}(\kappa)$ (by (16.7)), we obtain our claimed result without having to evaluate $\widetilde{\Xi}_\kappa(u)$ again. It is most convenient to establish (16.12) for $\kappa = 2.5, 3, 3.5, 4, 4.5$ by making specific calculations. For all $\kappa \geq 5$ we use just one calculation, at $\kappa = 5$, and a monotonicity relation.

For $2.5 \leq \kappa \leq 4.5$ with $2\kappa \in \mathbb{N}$, we first find the polynomial q_κ using equation (15.9). We then calculate ρ_κ by computer algebra and take

16.2 Solution of some Π and Ξ equations

Table 16.1. *Bounds on* $\widetilde{\Pi}_\kappa(v_\kappa)$

κ	$\rho_\kappa + 1$	v_κ	$\widehat{\widetilde{\Pi}}_\kappa(v_\kappa)$
2.5	6.35813 ...	6.358	1.9968
3.0	7.91907 ...	7.919	1.7895
3.5	9.50637 ...	9.506	1.6538
4.0	11.11367 ...	11.113	1.5590
4.5	12.73677 ...	12.736	1.4898

v_κ to be a convenient lower approximation for $\rho_\kappa + 1$. We show that $\widetilde{\Pi}_\kappa(v_\kappa) < 2$ using the upper bound from (16.9) as we did for $\widetilde{\Pi}_2(5.28)$.

Table 16.1 summarizes our results, with $\widehat{\widetilde{\Pi}}(u)$ denoting the right side of (16.9). Since in each case $\widetilde{\Pi}_\kappa(v_\kappa) < 2$ and $\widehat{\widetilde{\Pi}}_\kappa(u)$ is decreasing in u, we conclude that $z_{\widetilde{\Pi}}(\kappa) < v_\kappa < \rho + 1$ holds for $\kappa = 2.5, 3, 3.5, 4, 4.5$.

We complete the proof for the cases $\kappa \geq 5$ with the aid of the inequality $\rho_\kappa + 1 > 2.7\kappa$ established in Lemma 15.6. We have $\widetilde{\Pi}_\kappa(\rho+1) < \widetilde{\Pi}_\kappa(2.7\kappa)$; thus it suffices to show the last expression to be less than 2 for all $\kappa \geq 5$. We estimate $\widetilde{\Pi}_\kappa(2.7\kappa)$ from above by a cruder version of (16.9) using two intervals for the trapezoidal estimate and upper bounds for p and $1/\sigma$. We show the resulting expression to be a decreasing function of κ, and hence it suffices to show that $\widetilde{\Pi}_5(2.7 \cdot 5) < 2$. We begin by studying the upper estimate for p given in (16.10).

Lemma 16.4. *Let* $a > 0$, $b \geq 0$ *and* $a\kappa - b > 0$. *Then*

$$\kappa p_\kappa^*(a\kappa - b) = \kappa \left\{ a\kappa - b + \kappa(a\kappa - b)\log\left(1 + \frac{1}{a\kappa - b}\right) \right\}^{-1}$$

is a decreasing function of κ.

Proof. Consider the reciprocal expression

$$a - b/\kappa + (a\kappa - b)\log\{1 + 1/(a\kappa - b)\}.$$

Clearly, $-b/\kappa$ is increasing and, with $a\kappa - b =: x > 0$, we have

$$\frac{d}{dx}\left\{x \log\left(1 + \frac{1}{x}\right)\right\} = \log\left(1 + \frac{1}{x}\right) - \frac{1}{x+1} > 0. \qquad \square$$

This lemma does not apply to the expression $p_\kappa^*(2.7\kappa + 1)$, which occurs in our estimate of $\widetilde{\Pi}_\kappa(2.7\kappa)$. Here we have the simple bound

(16.13) $\qquad \kappa p_\kappa^*(2.7\kappa + 1) < 1/3.7, \quad \kappa > 0.$

Indeed, by (16.10) and the inequality $\log(1+1/x) > 1/(x+\tfrac{1}{2})$ for $x > 0$, we have

$$\{\kappa p_\kappa^*(2.7\kappa + 1)\}^{-1} > \kappa^{-1}\{2.7\kappa + 1 + \kappa(2.7\kappa + 1)/(2.7\kappa + 3/2)\}$$
$$> 2.7 + \frac{1}{\kappa} + 1 - \frac{1/2}{2.7\kappa + 3/2} > 3.7.$$

For $u \geq 2\kappa$ set

$$\sigma_\kappa^*(u) := 1 - \frac{1}{2}\left(\frac{u}{2\kappa}\right)^{2\kappa} e^{2\kappa - u}.$$

Recall from Theorem 14.1 that $\sigma_\kappa(u) > \sigma_\kappa^*(u)$ for all $u \geq 2\kappa$. Here we establish a monotonicity for σ^*.

Lemma 16.5. *If* $(a-2)\kappa > |b|$, *then* $\sigma_\kappa^*(a\kappa - b)$ *increases with* κ.

Proof. It suffices to show that

$$f(\kappa) := 2\kappa \log \frac{a\kappa - b}{2\kappa} + (2-a)\kappa$$

is decreasing in κ. Indeed,

$$f'(\kappa) = 2\{\log(a\kappa - b) - \log(2\kappa)\} + \frac{2b}{a\kappa - b} + 2 - a$$
$$\leq (a\kappa - b - 2\kappa)\left\{\frac{1}{a\kappa - b} + \frac{1}{2\kappa}\right\} - \frac{(a\kappa - b - 2\kappa)a}{a\kappa - b}$$

(using the trapezoidal estimate for $\log(a\kappa - b) - \log(2\kappa)$). Thus

$$f'(\kappa) \leq -\frac{\{(a-2)\kappa - b\}\{(a-2)\kappa + b\}}{2\kappa(a\kappa - b)}$$
$$= -\frac{\{(a-2)\kappa\}^2 - b^2}{2\kappa(a\kappa - b)} < 0. \qquad \square$$

16.3 Estimation of $\widetilde{\Pi}(2.7\kappa)$

Since the integrand occurring in $\widetilde{\Pi}_\kappa(u)$ is convex, we have (cf. (16.9))

$$\widetilde{\Pi}_\kappa(u) \leq \frac{u p_\kappa(u)}{\sigma_\kappa(u)} + \frac{\kappa}{2}\left\{\frac{p_\kappa(u+1)}{\sigma_\kappa(u)} + \frac{2p_\kappa(u)}{\sigma_\kappa(u-1)} + \frac{p_\kappa(u-1)}{\sigma_\kappa(u-2)}\right\}.$$

If we estimate p and σ by p^* and σ^* respectively, set $u = 2.7\kappa$ and suppress κ subscripts, we obtain

$$\widetilde{\Pi}(2.7\kappa) < \frac{2.7\kappa p^*(2.7\kappa)}{\sigma^*(2.7\kappa)} + \frac{\kappa p^*(2.7\kappa + 1)}{2\sigma^*(2.7\kappa)} + \frac{\kappa p^*(2.7\kappa)}{\sigma^*(2.7\kappa - 1)} + \frac{\kappa p^*(2.7\kappa - 1)}{2\sigma^*(2.7\kappa - 2)}.$$

16.3 Estimation of $\widetilde{\Pi}(2.7\kappa)$

We have $\kappa p^*(2.7\kappa + 1) < 1/3.7$ (by (16.13)) and use monotonicity of the other p^*'s and σ^*'s from the preceding lemmas. For $\kappa \geq 5$ we obtain

$$\widetilde{\Pi}_\kappa(2.7\kappa) < \frac{13.5 p_5^*(13.5)}{\sigma_5^*(13.5)} + \frac{1/3.7}{2\sigma_5^*(13.5)} + \frac{5 p_5^*(13.5)}{\sigma_5^*(12.5)} + \frac{5 p_5^*(12.5)}{2\sigma_5^*(11.5)}$$
$$< 1.957 < 2.$$

Thus $z_{\widetilde{\Pi}}(\kappa) < 2.7\kappa < \rho_\kappa + 1$ for all $\kappa \geq 5$, and so (16.12) holds in this case also.

In (16.7) we showed that $\rho_\kappa + 1 < z_{\widetilde{\Xi}}(\kappa)$, and combining this inequality with the preceding one completes the proof of Proposition 16.3. □

17
The parameters α_κ and β_κ

17.1 The cases $\kappa = 1, 1.5$

When we introduced Lemma 13.1, we said that we would use the formulas of that result to find α_κ and β_κ; the time has come to do so. This is the second step of the program laid out at the end of Chapter 13.

We begin by studying the cases $\kappa = 1, 1.5$, where we showed in Proposition 16.2 that $z_\Xi(\kappa) < z_{\widetilde{\Pi}}(\kappa)$ holds, and solve the first pair of equations from Lemma 13.1.

Theorem 17.1. *For $\kappa = 1$ the simultaneous equations*

(17.1) $$\Pi_\kappa(y, x) = 2, \quad \Xi_\kappa(y, x) = 0$$

possess a solution $y = \alpha_1 = 2$, $x = \beta_1 = 2$ that is unique among all solutions with $y \geq x \geq 2$.

For $\kappa = 1.5$ the equations (17.1) *possess a solution*

(17.2) $$y = \alpha_{1.5} = 3.91148\ldots, \quad x = \beta_{1.5} = 3.11582\ldots$$

that is unique among all solutions with

$$z_\Pi(1.5) \leq x \leq z_\Xi(1.5).$$

In each case, the solutions satisfy the conditions $\beta_\kappa \leq \alpha_\kappa \leq \beta_\kappa + 1$.

Proof. We begin by restating (17.1) in terms of $z_\Pi(\kappa)$ and $z_\Xi(\kappa)$. Recall that we have defined, for each value of $\kappa \geq 1$,

$$\Pi_\kappa(u) = \Pi(u) := \Pi_\kappa(u, u), \quad \Xi_\kappa(u) = \Xi(u) := \Xi_\kappa(u, u)$$

and showed in the last chapter that the equations $\Pi(u) = 2$, $\Xi(u) = 0$ have unique solutions, which we have called $z_\Pi(\kappa)$ and $z_\Xi(\kappa)$ respectively.

Rewrite the equation $\Pi(y,x) = 2$ in the equivalent form (cf. (12.24))

$$\kappa \int_{z_\Pi}^y \frac{p(t)\sigma(t-2)}{\sigma^2(t)}\,dt = 2 - \frac{(x-1)p(x-1)}{\sigma(x-1)} - \kappa \int_{x-1}^{z_\Pi} \frac{p(t)\sigma(t-2)}{\sigma^2(t)}\,dt$$

$$= 2 - \Pi(z_\Pi, x)$$

$$= 2 - \Pi(z_\Pi) - \kappa \int_{x-1}^{z_\Pi - 1} \frac{p(t+1)}{\sigma(t)}\,dt$$

$$= \kappa \int_{z_\Pi}^x \frac{p(t)}{\sigma(t-1)}\,dt$$

(since $\Pi(z_\Pi) = 2$), i.e., we have

(17.3) $$\int_{z_\Pi}^y \frac{p(t)\sigma(t-2)}{\sigma^2(t)}\,dt = \int_{z_\Pi}^x \frac{p(t)}{\sigma(t-1)}\,dt.$$

Also, rewrite $\Xi(y,x) = 0$ in the equivalent form (cf. (12.25))

$$\kappa \int_{z_\Xi}^y \frac{q(t)\sigma(t-2)}{\sigma^2(t)}\,dt = -\frac{(x-1)q(x-1)}{\sigma(x-1)} - \kappa \int_{x-1}^{z_\Xi} \frac{q(t)\sigma(t-2)}{\sigma^2(t)}\,dt,$$

$$= -\Xi(z_\Xi, x)$$

$$= -\Xi(z_\Xi) + \kappa \int_{x-1}^{z_\Xi - 1} \frac{q(t+1)}{\sigma(t)}\,dt$$

$$= \kappa \int_x^{z_\Xi} \frac{q(t)}{\sigma(t-1)}\,dt,$$

i.e.,

(17.4) $$\int_{z_\Xi}^y \frac{q(t)\sigma(t-2)}{\sigma^2(t)}\,dt = \int_x^{z_\Xi} \frac{q(t)}{\sigma(t-1)}\,dt.$$

We now proceed to the case $\kappa = 1$. Here

$$\Pi_1(2) = \Pi_1(2,2) = \frac{p_1(1)}{\sigma_1(1)} = \frac{e^{-\gamma}}{j_1(1/2)} = 2$$

by (12.24), (15.4), and (14.4); and by (12.25),

$$\Xi_1(2) = \Xi_1(2,2) = q_1(1)/\sigma_1(1) = 0,$$

since $q_1(u) = u - 1$. Thus $z_\Pi(1) = 2$ and $z_\Xi(1) = 2$ and $\alpha_1 = 2 = \beta_1$ are solutions of the equation (17.1) at $\kappa = 1$ and obviously satisfy $\beta_1 = \alpha_1 < \beta_1 + 1$. To see the uniqueness of the solution (2,2), note that if either of $x, y > 2$, then by (17.3) the other one satisfies this inequality as well, and this, in turn, violates (17.4).

17.1 The cases $\kappa = 1, 1.5$

We comment that the left side of each of (17.3) and (17.4) is 0 if $y < 2$, so each pair $(2, y)$ with $y < 2$ also provides a solution of (17.1); such solutions are not interesting for our purposes, since it is known (see [DHR90a]) that the relation $\alpha_\kappa \geq \beta_\kappa$ necessarily holds for our sieve.

Next, we show that (17.1) has a solution for $\kappa = 1.5$. We have here (by a computer calculation)

$$z_\Pi(1.5) = 2.98685\ldots \quad \text{and} \quad z_\Xi(1.5) = 3.30788\ldots.$$

Consider (17.3) for $z_\Pi(1.5) < x \leq z_\Xi(1.5)$. The integral on the left is 0 at $y = z_\Pi$, increasing in y, and divergent as $y \to \infty$, since the integrand is positive, $p(t) \sim 1/t$ and $\sigma(t) \sim 1$ as $t \to \infty$. Also, the integral on the right is positive valued here. Hence, for any given $x \in [z_\Pi, z_\Xi]$, the equation has a unique solution $y = \alpha(x)$, and $\alpha(x)$ is continuous and strictly increasing on this interval. Clearly,

$$\alpha(z_\Pi) = z_\Pi, \quad \alpha(z_\Xi) > z_\Pi;$$

indeed, the latter statement can be sharpened, for

$$\sigma(t-2)/\sigma^2(t) < 1/\sigma(t-1)$$

and therefore

$$\kappa \int_{z_\Pi}^{y} \frac{p(t)}{\sigma(t-1)} dt > \kappa \int_{z_\Pi}^{y} \frac{p(t)\sigma(t-2)}{\sigma^2(t)} dt = \kappa \int_{z_\Pi}^{x} \frac{p(t)}{\sigma(t-1)} dt,$$

and this means that

$$\alpha(x) > x, \quad z_\Pi < x \leq z_\Xi,$$

and, in particular, $\alpha(z_\Xi) > z_\Xi$.

Now consider equation (17.4) in the semi-infinite strip $z_\Pi \leq x \leq z_\Xi$, $z_\Xi \leq y$. Again, the value of the integral on the left is 0 at $y = z_\Xi$ and increasing to $+\infty$ as $y \to \infty$, and the value of the integral on the right is nonnegative, since $x \geq z_\Pi > \rho_{1.5} = 2.3660\ldots$. Hence, for each $x \in [z_\Pi, z_\Xi]$, there exists a unique solution $y = a^*(x)$, again a continuous function of x; but now $\hat{a}(x)$ is strictly decreasing in x on the interval, with

$$a^*(z_\Pi) > z_\Xi \quad \text{and} \quad a^*(z_\Xi) = z_\Xi.$$

We conclude that the curves $y = \alpha(x)$, $y = a^*(x)$ have a unique point of intersection in the range $z_\Pi \leq x \leq z_\Xi$, call it (β, α), and that

$$z_\Pi < \beta < z_\Xi < \alpha.$$

We find by computation the values of $\alpha_{1.5}$ and $\beta_{1.5}$ given in (17.2); then, by inspection, $\beta_{1.5} < \alpha_{1.5} < \beta_{1.5} + 1$. □

17.2 The cases $\kappa = 2, 2.5, 3, \ldots$

Now consider the cases $\kappa = 2, 2.5, 3, \ldots$, where we showed in Proposition 16.3 that

(17.5) $$z_{\widetilde{\Pi}}(\kappa) < z_{\widetilde{\Xi}}(\kappa)$$

holds. We shall solve the second pair of equations of Lemma 13.1 for such κ. For $y > 2$ and $x > 1$, define $f_\kappa(y-1, x)$ by

$$(y-1)^\kappa f_\kappa(y-1, x) := \kappa \int_x^{y-1} \frac{t^{\kappa-1}}{\sigma_\kappa(t-1)} dt.$$

In this section we prove

Theorem 17.2. *For each integer or half integer $\kappa \geq 2$, the simultaneous equations*

(17.6) $$\begin{cases} \widetilde{\Pi}_\kappa(y) + (y-1)p_\kappa(y-1)f_\kappa(y-1, x) = 2 \\ \widetilde{\Xi}_\kappa(y) - (y-1)q_\kappa(y-1)f_\kappa(y-1, x) = 0 \end{cases}$$

have a solution $y = \alpha_\kappa$, $x = \beta_\kappa$ that satisfies

(17.7) $$\alpha_\kappa > \max(\beta_\kappa + 1, z_{\widetilde{\Xi}}(\kappa)).$$

Elimination of f between the two equations (17.6) yields

(17.8) $$l(y) = l_\kappa(y) := q(y-1)(\widetilde{\Pi}(y) - 2) + p(y-1)\widetilde{\Xi}(y) = 0$$

for y (as usual, we have omitted subscripts κ, since no confusion is possible here). The proof of Theorem 17.2 depends on

Proposition 17.3. *For each integer or half integer $\kappa \geq 2$, the equation $l_\kappa(y) = 0$ has a solution $y \in (\rho_\kappa + 1, 3.75\kappa)$.*

Let us assume Proposition 17.3 for the moment and proceed to the proof of Theorem 17.2. If $l(y) = 0$ has more than one solution $y > \rho+1$, let α be the *least* such solution. We show that α satisfies the sharper inequality

(17.9) $$\alpha_\kappa > z_{\widetilde{\Xi}}(\kappa).$$

17.2 The cases $\kappa = 2, 2.5, 3, \ldots$

We showed in (16.7) and (16.12) that

$$z_{\widetilde{\Pi}}(\kappa) < \rho_\kappa + 1 < z_{\widetilde{\Xi}}(\kappa) \qquad \kappa = 2.5, 3, 3.5, \ldots.$$

We deal with this case first. By the Proposition and the last inequality, $\alpha > z_{\widetilde{\Pi}}$, and hence $\widetilde{\Pi}_\kappa(\alpha) < 2$ (by (16.5)). Also, $q(\alpha - 1) > 0$ because $\alpha - 1 > \rho$. Since $l(\alpha) = 0$, it now follows that $p(\alpha - 1)\widetilde{\Xi}(\alpha) \geq 0$ or $\widetilde{\Xi}(\alpha) \geq 0$. This proves (17.9) for $\kappa = 2.5, 3, 3.5, \ldots$, since $\widetilde{\Xi} \uparrow$ (by (16.6)). There remains the case $\kappa = 2$, where

$$z_{\widetilde{\Pi}}(2) = 5.2405\ldots > \rho_2 + 1 = 4.83398\ldots.$$

If $\alpha > z_{\widetilde{\Pi}}(2)$, then the preceding argument applies; we conclude the proof of (17.9) by showing that, in fact, it is impossible to have $\alpha_2 \leq z_{\widetilde{\Pi}}(2)$.

Lemma 17.4. *Let l_κ be as defined above. The equation $l_2(y) = 0$ has no solutions $y \in [\rho_2 + 1, z_{\widetilde{\Pi}}(2)]$.*

Proof. By (17.5) and (16.6), we have $l(z_{\widetilde{\Pi}}) = p(z_{\widetilde{\Pi}} - 1)\widetilde{\Xi}(z_{\widetilde{\Pi}}) < 0$. Let

$$L(y) := (y-1)^{-1} l(y),$$

so that also $L(z_{\widetilde{\Pi}}) < 0$. To show that $L < 0$ throughout the interval, it suffices to show that

(17.10) $\qquad L'(y) > 0 \text{ for } \rho + 1 \leq y \leq z_{\widetilde{\Pi}}.$

Now, by the differential equations (15.1) and (15.6) for p and q (with $\kappa = 2$),

$$L'(y) = 2(y-1)^{-2} q(y)(\widetilde{\Pi}(y) - 2) - 2(y-1)^{-2} p(y)\widetilde{\Xi}(y)$$
$$+ (y-1)^{-1} q(y-1)\widetilde{\Pi}'(y) + (y-1)^{-1} p(y-1)\widetilde{\Xi}'(y),$$

so that, by (16.5) and (16.6),

$$\frac{1}{2}(y-1)^2 L'(y) = q(y)(\widetilde{\Pi}(y) - 2) - p(y)\widetilde{\Xi}(y)$$
$$+ (y-1)\frac{\sigma(y-2)}{\sigma^2(y)}\{q(y-1)p(y) + q(y)p(y-1)\}.$$

Since $y \geq \rho + 1$, both $q(y-1)$ and $q(y)$ are nonnegative, and since $y \leq z_{\widetilde{\Pi}} < z_{\widetilde{\Xi}}$ by (17.5) and the condition in (17.10), $\widetilde{\Pi}(y) - 2 \geq 0$ and $\widetilde{\Xi}(y) < 0$. Hence $L'(y) > 0$ for $\rho + 1 \leq y \leq z_{\widetilde{\Pi}}$, and so $l_2(y) < 0$ throughout the interval. □

We have now proved (17.9), whence $\widetilde{\Xi}(\alpha) > 0$. Using Lemma 16.1, we see that the equation in x,

(17.11) $\quad \widetilde{\Xi}(\alpha) - (\alpha-1)^{1-\kappa} q(\alpha-1)\kappa \int_x^{\alpha-1} \frac{t^{\kappa-1}}{\sigma(t-1)} dt = 0$

has a unique solution $x = \beta$ satisfying

(17.12) $\quad 1 < \beta < \alpha - 1.$

This inequality and (17.9) together establish (17.7). Finally, $l(\alpha) = 0$ and (17.11) combine to give

$$\widetilde{\Pi}_\kappa(\alpha_\kappa) + (\alpha_\kappa - 1)p_\kappa(\alpha_\kappa - 1)f_\kappa(\alpha_\kappa - 1, \beta) = 2$$

and the proof of Theorem 17.2 is now complete, apart from establishing Proposition 17.3.

17.3 Proof of Proposition 17.3

Since $l(\rho+1) = p(\rho)\widetilde{\Xi}(\rho+1) < 0$, it suffices to prove that $l(u) > 0$ for some $u > \rho + 1$. Actually, we shall prove that, for all $\kappa \geq 2$,

(17.13) $\quad l(u) > 0 \text{ at } u = 3.75\kappa,$

which is greater than $\rho + 1$ by Proposition 15.7.

From the definition (17.8) of $l(u)$, after some simple rearrangement,

$l(u)/\{p(u-1)q(u-1)\}$
$= -\kappa \int_{u-2}^u \left(\frac{q(t+1)}{q(u-1)} - \frac{p(t+1)}{p(u-1)}\right)\left(\frac{1}{\sigma(t)} - \frac{1}{\sigma(u)}\right) dt$
$\quad - \frac{1}{\sigma(u)}\left\{\frac{\kappa}{q(u-1)}\int_{u-2}^u q(t+1)dt - \frac{\kappa}{p(u-1)}\int_{u-2}^u p(t+1)dt\right\}$
$\quad + \frac{1}{\sigma(u)}\left(\frac{up(u)}{p(u-1)} + \frac{uq(u)}{q(u-1)}\right) - \frac{2}{p(u-1)}, \quad u > \rho + 1.$

We apply (12.17) in the second expression on the right and also

$$\kappa \int_{u-2}^u p(t+1)dt = 2 - (u-1)p(u-1) - up(u),$$

which comes from two applications of (12.17), to obtain

(17.14) $\quad \dfrac{l(u)}{p(u-1)q(u-1)} = \dfrac{2}{p(u-1)}\left(\dfrac{1}{\sigma(u)} - 1\right) - I(u), \quad u > \rho + 1,$

where

(17.15) $$I(u) := \kappa \int_{u-2}^{u} \left(\frac{q(t+1)}{q(u-1)} - \frac{p(t+1)}{p(u-1)}\right)\left(\frac{1}{\sigma(t)} - \frac{1}{\sigma(u)}\right)dt.$$

By (17.14), (17.13) is equivalent to proving that

(17.16) $$I(u) < \frac{2}{p(u-1)}\left(\frac{1}{\sigma(u)} - 1\right) \quad \text{for } u = 3.75\kappa.$$

Since $\sigma(t)$ is strictly increasing in t, we see from (17.15) that

$$I(u) < \frac{\kappa}{\sigma(u-2)\sigma(u)} \int_{u-2}^{u} \left(\frac{q(t+1)}{q(u-1)} - \frac{p(t+1)}{p(u-1)}\right)(\sigma(u) - \sigma(t))dt.$$

Write

$$\sigma(u) - \sigma(t) = \int_{t}^{u} \sigma'(s)ds, \quad u-2 \leq t \leq u.$$

By Lemma 14.4, $\sigma'(s)$ is strictly decreasing in s for $s \geq 2\kappa$ and therefore the average

$$\frac{1}{u-t} \int_{t}^{u} \sigma'(s)ds$$

decreases as t increases from $u-2$ to u, provided that

(17.17) $$u \geq 2\kappa + 2.$$

Hence

$$\frac{1}{u-t} \int_{t}^{u} \sigma'(s)ds \leq \frac{1}{2} \int_{u-2}^{u} \sigma'(s)ds = \frac{\sigma(u) - \sigma(u-2)}{2} = \frac{u\sigma'(u)}{2\kappa}$$

for $2\kappa \leq u-2 \leq t \leq u$ (the last formula is (6.8')), and therefore

$$I(u) < \frac{u\sigma'(u)}{2\sigma(u-2)\sigma(u)} \int_{u-2}^{u} \left(\frac{q(t+1)}{q(u-1)} - \frac{p(t+1)}{p(u-1)}\right)(u-t)dt$$

for $u > \max(2\kappa + 2, \rho + 1)$. We write this inequality in the form

(17.18) $$I(u) < \frac{u\sigma'(u)}{2\sigma(u-2)\sigma(u)}(I_1(u) + I_2(u)), \quad u > \max(2\kappa + 2, \rho + 1),$$

where

$$I_1(u) = \int_{u-2}^{u} \left(\frac{q(t+1)}{q(u-1)} - 1\right)(u-t)dt$$

and

$$I_2(u) = \int_{u-2}^{u} \left(1 - \frac{p(t+1)}{p(u-1)}\right)(u-t)dt.$$

The integral $I_2(u)$ is rather easy to estimate with adequate precision. Since $up(u)$ increases with u, as we noted in Section 15.1,

$$(u-1)p(u-1) \leq (t+1)p(t+1), \quad u-2 \leq t,$$

and therefore

$$I_2(u) \leq \int_{u-2}^{u} \left(1 - \frac{u-1}{t+1}\right)(u-t)dt$$

$$\leq \frac{1}{u-1} \int_{u-2}^{u} (t-u+2)(u-t)dt = \frac{4/3}{u-1}.$$

To estimate $I_1(u)$, we begin with the remark that

$$\frac{q(t+1)}{q(u-1)} = \exp \int_{u-1}^{t+1} \frac{q'(s)}{q(s)} ds, \quad t+1 \geq u-1 > \rho.$$

Since $(q'/q)(s)$ is decreasing in $s > \rho$ (Lemma 15.4), we have at once that

(17.19) $$\frac{q(t+1)}{q(u-1)} \leq \exp\left\{(t+2-u)\frac{q'}{q}(u-1)\right\}, \quad t \geq u-2.$$

Actually, even $s(q'/q)(s)$ is decreasing in $s > \rho$, by Corollary 15.5; therefore, since $s(q'/q)(s) \to +\infty$ as $s \to \rho + 0$, it follows that $(q'/q)(s)$ is strictly decreasing to 0 as $s \to \infty$ and assumes every positive value exactly once in (ρ, ∞). In particular, there exists a unique number $s_0 > \rho$ such that $(q'/q)(s_0) = 1$ and

$$\frac{q'}{q}(s) \leq 1, \quad s \geq s_0 \,(>\rho).$$

But, by (15.5),

$$s\frac{q'}{q}(s) = \kappa - 1 + \kappa \frac{q(s+1)}{q(s)} = \kappa - 1 + \kappa \exp \int_{s}^{s+1} \frac{q'}{q}(t)dt$$

$$< \kappa - 1 + \kappa \exp \frac{q'}{q}(s),$$

whence $s_0 < \kappa - 1 + \kappa e$, and hence

$$\frac{q'}{q}(s) < 1 \quad \text{if} \quad s \geq (e+1)\kappa - 1.$$

It follows from (17.19) that

$$\frac{q(t+1)}{q(u-1)} \leq e^{t+2-u}, \quad t \geq u-2,$$

provided that

(17.20) $$u - 1 \geq (e+1)\kappa - 1.$$

Then, from the definition of I_1, we have

$$I_1(u) \leq \int_{u-2}^{u} (e^{t+2-u} - 1)(u-t)dt = \int_0^2 (e^x - 1)(2-x)dx$$
$$= e^2 - 5 \leq 2.390, \quad u \geq \kappa(e+1).$$

Now we take

$$u := u^* = 3.75\,\kappa > (e+1)\,\kappa,$$

which satisfies (17.20). Further, from the estimate $\rho_\kappa < 3.6\kappa - 1.56$ (Proposition 15.7), we get

$$3.75\,\kappa > \max(2\kappa + 2, \rho + 1) \quad \text{for all} \quad \kappa \geq 2;$$

thus the concavity condition (17.17) and the hypothesis of the proposition are satisfied for $u \geq u^*$.

Combining the estimates of I_1 and I_2 with (17.18), we find that

$$I(u) < \frac{u\sigma'(u)}{2\sigma(u-2)\,\sigma(u)} \left(2.390 + \frac{4/3}{u-1} \right), \quad u \geq u^*,$$

and hence, from (17.16), that (17.13) follows if we can prove that

(17.21) $$\frac{up(u-1)}{4} \frac{\sigma'(u)}{1-\sigma(u)} \left(2.390 + \frac{4/3}{u-1} \right) \frac{1}{\sigma(u-2)} < 1$$

for $u = u^*$.

With this choice of u and $\kappa \geq 2$, we have by Corollary 15.2,

$$up(u-1) < \frac{u}{u-1+\kappa-1/2} = \frac{3.75\kappa}{4.75\kappa - 1.5}$$
$$= \frac{3.75}{4.75 - 1.5/\kappa} \leq \frac{3.75}{4.75 - .75} < 0.938.$$

Our second factor is

$$\frac{\sigma'(u)}{1-\sigma(u)} \leq 0.64 + 0.57 \log \frac{3.75\kappa + 2}{2\kappa} \leq 1.134$$

by Corollary 14.17. (Note that we have treated here the case $3.75\,\kappa > 4.6\kappa - 2$, which applies for $\kappa = 2$; the estimate 1.12, which holds for all $\kappa \geq 2.5$, is even smaller.)

For the third factor, we have

$$2.390 + \frac{4/3}{u-1} \le 2.390 + \frac{4/3}{3.75 \cdot 2 - 1} \le 2.596.$$

For the last factor, we need to handle the cases $\kappa = 2, 2.5$ by evaluating $\sigma_\kappa(3.75\kappa - 2)$ directly. Using the package described in the Appendix, we find that

$$\sigma_2(5.5) > 0.7865, \qquad \sigma_{2.5}(7.375) > 0.8543.$$

For $\kappa \ge 3$, we use the first form of Proposition 14.18:

$$1 - \sigma_\kappa(3.75\kappa - 2) = 1 - j_\kappa(1.875\kappa - 1)$$
$$< \frac{1}{2} \exp\Big\{ - \kappa \int_1^{1.875 - 1/\kappa} (\log v + \log(1 + \log v))\, dv \Big\}.$$

For $\kappa \ge 3$,

$$1.875 - 1/\kappa \ge 1.875 - 1/3 > 1.541,$$

and thus (with the help of numerical integration)

$$1 - \sigma_\kappa(3.75\kappa - 2) < \exp\Big\{ - \kappa \int_1^{1.541} (\log v + \log(1 + \log v))\, dv \Big\}/2$$
$$< \exp\{-0.2353\kappa\}/2 < 0.247,$$

giving the bound

$$\sigma_\kappa(3.75\kappa - 2) > 1 - 0.247 = 0.753.$$

Thus, in all cases, $\sigma_\kappa(3.75\kappa - 2) > 0.753$.

Combining the preceding estimates, we see that the left side of (17.21) is less than $0.917 < 1$, and hence (17.13) is true for $\kappa \ge 2$.

With Proposition 17.3 established, we have proved Theorem 17.2 unconditionally. Note that the restriction to $\kappa \ge 2$ arises only indirectly via (17.5). $\qquad\square$

Once α_κ has been determined, the sieving limit β_κ is given implicitly by (17.6): we have

$$\widetilde{\Xi}_\kappa(\alpha_\kappa)/\{(\alpha_\kappa - 1) q(\alpha_\kappa - 1)\} = f(\alpha_\kappa - 1, \beta_\kappa),$$

so that

$$\int_{\beta_\kappa}^{\alpha_\kappa - 1} \frac{t^{\kappa - 1}}{\sigma_\kappa(t-1)}\, dt = \frac{(\alpha_\kappa - 1)^{\kappa - 1}}{\alpha_\kappa\, q(\alpha_\kappa - 1)} \widetilde{\Xi}_\kappa(\alpha_\kappa).$$

Table 17.1. *Values of* α_κ, β_κ, *and* $\rho_\kappa + 1$

κ	α_κ	β_κ	$\rho_\kappa + 1$
1.0	2.000000	2.000000	2.000000
1.5	3.911481	3.115821	3.366025
2.0	5.357727	4.266450	4.833987
2.5	6.839998	5.444068	6.358138
3.0	8.371931	6.640859	7.919077
3.5	9.938884	7.851463	9.506376
4.0	11.531799	9.072248	11.113677
4.5	13.144726	10.300628	12.736774
5.0	14.773560	11.534709	14.372723
5.5	16.415350	12.773074	16.019369
6.0	18.067899	14.014644	17.675086
6.5	19.729536	15.258588	19.338609
7.0	21.398952	16.504258	21.008934
7.5	23.075103	17.751146	22.685249
8.0	24.757152	18.998853	24.366886
8.5	26.444401	20.247056	26.053290
9.0	28.136280	21.495510	27.743992
9.5	29.832306	22.744013	29.438593
10.0	31.532074	23.992408	31.136752

Table 17.1 lists values of α_κ and β_κ, along with $\rho_\kappa + 1$ values added for purposes of comparison, for $\kappa \leq 10$. For $\kappa \leq 7$, the α_κ and β_κ values were computed using a software package developed by Wheeler. Compared with ones produced using the package described in Appendix 1, they were found to differ by at most 1 in the last displayed decimal place.

17.4 Notes on Chapter 17

Early in Chapter 13 we spoke of α_κ and β_κ as being "probably unique." Now once α_κ is known, β_κ is indeed uniquely determined; but all that we have proved here about α_κ, the least positive zero of the transcendental equation $l_\kappa(y) = 0$, is that it lies in $(\rho_\kappa + 1, 3.75\kappa)$. (Recall from Section 15.6 that $\rho + 1 \sim r_c \kappa$ as $\kappa \to \infty$, with $r_c \approx 3.591121$.) Since α_κ marks the point at which our upper sieve estimate starts to improve on that of Ankeny–Onishi, it is clear that we want α_κ to be as small as possible, which attaches an element of uniqueness to this parameter. However, we believe that the equation $l_\kappa(y) = 0$ has in fact only the one positive root. In [DH97b], Theorem 1, we have proved that

$$\rho_\kappa + 1 < \alpha_\kappa < \rho_\kappa + 2.5, \quad \kappa \geq 200,$$

and for all smaller values of κ ($1 < \kappa \leq 10$) that have been computed, α_κ satisfies these inequalities too.

We proved in [DHR93a], Proposition 7.3, that $\beta_\kappa > 3$ for $\kappa \geq 2$ and $\beta_\kappa > 2\kappa$ for $\kappa \geq 4.6$; that $\beta_\kappa > z_{\widetilde{\Pi}}(\kappa) - 1$ when $\kappa \geq 200$ in [DH01], Corollary 4.2; and in [DH97b] we showed that β_κ is always less than the Ankeny–Onishi sieving limit (which is referred to there as ν_κ). It is shown in [AO65] that $\nu_\kappa \sim c\kappa$, with $c \approx 2.445$, as $\kappa \to \infty$.

In [Grv01], §7.3.2, a description is given of Selberg's argument (see also [Sel91], Chapter 45, Section 14) that the "right" sieving limit should be asymptotic to 2κ as $\kappa \to \infty$. For this he set out from a type of lower bound χ^- as described at the end of Section 3.3.

Table 17.1 suggests that it is likely that $\beta_\kappa/\kappa \uparrow$, and perhaps to the same limit, $2.445\ldots$, as the Ankeny–Onishi sieving limit.

18
Properties of F_κ and f_κ

18.1 F_κ and $f_\kappa \to 1$ at ∞

In the preceding two chapters we determined numbers α_κ and β_κ which satisfy the equations occurring in Lemma 13.1. Equipped with these parameters, we can at last define F_κ and f_κ on $(0, \infty)$: we define F_κ and f_κ by (6.1) and (6.2) for the initial intervals $(0, \alpha_\kappa]$ and $[0, \beta_\kappa]$ respectively, and we continue the functions forward by recursively applying formulas (6.3) and (6.4). Details of our computer calculation are described in Section A1.5 in the Appendix.

It remains to carry out the last steps in the program described in Chapter 13: to show that this choice satisfies condition (6.5), that $F > f$, and to establish the monotonicity of F and f as asserted in Theorem 6.1. (The κ-subscripts are again generally suppressed.) We can use $P = F+f$ and $Q = F - f$ in place of F and f, since the conditions (6.1)–(6.4) are equivalent to (12.1)–(12.6).

Upon setting $\alpha = \alpha_\kappa$ and $\beta = \beta_\kappa$, the functions P and Q satisfy the Iwaniec inner product formulas (12.18) and (12.19). The results of Section 12.2 apply, and they yield

$$P(u) - 2 \ll e^{-u}, \quad Q(u) \ll e^{-u}.$$

Thus F and f converge to 1 at infinity at the rate claimed in (6.5).

18.2 $Q_\kappa(u) > 0$ for $u > 0$

We asserted in (12.8) that Q is positive on $(0, \infty)$. This property is crucial for our theory, since Q is the difference of purported upper and lower bound functions. We now prove this result.

By (12.2), $Q(u) = 1/\sigma(u) > 0$ for $0 < u \leq \beta$. When $\beta < u \leq \alpha$, (12.4) implies that

$$(u^\kappa Q(u))' = \left(\frac{u^\kappa}{\sigma(u)}\right)' - \kappa \frac{u^{\kappa-1}}{\sigma(u-1)}$$

$$= \kappa u^{\kappa-1}\left\{\frac{\sigma(u-2)}{\sigma^2(u)} - \frac{1}{\sigma(u-1)}\right\} < 0$$

(by formula (6.8) and the monotonicity of σ). Thus $u^\kappa Q(u)$ is strictly decreasing for $\beta < u \leq \alpha$, and so $Q(u) > 0$ for $\beta < u \leq \alpha$ if $Q(\alpha) > 0$.

Indeed, the truth of the last inequality implies also that $Q(u) > 0$ for all $u > \alpha$. For suppose on the contrary that $Q(u)$ changes sign beyond α and that $v > \alpha$ is the least zero of $Q(u)$. By (12.19),

$$0 = \int_{v-1}^{v} Q(t)q(t+1)dt;$$

but this is impossible, since $Q(t) > 0$ for $v - 1 \leq t < v$ and $q(t+1)$ is positive because $t+1 \geq v > \alpha > \rho$. To see the last inequality, note that $\alpha_1 = 2$, $\rho_1 = 1$; $\alpha_{3/2} = 3.91148\ldots$, $\rho_{3/2} = 2.36602\ldots$; and for larger values of κ we have $\alpha_\kappa > z_{\Xi}(\kappa) > \rho_\kappa + 1$ by Theorem 17.2 and (16.7). It follows that $Q(u) > 0$ for all $u > 0$ if $Q(\alpha) > 0$.

Lemma 18.1. $Q_\kappa(\alpha_\kappa) > 0$, $\kappa \geq 1$.

Proof. (i) The case $\kappa = 1$ is trivial, for $\alpha_1 = \beta_1 = 2$, and it follows that $Q_1(2) = F_1(2) - f_1(2) = 1/\sigma_1(2) = e^\gamma > 0$.

For the remaining cases, we apply (12.19) at $u = \alpha$,

(18.1) $$\alpha q(\alpha) Q(\alpha) = \kappa \int_{\alpha-1}^{\alpha} Q(t)q(t+1)dt.$$

We have just noted that $\alpha > \rho + 1$ for all $\kappa \geq 3/2$, so that $q(\alpha) > 0$ and $q(t+1) > 0$ for $\alpha - 1 \leq t \leq \alpha$.

(ii) $\kappa = 3/2$. Here $\beta < \alpha < \beta + 1$ and, since $t^{3/2}Q(t)$ is decreasing in $[\beta, \alpha]$, we have, by (18.1),

$$\alpha q(\alpha) Q(\alpha) = \frac{3}{2}\int_{\alpha-1}^{\beta} \frac{q(t+1)}{\sigma(t)}dt + \frac{3}{2}\int_{\beta}^{\alpha} t^{3/2}Q(t)t^{-3/2}q(t+1)dt$$

$$\geq \frac{3}{2}\int_{\alpha-1}^{\beta} \frac{q(t+1)}{\sigma(t)}dt + \frac{3}{2}\alpha^{3/2}Q(\alpha)\int_{\beta}^{\alpha} t^{-3/2}q(t+1)dt$$

$$= \frac{3}{2}\int_{\alpha-1}^{\beta} \frac{q(t+1)}{\sigma(t)}dt + \alpha^{3/2}Q(\alpha)\{\alpha^{-1/2}q(\alpha) - \beta^{-1/2}q(\beta)\},$$

the last by (15.7). Hence

$$\alpha^{3/2}\beta^{-1/2}q(\beta)Q(\alpha) \geq \frac{3}{2}\int_{\alpha-1}^{\beta}\frac{q(t+1)}{\sigma(t)}dt > 0.$$

Now $\beta_{3/2} > 3.115 > 2.367 > \rho_{3/2}$, so $q(\beta) > 0$ and $Q(\alpha) > 0$ follows.

(iii) $\kappa = 2, 2.5, 3, \ldots$. Here $\alpha - 1 > \beta$ and $\alpha - 1 > z_\Xi - 1 > \rho$ by Theorem 17.2 and (16.7). The argument is similar to that of (ii) but simpler. This time, since $[\alpha - 1, \alpha] \subset [\beta, \alpha]$, $t^\kappa Q(t)$ is strictly decreasing in $[\alpha - 1, \alpha]$ and therefore, by (12.18),

$$\alpha q(\alpha)Q(\alpha) > \alpha^\kappa Q(\alpha) \int_{\alpha-1}^{\alpha} \kappa t^{-\kappa} q(t+1) dt$$
$$= \alpha^\kappa Q(\alpha)\{\alpha^{1-\kappa}q(\alpha) - (\alpha-1)^{1-\kappa}q(\alpha-1)\}.$$

Hence $\alpha^\kappa(\alpha-1)^{1-\kappa}q(\alpha-1)Q(\alpha) > 0$, and therefore, $Q(\alpha) > 0$. $\square$

Summarizing progress to this point, we have now proved that the boundary value problem specified by (12.2)–(12.7) is solvable in α_κ, β_κ, P, and Q; and that $Q_\kappa(u) > 0$ for all $u > 0$ and all $\kappa \geq 1$. We conclude the proof of Theorem 6.1 by showing that $F(u)$ decreases monotonically for $u > 0$ and $f(u)$ increases monotonically for $u > \beta$.

Lemma 18.2. *If $\kappa \geq 1$, then*

(18.2) $$f'_\kappa(u) > 0, \quad u > \beta_\kappa,$$
(18.3) $$F'_\kappa(u) < 0, \quad u > 0.$$

Proof. Since $Q > 0$, we have

(18.4) $$F(u) > f(u), \quad u > 0.$$

We now restate formulas (6.3) and (6.4) in the form

(18.5) $$\frac{u}{\kappa}F'(u) = f(u-1) - F(u), \quad u > \alpha,$$
(18.6) $$\frac{u}{\kappa}f'(u) = F(u-1) - f(u), \quad u > \beta.$$

If $\beta < u \leq \alpha$, by (18.6) and the condition that $F(u) = 1/\sigma(u)$, we have

$$\frac{u}{\kappa}f'(u) = \frac{1}{\sigma(u-1)} - f(u) > \frac{1}{\sigma(u)} - f(u) = F(u) - f(u) > 0$$

by (18.4), so that (18.2) holds for $\beta < u \leq \alpha$, and by continuity of F and f, for some distance to the right of α.

We now turn to F. Again, $F(u) = 1/\sigma(u)$ if $0 < u \le \alpha$, and therefore
$$F'(u) = -\sigma'(u)/\sigma^2(u) < 0, \quad 0 < u \le \alpha-;$$
since $F'(u)$ need not be continuous at $u = \alpha$, we give a separate argument to show $F'(\alpha+) < 0$. By (18.5),
$$\alpha F'(\alpha+)/\kappa = f(\alpha-1) - 1/\sigma(\alpha).$$
If $\kappa = 1, 1.5$, then $\alpha \le \beta+1$ and $f(\alpha-1) = 0$, so
$$F'(\alpha+) = -\kappa/\{\alpha\sigma(\alpha)\} < 0.$$
For the cases $\kappa = 2, 2.5, 3, \ldots$, we have $\alpha > \beta+1$, and so (18.5) and Theorem 17.2 (with $f(\alpha-1,\beta) = f(\alpha-1)$) yield
$$\frac{\alpha}{\kappa}F'(\alpha+) = \frac{\widetilde{\Xi}(\alpha)}{(\alpha-1)q(\alpha-1)} - \frac{1}{\sigma(\alpha)}.$$
By the fact that $\alpha > \rho+1$ (which ensures that $q(t+1) > 0$ for $t \ge \alpha-2$), the strict monotonicity of σ, and (12.17), we get
$$\widetilde{\Xi}(\alpha) = \alpha\frac{q(\alpha)}{\sigma(\alpha)} - \kappa\int_{\alpha-2}^{\alpha}\frac{q(t+1)}{\sigma(t)}dt$$
$$< \frac{1}{\sigma(\alpha)}\{\alpha q(\alpha) - \kappa\int_{\alpha-2}^{\alpha}q(t+1)dt\} = \frac{(\alpha-1)q(\alpha-1)}{\sigma(\alpha)}.$$

It follows that $F'(\alpha+) < 0$ in these cases too. Since F' is continuous on (α, ∞), $F'(u) < 0$ for some u interval beyond α.

Suppose now that (18.2) or (18.3) is false. The first failure has to occur at a value $u = v$, where $v > \alpha$. Suppose that (18.2) fails first. Then $f'(v) = 0$ and $F'(u) < 0$ for $0 < u < v$. By (18.6), $f(v) = F(v-1)$ and therefore
$$Q(v) = F(v) - f(v) = F(v) - F(v-1) = \int_{v-1}^{v}F'(t)dt < 0,$$
contradicting (18.4). Hence, if there is a failure at v, it must be (18.3) that fails there. Now we have $F'(v) = 0$, whereas $f'(u) > 0$, $\beta < u < v$. Hence, by (18.5), $F(v) = f(v-1)$ and
$$Q(v) = F(v) - f(v) = f(v-1) - f(v) = -\int_{v-1}^{v}f'(t)dt < 0,$$
since $v > \beta$, again contradicting (18.4). It follows that (18.2) and (18.3) both hold.

Thus F_κ and f_κ have all the properties claimed in Theorem 6.1. □

Appendix 1
Procedures for computing sieve functions

In this appendix we explain some of the techniques used in a *Mathematica*® 5.2† package developed for computing many of the functions described in previous chapters. All of the figures and many of the tables in this book were generated using this software. Although the results of this book were proved only for $\kappa \geq 1$, $2\kappa \in \mathbb{Z}$, we support computation for $\kappa > 0$, $\kappa \in \mathbb{R}$ whenever it is convenient to do so.

Other systems developed for computing sieve functions include those of te Riele [teR80], Bradley [Brd96], and Wheeler [Whe88]. Ferrell Wheeler provided some advice on the implementation described here.

Some of the software developed by others was designed to guarantee results of a given accuracy. However, our package was written with an eye towards generality and convenience of design, and we do not guarantee the accuracy of our results. Also, the computational methods described here differ from some of the methods described in the main body of this book.

As mentioned in the preface, we will maintain a copy of our package at www.math.uiuc.edu/SieveTheoryBook, kept current as it goes through further revision. At that site interested readers also can find fuller and more current documentation on its implementation.

We begin by outlining our approach to computing $F_\kappa(u)$ and $f_\kappa(u)$, following much the same plan as was used in Part II to prove Theorem 6.1. While summarizing our approach we will introduce new terminology—some of which is not standard.

---

† *Mathematica* is a registered trademark of Wolfram Research, Inc.

A1.1 DDEs and the Iwaniec inner product

Given fixed real numbers a, b, and u_0, suppose a function $G(u)$ satisfies the difference differential equation (or "DDE")

$$(\text{A1.1}) \qquad u\frac{d}{du}G(u) = -a\,G(u) - b\,G(u-1), \qquad u > u_0.$$

We call equation (A1.1) a *delay differential equation of type (a,b)* and say that $G(u)$ is of type (a,b) to mean that $G(u)$ satisfies equation (A1.1) for some $u_0 \in \mathbb{R}$. For example, recalling from equation (14.5) that

$$uj'_\kappa(u) = \kappa j_\kappa(u) - \kappa j_\kappa(u-1), \qquad u > 1,$$

we see that $j_\kappa(u)$ is of type $(-\kappa, \kappa)$. Equation (A1.1) is equivalent to (and is often stated as)

$$\frac{d}{du}\left(u^a G(u)\right) = -bu^{a-1}G(u-1), \qquad u > u_0.$$

We call a function $g(u)$ an *adjoint* (of type (a,b)) to $G(u)$ provided $g(u)$ satisfies the advanced argument differential equation

$$(\text{A1.2}) \qquad \frac{d}{du}\left(ug(u)\right) = a\,g(u) + b\,g(u+1), \qquad u > u_0.$$

We introduce adjoints because, as noted in Section 12.4, they have better analytic properties than their associated "original" functions, and they can be used to analyze the original functions via the *Iwaniec inner product*. This inner product depends on the quantity b, and, for arbitrary piecewise continuous $G(u)$ and $g(u)$, is defined as the function

$$(\text{A1.3}) \qquad \langle G, g\rangle_b(u) := uG(u)g(u) - b\int_{u-1}^{u} G(t)g(t+1)\,dt.$$

Note that $\langle G, g\rangle_b$ is *not* an inner product over a vector space. For example, it does not yield a "scalar" (an element of $\mathbb{R}$), but rather a function (which may or may not be constant). Also, although $\langle G, g\rangle_b$ is bilinear in G and g it is not symmetric: we do not generally have $\langle G, g\rangle_b = \langle g, G\rangle_b$.

As noted earlier in some specific cases, differentiation of (A1.3) gives

$$\frac{d}{du}\langle G, g\rangle_b(u) = G'(u)\,u\,g(u) + G(u)\frac{d}{du}\left(u\,g(u)\right)$$
$$- b\,G(u)g(u+1) + b\,G(u-1)g(u).$$

Furthermore, if $G(u)$ is of type (a, b) and $g(u)$ is adjoint to $G(u)$ then by equations (A1.1) and (A1.2), we find that for $u > u_0$ this is

$$= -aG(u)g(u) - bG(u-1)g(u) + aG(u)g(u)$$
$$+ bG(u)g(u+1) - bG(u)g(u+1) + bG(u-1)g(u)$$
$$= 0.$$

Thus, given $G(u)$ of type (a, b) with adjoint $g(u)$, we have

(A1.4) $$\langle G, g \rangle_b(u) = C, \qquad u \geq u_0,$$

for some constant C.

A1.2 The upper and lower bound sieve functions

With this notation established, we now restate the defining equations for $F_\kappa(u)$ and $f_\kappa(u)$ (equations (6.1)–(6.4)), making the roles of the parameters α and β more explicit. Recall that $\sigma_\kappa(u)$ was first defined in Section 5.3. Given $\kappa \geq 1$ and $2 \leq \beta_\kappa \leq \alpha_\kappa$, let

(A1.5) $$F_\kappa(u; \alpha, \beta) = 1/\sigma_\kappa(u), \qquad 0 < u \leq \alpha,$$

(A1.6) $$f_\kappa(u; \alpha, \beta) = 0, \qquad 0 < u \leq \beta,$$

(A1.7) $$\frac{d}{du}\left(u^\kappa F_\kappa(u; \alpha, \beta)\right) = \kappa u^{\kappa-1} f_\kappa(u-1; \alpha, \beta), \qquad \alpha < u,$$

(A1.8) $$\frac{d}{du}\left(u^\kappa f_\kappa(u; \alpha, \beta)\right) = \kappa u^{\kappa-1} F_\kappa(u-1; \alpha, \beta), \qquad \beta < u.$$

Theorem 6.1 states that there exist $\alpha = \alpha_\kappa$ and $\beta = \beta_\kappa$ such that

(A1.9) $$F_\kappa(u; \alpha_\kappa, \beta_\kappa) = 1 + O(e^{-u}) \quad \text{as } u \to \infty,$$
(A1.10) $$f_\kappa(u; \alpha_\kappa, \beta_\kappa) = 1 + O(e^{-u}) \quad \text{as } u \to \infty.$$

In contrast to an arbitrary choice of α, β, we think of a pair α_κ, β_κ satisfying (A1.9) and (A1.10) as being "correct values." For any such pair α_κ, β_κ we write

$$F_\kappa(u) := F_\kappa(u; \alpha_\kappa, \beta_\kappa),$$
$$f_\kappa(u) := f_\kappa(u; \alpha_\kappa, \beta_\kappa).$$

(Note how the monotonicities of $F_\kappa(u; \alpha, \beta)$ and $f_\kappa(u; \alpha, \beta)$, which hold by Theorem 6.1 when $\alpha = \alpha_\kappa$ and $\beta = \beta_\kappa$, fail to hold in Figure A1.1(a).)

Figure A1.2 shows some computed values of α_κ and β_κ. (These values are tabulated more precisely in Table 17.1.)

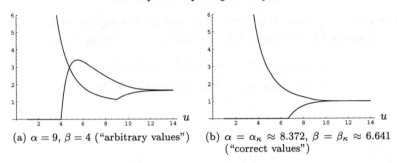

(a) $\alpha = 9$, $\beta = 4$ ("arbitrary values")   (b) $\alpha = \alpha_\kappa \approx 8.372$, $\beta = \beta_\kappa \approx 6.641$ ("correct values")

Fig. A1.1. $F_\kappa(u; \alpha, \beta)$ and $f_\kappa(u; \alpha, \beta)$ for two choices of α and β, $\kappa = 3$

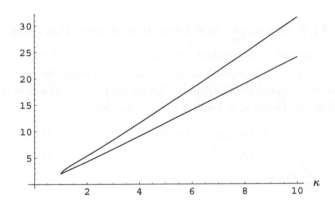

Fig. A1.2. α_κ (upper curve) and β_κ (lower curve)

A1.3 Using the Iwaniec inner product

The criteria given by (A1.9) and (A1.10) are not computationally effective, since they would require computing the limiting values of $F_\kappa(u; \alpha, \beta)$ and $f_\kappa(u; \alpha, \beta)$ as $u \to \infty$. However, as detailed below, the Iwaniec inner product gives a computationally effective method for determining that limiting behavior.

The coupled delay differential equations that define $F_\kappa(u)$ and $f_\kappa(u)$ are not of the form (A1.1), but recall that

(A1.11) $\qquad P_\kappa(u; \alpha, \beta) := F_\kappa(u; \alpha, \beta) + f_\kappa(u; \alpha, \beta),$

(A1.12) $\qquad Q_\kappa(u; \alpha, \beta) := F_\kappa(u; \alpha, \beta) - f_\kappa(u; \alpha, \beta),$

A1.3 Using the Iwaniec inner product

and that $P_\kappa(u; \alpha, \beta)$ is of type $(\kappa, -\kappa)$ and $Q_\kappa(u; \alpha, \beta)$ is of type (κ, κ):

$$\frac{d}{du}\left(u^\kappa P_\kappa(u; \alpha, \beta)\right) = \kappa u^{\kappa-1} P_\kappa(u-1; \alpha, \beta), \qquad u > \alpha,$$

$$\frac{d}{du}\left(u^\kappa Q_\kappa(u; \alpha, \beta)\right) = -\kappa u^{\kappa-1} Q_\kappa(u-1; \alpha, \beta), \qquad u > \alpha.$$

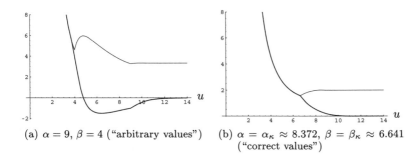

(a) $\alpha = 9$, $\beta = 4$ ("arbitrary values")   (b) $\alpha = \alpha_\kappa \approx 8.372$, $\beta = \beta_\kappa \approx 6.641$ ("correct values")

Fig. A1.3. $P_\kappa(u; \alpha, \beta)$ (thin line) and $Q_\kappa(u; \alpha, \beta)$ (thick line) for two choices of α and β, $\kappa = 3$

Recasting the criteria (A1.9) and (A1.10) in terms of $P_\kappa(u; \alpha, \beta)$ and $Q_\kappa(u; \alpha, \beta)$ gives the equivalent conditions

(A1.13) $\qquad P_\kappa(u; \alpha, \beta) = 2 + O(e^{-u}) \quad$ as $u \to \infty$,

(A1.14) $\qquad Q_\kappa(u; \alpha, \beta) = O(e^{-u}) \quad$ as $u \to \infty$.

Note in Figure A1.3(a) how condition (A1.13) is violated—as is the property that $Q_\kappa(u; \alpha, \beta) > 0$.

Recall from Section 12.1 that $p_\kappa(u)$ is adjoint to $P_\kappa(u; \alpha, \beta)$ and $q_\kappa(u)$ is adjoint to $Q_\kappa(u; \alpha, \beta)$:

$$\frac{d}{du}\left(up_\kappa(u)\right) = \kappa p_\kappa(u) - \kappa p_\kappa(u+1), \qquad u > 0,$$

$$\frac{d}{du}\left(uq_\kappa(u)\right) = \kappa q_\kappa(u) + \kappa q_\kappa(u+1), \qquad u > 0,$$

and they satisfy $p_\kappa(u) \sim u^{-1}$, $q_\kappa(u) \sim u^{2\kappa-1}$ as $u \to \infty$. (See Figures A1.4 and A1.5.) These two functions are independent of the parameters α and β.

From the definition of $\langle G, g \rangle_b(u)$ and using equation (A1.4), we find

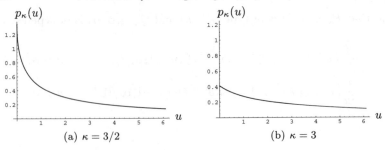

Fig. A1.4. $p_\kappa(u)$ for two values of κ

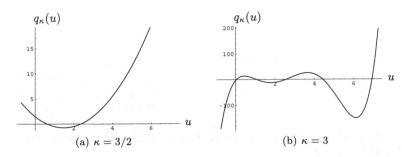

Fig. A1.5. $q_\kappa(u)$ for two values of κ

that for some constants C_P and C_Q we have, for $u \geq \alpha$,

$$\langle P, p \rangle_{-\kappa}(u) = u P_\kappa(u; \alpha, \beta) p_\kappa(u) + \kappa \int_{u-1}^{u} P_\kappa(t; \alpha, \beta) p_\kappa(t+1)\, dt = C_P,$$

$$\langle Q, q \rangle_\kappa(u) = u Q_\kappa(u; \alpha, \beta) q_\kappa(u) - \kappa \int_{u-1}^{u} Q_\kappa(t; \alpha, \beta) q_\kappa(t+1)\, dt = C_Q.$$

Figure A1.6 shows the behavior of $\langle Q, q \rangle_\kappa(u)$ for two choices of α, β.

First letting $u \to \infty$, and then setting $u = \alpha$ in these last equations, we find that if $\alpha = \alpha_\kappa$ and $\beta = \beta_\kappa$ then

(A1.15) $\qquad\qquad \langle P, p \rangle_{-\kappa}(\alpha) = 2,$ and

(A1.16) $\qquad\qquad \langle Q, q \rangle_\kappa(\alpha) = 0.$

Furthermore, the discussion following Lemma 12.1 demonstrates that equations (A1.15) and (A1.16) hold *only* if (A1.9) and (A1.10) hold. For a given κ, these results give a computationally effective method for approximating α_κ and β_κ, viz., we find solutions for α and β that, within an acceptable error tolerance, satisfy (A1.15) and (A1.16). (Note that β is implicitly a parameter in these equations, since it affects the values

A1.4 Some features of Mathematica

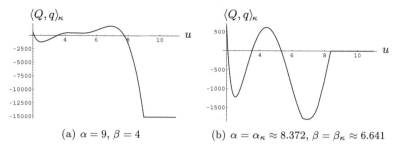

Fig. A1.6. $\langle Q, q \rangle_\kappa$ for two choices of α and β, $\kappa = 3$

of the functions P and Q.) After a short interlude (Section A1.4), we will detail how we go about approximating those solutions to (A1.15) and (A1.16).

A careful reader will note that we do not re-express equations (A1.15) and (A1.16) in terms of the functions $\Pi_\kappa(u, v)$ and $\Xi_\kappa(u, v)$ (defined in Section 12.3)—as was done in Lemma 13.1. The reason is that for our computational purposes it suffices to work directly with (A1.15) and (A1.16), although $\Pi_\kappa(u, v)$ and $\Xi_\kappa(u, v)$ are better suited for the analysis in Part II of this book.

A1.4 Some features of *Mathematica*

Our package, available at www.math.uiuc.edu/SieveTheoryBook, takes advantage of some design features of *Mathematica*, which we summarize here. Detailed information on *Mathematica* can be found in the documentation included with that system, or online at www.wolfram.com.

Arithmetic: Arithmetic can be performed on integers and rationals of arbitrary size or precision and is exact. For arithmetic on arbitrary real numbers we use the default floating-point representation, accurate to about 16 decimal-places of relative precision.

Contour plots: We used the ContourPlot command to create many of our figures. These plots illustrate a function of two variables, say x and y. Most of our contour plots, such as Figure A1.7(a), illustrate a real-valued function of $z = x + iy$. The contour lines are lines of constant height, and the shading indicates the approximate magnitude of the function, with darker shades indicating smaller values.

Re-using results: *Mathematica* provides a mechanism for saving the result of a computation for later re-use. (This technique is often called *caching* or *dynamic programming*.) Caching can save a great deal of time when the result of a long computation is used multiple times, although at the cost of using some additional storage.

Interpolating-functions: The InterpolatingFunction construct is defined by a set of data points, say $\mathcal{D} := \{(x_k, y_k)\}$ in the interval $x_0 \leq x_k \leq x_n$. An InterpolatingFunction acts like any other function in *Mathematica*—it returns y_k when applied to $x_k \in \mathcal{D}$ and otherwise, as the name suggests, it computes the returned value using an interpolating polynomial of low degree. When applied to an argument outside the interval $[x_0, x_n]$ it issues a warning message and returns the value extrapolated by the interpolating polynomial.

Differential equations: We use the NDSolve operator to find numerical solutions of differential equations. The solution is returned as an InterpolatingFunction, which we always save to avoid later recomputation.

Numerical integration: The NIntegrate operator implements numerical integration, with

$$\text{NIntegrate}[G(z), \{z, z_0, \ldots, z_n\}] \approx \sum_{k=1}^{n} \int_{z_{k-1}}^{z_k} G(z)\, dz.$$

When $z_k \in \mathbb{C}$ the integral is evaluated along a piecewise linear "polyline" path connecting the points. Whether or not $z_k \in \mathbb{C}$, each z_k is flagged as a potential singularity and NIntegrate adjusts its quadrature algorithm to use exceptional care near those points.

Those functions in this book which are defined by delay differential equations have bad differentiability properties at some points. For example: $j_\kappa(u)$ when $u \in \mathbb{Z}$ (as was demonstrated in Section 14.1), and $F_\kappa(u; \alpha, \beta)$ and $f_\kappa(u; \alpha, \beta)$ when $u \in \mathbb{Z} \cup (\alpha + \mathbb{Z}) \cup (\beta + \mathbb{Z})$. When such a function appears in an integrand we take care to flag potentially "bad" points for NIntegrate. Similarly, when using NDSolve to compute such a function over an interval, we "chop up" the interval into sections, so that no "bad" points lie within a section.

A1.5 Computing $F_\kappa(u)$ and $f_\kappa(u)$

Referring to equations (A1.5) and (A1.6), we see that for $0 < u \leq \beta$ the computation of $f_\kappa(u; \alpha, \beta)$ is trivial, while for $0 < u \leq \alpha$ the computation of $F_\kappa(u;\alpha,\beta)$ reduces to the computation of $j_\kappa(u) = \sigma_\kappa(2u)$. We describe methods for computing $j_\kappa(u)$ in Section A1.8.

To calculate $F_\kappa(u;\alpha,\beta)$ and $f_\kappa(u;\alpha,\beta)$ beyond their initial ranges of definition, we use the NDSolve operator to approximate the solutions to equations (A1.7) and (A1.8), i.e., the delay differential equations satisfied by $F_\kappa(u;\alpha,\beta)$ for $u > \alpha$ and by $f_\kappa(u;\alpha,\beta)$ for $u > \beta$. We find

$$P_\kappa(u;\alpha,\beta) := F_\kappa(u;\alpha,\beta) + f_\kappa(u;\alpha,\beta), \text{ and}$$
$$Q_\kappa(u;\alpha,\beta) := F_\kappa(u;\alpha,\beta) - f_\kappa(u;\alpha,\beta),$$

directly from the values of $F_\kappa(u;\alpha,\beta)$ and $f_\kappa(u;\alpha,\beta)$.

A1.6 The function Ein(z)

Several of the functions in our package are computed by integrating some function in which a factor of the form $e^{\pm\kappa\,\mathrm{Ein}(\pm z)}$ appears, where

$$\mathrm{Ein}(z) := \int_0^z (1-e^{-s})\frac{ds}{s}.$$

(The Ein function was first introduced in Section 12.1. A thorough survey of its properties may be found in [AS94, Chapter 5].)

Recall that $\mathrm{Ein}(z)$ is entire, with the power series expansion

$$\mathrm{Ein}(z) = \sum_{n=1}^\infty (-1)^{n-1} \frac{z^n}{n!\,n}.$$

However, for our purposes a more useful expression is

(A1.17) $\qquad \mathrm{Ein}(z) = \gamma + \log z + E_1(z), \quad \arg(z) < \pi,$

where

(A1.18) $\qquad E_1(z) := \int_z^\infty \frac{e^{-s}}{s}\,ds.$

(If the branch cuts for $\log z$ and $E_1(z)$ are chosen consistently in (A1.17) then the restriction that $\arg(z) < \pi$ is unnecessary.) Since $E_1(z)$ is included in the standard repertoire of functions of *Mathematica* 5, while $\mathrm{Ein}(z)$ is not, we use equation (A1.17) to implement $\mathrm{Ein}(z)$.

Equation (A1.17) and a well-known asymptotic expansion for $E_1(z)$ [AS94, Entry 5.1.51] give

(A1.19) $\mathrm{Ein}(z) = \gamma + \log z + \dfrac{e^{-z}}{z}(1 + O(1/z)), \quad |z| \to \infty,\ \arg(z) < \pi.$

It follows from this expansion that $\left|e^{\kappa\,\operatorname{Ein}(z)}\right|$ is on the rough order of $|z|^\kappa$ when $\Re(z)$ is sufficiently positive, while, for $\Re(z)$ sufficiently negative, it behaves in a "violently" oscillatory way as $\Im(z)$ varies. For example, as $\Re(z) \to -\infty$, with $\Im(z) = 2\pi m$ for some $m \in \mathbb{Z}$, we see that $\left|e^{\kappa\,\operatorname{Ein}(z)}\right|$ is on the rough order of $\exp(\kappa\,e^{|z|}/|z|)$; while with $\Im(z) = 2\pi m + 1$, $\left|e^{\kappa\,\operatorname{Ein}(z)}\right|$ is on the rough order of $\exp(-\kappa\,e^{|z|}/|z|)$.

A1.7 Computing the adjoint functions

As we will see below, the computation of $p_\kappa(u)$ is straightforward, while the computation of $q_\kappa(u)$ is simple when $2\kappa \in \mathbb{N}$ and more complicated otherwise.

Recall from equation (12.11) that $p_\kappa(u)$ can be written as the Laplace transform

$$(A1.20) \qquad p_\kappa(u) = \int_0^\infty e^{-uz - \kappa\,\operatorname{Ein}(z)}\,dz.$$

We approximate $p_\kappa(u)$ by applying *Mathematica*'s NIntegrate operator to equation (A1.20), letting NIntegrate determine the appropriate truncation point for the upper limit of the integral.

Although some of the functions in our package can be evaluated only for $\kappa \geq 1$, our implementation of $p_\kappa(u)$ is an exception, for which we allow $\kappa > 0$. However, by the remarks which follow equation (A1.19), we see that the integral of equation (A1.20) converges for all $u \geq 0$ provided $\kappa > 1$, while it converges only for $u > 0$ when $\kappa \leq 1$. In our *Mathematica* implementation, the approximation of the integral becomes problematic as $u \to 0+$ even when κ is slightly larger than 1, and we simply trust the user to avoid attempting computations in the "problematic range."

Turning to the computation of $q_\kappa(u)$, recall that when $2\kappa \in \mathbb{N}$ then $q_\kappa(u)$ is a polynomial in u of degree $2\kappa - 1$:

$$(A1.21) \qquad q_\kappa(u) = \sum_{n=0}^{2\kappa-1} (-1)^n \binom{2\kappa-1}{n} c_n(\kappa) u^{2\kappa-1-n},$$

where the coefficients $c_n(\kappa)$ are polynomials in κ satisfying

$$(A1.22) \qquad c_0(\kappa) = 1,$$

$$(A1.23) \qquad c_{n+1}(\kappa) = \kappa \sum_{m=0}^{n} \frac{(-1)^m}{m+1} \binom{n}{m} c_{n-m}(\kappa).$$

A1.7 Computing the adjoint functions

For a given value of κ with $2\kappa \in \mathbb{N}$, upon the first computation of $q_\kappa(u)$ we use (A1.22) and (A1.23) to compute, and store for later use, a table of $c_n(\kappa)$, $0 \le n \le 2\kappa - 1$. (Table 15.1 lists the first few polynomials.)

When $2\kappa \notin \mathbb{N}$ we approximate $q_\kappa(u)$ using its contour integral representation, valid for $u > 0$:

$$(A1.24) \qquad q_\kappa(u) = \frac{\Gamma(2\kappa)}{2\pi i} \int_{\supset} z^{-2\kappa} e^{uz + \kappa \operatorname{Ein}(-z)} \, dz,$$

where the path of integration traverses a "keyhole" shape, traveling from $-\infty$, wrapping about the negative real axis (and the origin) in the positive sense and returning to $-\infty$. We will spend quite some time explaining how we approximate this contour integral, although computation of $q_\kappa(u)$ when $2\kappa \notin \mathbb{N}$ is outside the general scope of this book. However, we use the same general ideas to compute $j_\kappa(u)$ as well as $q_\kappa(u)$, and this section serves to introduce the necessary concepts.

We approximate the right side of equation (A1.24) by deforming and truncating the original keyhole path into a polyline path designed to be well suited for numerical quadrature. Our choice of path is based on the saddle point method (or "method of steepest descent").

Before explaining this adaptation of the saddle point method, we note that if $G(z)$ denotes the integrand of equation (A1.24) then $G(z)$ is symmetric under complex conjugation, i.e., $G(\bar{z}) = \overline{G(z)}$. For any such integrand and a path $\mathcal{C}$ that is symmetric about the real axis we have

$$(A1.25) \qquad \int_\mathcal{C} G(z) \, dz = 2 \left(\Im \int_{\mathcal{C}_+} G(z) \, dz \right),$$

where $\mathcal{C}_+$ denotes the restriction of $\mathcal{C}$ to the upper half of the complex plane. We expect numerical quadrature of the right side of equation (A1.25) to take roughly half as much computation as would be needed to approximate the left side, and we recast all such contour integrals into the form of the right side before computing them.

The saddle point method is usually used to find asymptotic expansions of special functions (see [dBr81, Chapter 5] or [Hen91, §11.8]). The same ideas can be used to choose the path of a contour integral so that the integral is well-suited for numerical quadrature [Tem77]. We begin with an informal discussion of the method—skipping over some complicating issues which we will return to later.

Suppose we wish to evaluate an integral of the form

$$\text{(A1.26)} \qquad \int_C G(z)\,dz,$$

where the path C passes through a point z_s (a *saddle point*) in a neighborhood of which we have

$$\text{(A1.27)} \qquad \log G(z) = \sum_{m \geq 0} a_m (z - z_s)^m.$$

with $a_1 = 0$ and $a_2 \neq 0$. Saddle points are solutions of $G'(z)/G(z) = 0$, although at a solution it may be that $a_2 = 0$.

If C can be parametrized near z_s as $z = z_s + \vartheta t + O(t^2)$, then we say that the path travels in the direction $\vartheta/|\vartheta|$ at z_s. If ϑ satisfies

$$\text{(A1.28)} \qquad a_2 \vartheta^2 = -1,$$

then along this path we have

$$\text{(A1.29)} \qquad G(z) = \exp\left(a_0 - t^2\right)\left(1 + O(t^2)\right).$$

In the region where the error term in equation (A1.29) is small, the integrand is well-suited to numerical quadrature. This remains true when the path passes near z_s and travels only roughly in the optimal direction. When C passes through a well-chosen subset of the saddle points of $G(z)$ then it is likely that the vast bulk of the integrand is concentrated near those points, and we can truncate C to a relatively short range over which numerical quadrature yields an accurate approximation to (A1.26).

Returning to the specific problem of computing $q_\kappa(u)$, in Figure A1.7 we illustrate this method from two points of view. Figure A1.7(a) shows the magnitude of the integrand and the truncated path of integration used to approximate $q_\kappa(u)$ when $\kappa = 1.5$, $u = 7$. The two saddle points of the integrand are marked as white "x"s and the heavy dots show the sample points chosen by *Mathematica*'s `NIntegrate` operator.

For our second point of view, we parametrize $z = z(\varsigma) \in C$ by arc-length ς, let

$$\text{(A1.30)} \qquad H(z) := z^{-2\kappa} e^{uz + \kappa \, \text{Ein}(-z)} \frac{dz}{i\,d\varsigma},$$

and rewrite equation (A1.24) as

$$\text{(A1.31)} \qquad q_\kappa(u) = \frac{\Gamma(2\kappa)}{\pi} \int_0^\infty \Re H(z(\varsigma))\,d\varsigma.$$

A1.7 Computing the adjoint functions

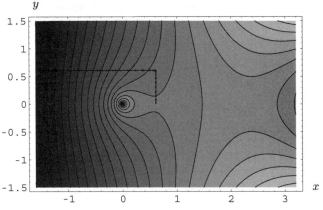

(a) $\log_{10} |G(z+iy)|$, where $G(z)$ is the integrand of equation (A1.24)

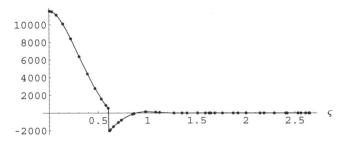

(b) $\Re H(z(\varsigma))$, where $H(z)$ is defined by equation (A1.30), and ς measures distance along the path of Figure A1.7(a)

Fig. A1.7. Two views of the integrand used to find $q_\kappa(u)$, $\kappa = 3$, $u = 14$

Figure A1.7(b) shows the integrand of equation (A1.31). (The discontinuity in the integrand arises from the corresponding discontinuity in the direction followed by the path shown in Figure A1.7(a).)

To locate the saddle points of $G(z)$, with

$$G(z) := z^{-2\kappa} e^{uz + \kappa \operatorname{Ein}(-z)},$$

we look for solutions to

(A1.32) $$0 = \frac{G'}{G}(z) = u - \frac{\kappa(1 + e^z)}{z}.$$

To solve (A1.32), we let $r := u/\kappa$ and instead solve

(A1.33) $$0 = 1 + e^z - rz =: K(r, z).$$

Focusing on real-valued solutions, equation (A1.33) is satisfied where a line of slope r intersects the curve $1+e^z$. There is a "critical" value of r, say r_c, for which the line is tangent to the curve, meeting it at a single point z_c. Thus $0 = 1 + e^{z_c} - r_c z_c$ and $r_c = e^{z_c}$, giving

$$r_c = r_c \log(r_c) - 1 \approx 3.59112, \text{ and}$$
(A1.34)
$$z_c = \log r_c \approx 1.27846.$$

Note that the constant r_c is the one occurring in Proposition 15.7.

To determine the saddle point $z =: z_s$ satisfying equation (A1.33) we start with a rough approximation to z_s and then, usually, refine it using Newton's method. Let $K'(r,z) := (d/dz) K(r,z)$. Provided z is a sufficiently accurate initial approximation to z_s, Newton's method sends z to a very accurate approximation by iterating the map

(A1.35) $$z \mapsto z - \frac{K(r,z)}{K'(r,z)} = \frac{(z-1)e^z - 1}{e^z - r}.$$

We use one of three initial rough approximations to z_s, with the choice of approximation depending on the relationship between r and r_c. Note that equation (A1.33) has two complex conjugate roots when $r < r_c$, a single real root when $r = r_c$, and two real roots when $r > r_c$.

The first rough approximation applies when $r < r_c$ is sufficiently small. Without quantifying the error, we find that as $r \to 0$

$$z_s \approx Z_0(r) := \frac{\pi^2}{2} r^2 + \pi(1 - r + r^2)i,$$

where z_s denotes the complex root of (A1.33) with positive real part.

Our second rough approximation is applicable for r in an "intermediate range" where $|r - r_c|$ is sufficiently small. In this case we expand $K(r,z)$ about $r = r_c$, $z = z_c$ to find that

$$K(r,z) = -z_c(r - r_c) - (r - r_c)(z - z_c) + \frac{r_c}{2}(z - z_c)^2 + \cdots.$$

Ignoring the higher-order terms and solving for the zeros of the resulting quadratic equation, we find that

(A1.36) $$z_s \approx Z_1(r) := z_c + \frac{r - r_c \pm \sqrt{(r - r_c)^2 + 2 z_c r_c (r - r_c)}}{r_c}.$$

Our third rough approximation is applicable for r sufficiently large. Letting z_s denote the smaller of the two real roots of (A1.33), we find that as $r \to \infty$ we have

$$z_s \approx Z_2(r) := \frac{2}{r} + \frac{2}{r^2}.$$

A1.7 Computing the adjoint functions

Our choice of transition points between these three approximations was determined experimentally. To summarize: writing z for our initial rough approximation to z_s, we use

$$\text{(A1.37)} \qquad z = \begin{cases} Z_0(r), & r < 0.517, \\ Z_1(r), & 0.517 \leq r \leq 5.3, \\ Z_2(r), & 5.3 < r. \end{cases}$$

Starting with z given by (A1.37) we find that 4 iterations of (A1.35) suffice to approximate z_s with an error-bound on the order of 10^{-5}. This is sufficient for our purposes since, as observed above, our path need only pass near the saddle point to be effective. Recalling that $r_c = e^{z_c}$, note that when $r = r_c$, $z = z_c$ the Newton iteration (A1.35) is ill-defined. For this reason, and since the approximation (A1.36) is very accurate as $r \to r_c$, we do not apply Newton's method when $|r - r_c| < 10^{-12}$.

The value of z_s determines our path of integration—which is a polyline running consecutively through $\{z_0, z_1, z_2\}$; where z_0 lies on the real axis, z_s lies on the segment connecting z_0 and z_1, and z_2 is chosen to bound the truncation error arising from using a finite path. Figures A1.7(a) and A1.8(a)–A1.8(c) show some of the resulting paths.

When $z_s \in \mathbb{R}$ we set $z_0 = z_s$ and $z_1 = (1+i)z_0$, otherwise we choose z_0 and z_1 so that z_s lies midway between them and so that the path travels in the direction determined by equation (A1.28).

Given an allowable truncation error of ε, we choose z_2 to satisfy

$$\text{(A1.38)} \qquad \left| \int_{z_2}^{z_2 - \infty} G(z)\, dz \right| \leq \varepsilon.$$

Letting $C := \gamma + e^{-1}$ and assuming that $\Re z_2 \leq -1$ we can show that

$$\left| \int_{z_2}^{z_2 - \infty} G(z)\, dz \right| \leq \exp\left(\kappa C + u \Re z_2 \right) / u.$$

Thus, to ensure that the bound (A1.38) holds, we set

$$\Re z_2 = \min\left(-1, \Re z_1, \frac{\log(\varepsilon u) - \kappa C}{u} \right).$$

When $z_s \in \mathbb{R}$ we set $\Im z_2 = \Im z_1$. However, when $z_s \notin \mathbb{R}$ it is possible for $\Im z_1$ to be arbitrarily small, and setting $\Im z_2 = \Im z_1$ could cause our path to pass uncomfortably close to the singularity at the origin. (See Figure A1.8(b).) For this reason, when $z_s \notin \mathbb{R}$ we set $\Im z_2 = \max(1, \Im z_1)$.

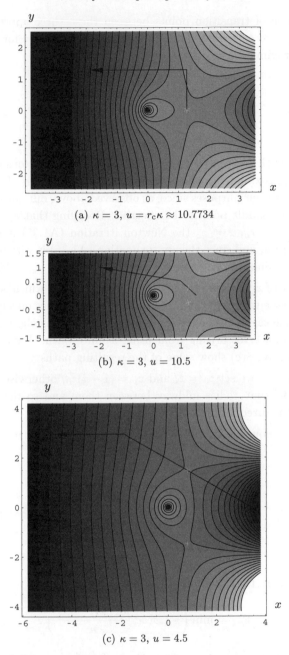

(a) $\kappa = 3$, $u = r_c\kappa \approx 10.7734$

(b) $\kappa = 3$, $u = 10.5$

(c) $\kappa = 3$, $u = 4.5$

Fig. A1.8. $\log_{10}|G(x+iy)|$, where $G(z)$ is the integrand of equation (A1.24), for three values of u/κ, contour interval is 1

A1.7 Computing the adjoint functions

As an aside, note that when $u/\kappa = r_c$ the Taylor series expansion of $\log G(z)$ about z_c (equation (A1.27) with $z_s = z_c$) has $a_2 = 0$ while $a_3 = -u/z_s$, and so the integrand does not have a "proper" saddle point but rather a three-lobed saddle point, indicated by the 6-pointed star in Figure A1.8(a). Since $a_2 = 0$, equation (A1.28) is not applicable for determining the direction in which the path should travel. Instead, the optimal path would be one which entered z_s in a direction $\vartheta/|\vartheta|$ with ϑ being one of the three roots to

$$(A1.39) \qquad a_3 \vartheta^3 = -1,$$

and then exited z_s along a path also satisfying (A1.39), but with a different choice of root. The same reasoning suggests that such a path would be better than our current choice if a_2 is sufficiently near 0. However, our less sophisticated path seems quite satisfactory in practice.

When $2\kappa \in \mathbb{N}$ and $u \in \mathbb{Q}$ we can measure the accuracy of our saddle-point approximation of $q_\kappa(u)$ by comparing it with the exact result given by the polynomial expansion (A1.21). Letting $\tilde{q}_\kappa(u)$ denote the approximation, we measure the relative precision of our approximation in "digits" as defined by

$$(A1.40) \qquad d(\kappa, u) := -\log_{10} \left| 1 - \frac{q_\kappa(u)}{\tilde{q}_\kappa(u)} \right|.$$

Table A1.7 illustrates the accuracy of our approximations to $q_\kappa([\alpha_\kappa])$.

Table A1.1. *Accuracy of approximations to $q_\kappa(u)$, measured in "digits" as given by $d(\kappa, u)$, defined in equation (A1.40)*

κ	$d(\kappa, [\alpha_\kappa])$	κ	$d(\kappa, [\alpha_\kappa])$
1.0	10.4	1.5	10.6
2.0	11.4	2.5	11.6
3.0	13.3	3.5	13.5
4.0	13.0	4.5	13.9
5.0	11.3	5.5	13.9
6.0	14.2	6.5	11.1
7.0	14.1	7.5	14.0
8.0	10.8	8.5	14.1
9.0	14.1	9.5	10.6
10.0	13.4		

A1.8 Computing $j_\kappa(u)$

In this section we will describe several methods for computing $j_\kappa(u)$, illustrated in Figure A1.9. We start by applying our saddle point method to two of its contour integral representations. At the end of this section we will explain how we choose which of these methods to use.

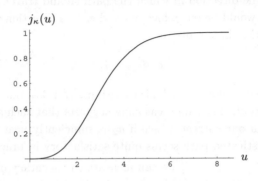

Fig. A1.9. $j_\kappa(u)$, $\kappa = 3$

As was shown in Lemma 14.2, the Laplace transform of $j_\kappa(u)$ is

$$\widehat{j}_\kappa(z) := \int_0^\infty e^{-uz} j_\kappa(u)\, du = \frac{1}{z} e^{-\kappa \operatorname{Ein}(z)}.$$

Taking the inverse Laplace transform of $\widehat{j}_\kappa(z)$ gives, for all $u \in \mathbb{R}$,

$$j_\kappa(u) = \frac{1}{2\pi i} \int_{1-i\infty}^{1+i\infty} \widehat{j}_\kappa(z) e^{uz}\, dz$$

(A1.41)
$$= \frac{1}{2\pi i} \int_{1-i\infty}^{1+i\infty} e^{uz - \kappa \operatorname{Ein}(z)} \frac{dz}{z}$$

(A1.42)
$$= 1 + \frac{1}{2\pi i} \int_{-1-i\infty}^{-1+i\infty} e^{uz - \kappa \operatorname{Ein}(z)} \frac{dz}{z},$$

where the identity (A1.42) follows from shifting the path of integration used in (A1.41) to the left, picking up a residue of 1 arising from the pole of the integrand at the origin.

As we did for $q_\kappa(u)$ in Section A1.7, we recast the original integral into the form of the right side of equation (A1.25), and deform and truncate the path of integration into a suitable polyline path to be passed to *Mathematica*'s `NIntegrate` operator.

Our integrand is now

$$G(z) := \frac{1}{z} e^{uz - \kappa \operatorname{Ein}(z)},$$

A1.8 Computing $j_\kappa(u)$

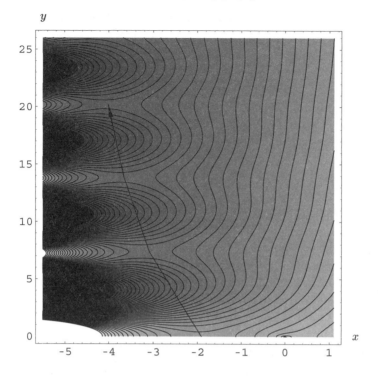

Fig. A1.10. $\log_{10} |G(x+iy)|$, where $G(z)$ is the integrand appearing in equation (A1.42), contour interval is 0.5, $\kappa = 3/2$, $u = 4$

with saddle points at the solutions to

(A1.43) $$0 = \frac{G'}{G}(z) = u - \frac{1}{z} - \frac{\kappa}{z}(1 - e^{-z})$$

or equivalently,

(A1.44) $$0 = uz - (1 + \kappa) + \kappa e^{-z}.$$

As can be seen in Figures A1.10 and A1.11, the saddle points of $G(z)$ have a more complicated structure than those analyzed in Section A1.7. Fortunately, the task of locating them is much easier in this section.

From this point on, we assume that $u > 0$. The design of our path is simpler for the integral in (A1.42), where the path lies to the left of the singularity at the origin, than it is for the integral in (A1.41), so we analyze (A1.42) first. (See Figure A1.10).

Equation (A1.44) has a family of solutions z_m, $m \in \mathbb{Z}$, satisfying

(A1.45) $\qquad z_m = 2\pi i m + \log(\kappa) - \log(1 + \kappa - u z_m), \ \Re z_m < 0.$

Note that $\Im z_0 = 0$. We can show that $\Im z_m = 2\pi m + O(1)$, with an absolute, and small, O-constant, and that $\Re z_m = -\log(|m|) + O(1)$ for $m \gg \kappa/u$. For each m, z_m is readily approximated by starting with the initial approximation $z = -\log(1 + 1/\kappa)$, and then iterating the map

$$z \mapsto 2\pi i m + \log(\kappa) - \log(1 + \kappa - uz)$$

until z fails to change by more than our desired error tolerance.

We evaluate the integral of (A1.42) along the path $\{z_0, \ldots, z_M\}$, where the terminal saddle point, z_M, is chosen to bound the truncation error arising from using a finite path. We do not ensure that the path passes through these saddle points in the "optimal" direction as given by equation (A1.28). This casual approach is justified by the remarks following equation (A1.45), which suggest that the saddle points lie along a nearly vertical path; so in going straight from one saddle point to the next the direction followed is not too far off the mark.

Before discussing the choice of z_M, we turn to evaluation of the integral in equation (A1.41), where the path lies to the right of the singularity at the origin (see Figure A1.11). This proceeds almost precisely as when evaluating (A1.42), but with a different choice for the first few points defining the path.

In addition to the saddle points z_m, $G(z)$ also has a saddle point at $z = z_+$, where z_+ denotes the positive real solution of equation (A1.44). We can easily show that $1/u < z_+ < (\kappa + 1)/u$, and we compute z_+ by applying *Mathematica*'s `FindRoot` operator to equation (A1.44) with the initial approximation of $z = (\kappa + 1)/u$. Letting $z_{0+} := \max(10, z_+)$, we use the three points

$$z_{k+} := e^{\pi i k/8} z_{0+}, \quad 0 \le k \le 2,$$

as the first points along the path used when evaluating (A1.41). (We arbitrarily bound z_{0+} by 10 to prevent the path from lying too far to the right when u is very near 0.) Letting $M_0 := \lceil z_{0+} \rceil / (2\pi)$, we use $\{z_{M_0}, \ldots, z_M\}$ as the remaining points along our path. To summarize, we use the path $\{z_{0+}, z_{1+}, z_{2+}, z_{M_0}, \ldots, z_M\}$ when evaluating (A1.41).

To find our terminal point z_M, without *any* rigorous error analysis we

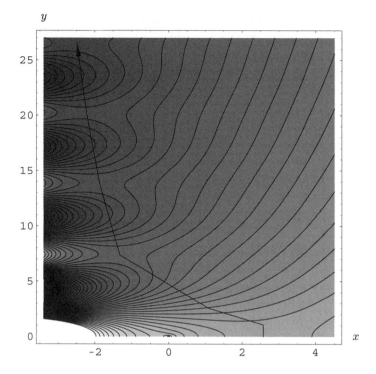

Fig. A1.11. $\log_{10} |G(x+iy)|$, where $G(z)$ is the integrand appearing in equation (A1.41), contour interval is 0.5, $\kappa = 9/2$, $u = 2$

observe that $|G(z_m)|$ decreases nicely as m increases, and that

$$\left| \int_{z_m}^{z_m+i\infty} G(z)\, dz \right| \ll |G(z_m)|\, |z_m|^\kappa.$$

On this basis, for a given truncation-error tolerance ε we simply choose M to satisfy $|G(z_M)| < \varepsilon$. (Note that $M = 3$ in Figure A1.10, while $M = 4$ in Figure A1.11.)

To choose between equations (A1.41) and (A1.42) as approximations for $j_\kappa(u)$, we recall that

(A1.46) $\quad j_\kappa(u) = e^{-\gamma \kappa} u^\kappa / \Gamma(\kappa + 1), \qquad 0 < u \leq 1,$

(A1.47) $\quad u \dfrac{d}{du} j_\kappa(u) = \kappa j_\kappa(u) - \kappa j_\kappa(u-1) \qquad u > 1.$

From equation (A1.46) we see that $j_\kappa(u)$ is very small when $u \leq 1$ and κ is large. In Chapter 14 we showed that $j_\kappa(u)$ increases to 1 as $u \to \infty$, and equation (14.41) implies that $j_\kappa(\kappa) \approx 1/2$.

It follows that when u is much smaller than κ then $j_\kappa(u)$ is small—very small when κ is large. In this case the two terms in equation (A1.42) must nearly cancel, and we would need to evaluate the integral in (A1.42) to very high accuracy to get an acceptable relative-error bound in our approximation of $j_\kappa(u)$.

On the other hand, when u is much larger than κ then the integral term in (A1.42) is near 0, and need only be found with modest accuracy to estimate $j_\kappa(u)$ accurately. For these reasons, we use equation (A1.41) when $\kappa \leq u$ and equation (A1.42) when $\kappa > u$. Since it is readily computed, we always use equation (A1.46) to find $j_\kappa(u)$ when $u \leq 1$, except when, for testing purposes, we wish to compare the results from our contour integral representations against (A1.46).

Instead of our contour integral formulas, when $u > 1$ we may use NDSolve to solve equation (A1.47): the defining delay differential equation for $j_\kappa(u)$. Since we save the InterpolatingFunction returned by NDSolve, this approach has the advantage that, once $j_\kappa(u_1)$ has been computed, given $1 \leq u \leq u_1$ we can find $j_\kappa(u)$ more rapidly than by evaluating a contour integral.

However, we find that the solution returned by NDSolve tends to drift away from the correct value when $u > \kappa$, particularly when κ is large. For this reason, to balance the goals of speed and accuracy, only when $u \in \mathbb{Z}$ do we use a contour integral formula to find $j_\kappa(u)$. Otherwise, we apply NDSolve to compute $j_\kappa(u)$ over the interval $[u] < u < [u]+1$. Note that we use the contour integral value for $j_\kappa([u])$, and this initial condition for NDSolve reduces the potential for drift in our solution.

A1.9 Computing α_κ and β_κ

To compute α_κ and β_κ we use *Mathematica*'s FindRoot operator to solve the system of equations

(A1.48)
$$\langle P, p\rangle_{-\kappa}(\alpha) - 2 = 0,$$
$$\alpha^{1-2\kappa}\langle Q, q\rangle_\kappa(\alpha) = 0,$$

where $\langle P, p\rangle_{-\kappa}(u)$ and $\langle Q, q\rangle_\kappa(u)$ are approximated by applying *Mathematica*'s NIntegrate operator to their defining expressions as implied by equation (A1.3). FindRoot requires initial approximations to α_κ and β_κ, which we interpolate (or extrapolate) from an InterpolatingFunction which was "bootstrapped" from a few values calculated using the software developed by Ferrell Wheeler [Whe88].

Equation (A1.48) includes the scaling factor $\alpha^{1-2\kappa}$ since FindRoot is not well suited to solving two equations in two unknowns when the magnitude of acceptable error differs greatly between the two equations. For example, with $\kappa = 10$, we calculate that

$$\langle P, p \rangle_{-\kappa}(\alpha_\kappa) - 2 \approx 5.32463 \cdot 10^{-13},$$
$$\alpha_\kappa^{1-2\kappa} \langle Q, q \rangle_\kappa(\alpha_\kappa) \approx 2.99870 \cdot 10^{-15},$$

while the latter inner product when unscaled is computed as

$$\langle Q, q \rangle_\kappa(\alpha_\kappa) \approx 8.97906 \cdot 10^{13}.$$

A1.10 Weighted-sieve computations

In Chapter 11 we used a weighted sieve to put lower bounds on r for which almost-prime numbers, with at most r prime factors, occur frequently in a sequence $\mathcal{A}$. In Theorem 11.1 we introduced the function $N(u, v; \kappa, \mu_0, \tau)$ which serves as such a lower bound.

To implement this function, we change variables in the integral of equation (11.16), which defines $N(u, v; \kappa, \mu_0, \tau)$. Letting $s \mapsto 1/t$ gives

(A1.49)
$$N(u, v; \kappa, \mu_0, \tau)$$
$$= \mu_0 \tau u - 1 + \frac{\kappa}{f_\kappa(\tau v)} \int_1^{v/u} F_\kappa(\tau v - t) \left(1 - \frac{u}{v} t\right) \frac{dt}{t}.$$

We use NIntegrate to approximate the right side of (A1.49), taking care not to integrate across the points t where

$$\tau v - t \in \mathbb{Z} \cup (\alpha_\kappa + \mathbb{Z}) \cup (\beta_\kappa + \mathbb{Z}),$$

at which the integrand has bad differentiability properties. (See the discussion at the end of Section A1.4.)

Of course, to get the best possible lower bound in Theorem 11.1, we want to approximate the minimal value achieved by $N(u, v; \kappa, \mu_0, \tau)$ when u and v are free to vary (subject to the conditions imposed by that theorem). That is, we want to approximate

(A1.50) $$N_{\min}(\kappa, \mu_0, \tau) := \min_{\substack{u,v \\ v > \beta_\kappa/\tau \\ 1/\tau < u < v}} N(u, v; \kappa, \mu_0, \tau).$$

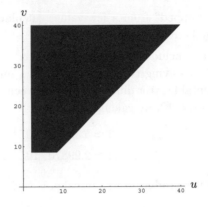

Fig. A1.12. Region determined by the constraints of Theorem 11.1; truncated at $v = 40$. $\kappa = 2$, $\mu_0 = 6$, $\tau = 1/2$

The constraints on u, v in equation (A1.50) define a triangular region, truncated to satisfy $v > \beta_\kappa/\tau$ and unbounded as $v \to \infty$. (See Figure A1.12.) Since *Mathematica* is not well-suited to search non-rectangular regions for a root or minimum, we reparametrize u in terms of a variable ν, setting

$$(\text{A1.51}) \qquad u = u(\nu) := \frac{1}{\tau} + e^{-\nu}(v - \frac{1}{\tau}),$$

so that $\nu = \log(v - 1/\tau) - \log(u - 1/\tau)$. Figure A1.13 shows a contour plot of $N(u(\nu), v; \kappa, \mu_0, \tau)$.

With this change of variable, we can restate equation (A1.50) as

$$(\text{A1.52}) \qquad N_{\min}(\kappa, \mu_0, \tau) = \min_{v > \beta_\kappa/\tau} \min_{0 < \nu < \infty} N(u(\nu), v; \kappa, \mu_0, \tau).$$

To approximate the values of ν and v that achieve the minimum in equation (A1.52) we apply *Mathematica*'s `FindMinimum` operator, using some simple estimates to bound the region searched and to find initial estimates for the solutions. As of this writing, our implementation of $N_{\min}$ is under development, so we refer the reader to the user's guide that accompanies our package for further details on our procedure.

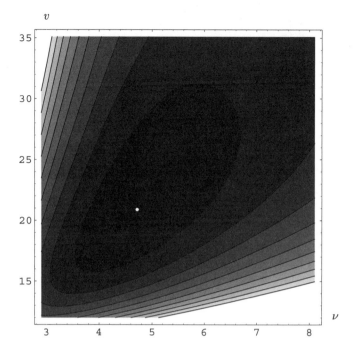

Fig. A1.13. $N(u(\nu), v; \kappa, \mu_0, \tau)$, where $u(\nu)$ satisfies equation (A1.51). The white dot indicates the approximate location where $N(u(\nu), v; \kappa, \mu_0, \tau)$ is minimal. $\kappa = 2$, $\mu_0 = 6$, $\tau = 1/2$

Bibliography

[AS94] M. Abramowitz and I. A. Stegun (eds.), *Handbook of mathematical functions with formulas, graphs, and mathematical tables*, Dover Publications Inc., New York, 1992, Reprint of the 1972 edition. MR 1225604 (94b:00012)

[AO65] N. C. Ankeny and H. Onishi, *The general sieve*, Acta Arith. **10** (1964/1965), 31–62. MR 0167467 (29 #4740)

[Bal85] A. Balog, *On sums over primes*, Elementary and analytic theory of numbers (Warsaw, 1982), Banach Center Publ., vol. 17, PWN, Warsaw, 1985, pp. 9–19. MR 840470 (87i:11119)

[BaD04] P. T. Bateman and H. G. Diamond, *Analytic number theory, an introductory course*, World Scientific Publishing Co. Pte. Ltd., Hackensack, NJ, 2004. MR 2111739 (2005h:11208)

[BH62] P. T. Bateman and R. A. Horn, *A heuristic asymptotic formula concerning the distribution of prime numbers*, Math. Comp. **16** (1962), 363–367. MR 0148632 (26 #6139)

[Bom65] E. Bombieri, *On the large sieve*, Mathematika **12** (1965), 201–225. MR 0197425 (33 #5590)

[Brd96] D. M. Bradley, *A sieve auxiliary function*, Analytic number theory, Vol. 1 (Allerton Park, IL, 1995), Progr. Math., vol. 138, Birkhäuser Boston, Boston, MA, 1996, pp. 173–210. MR 1399338 (97h:11099)

[BrD97] D. M. Bradley and H. G. Diamond, *A difference differential equation of Euler–Cauchy type*, J. Differential Equations **138** (1997), 267–300. MR 1462269 (99a:34172)

[BrF94] J. Brüdern and É. Fouvry, *Lagrange's four squares theorem with almost prime variables*, J. reine angew. Math. **454** (1994), 59–96. MR 1288679 (96e:11125)

[BrF96] J. Brüdern and E. Fouvry, *Le crible à vecteurs*, Compositio Math. **102** (1996), 337–355. MR 1401427 (97f:11079)

[Buc37] A. A. Buchstab, *Asymptotic estimates of a general number-theoretic function*, Math. Sbornik **44** (old series) **2** (new series) (1937), 1239–1246 (Russian, German summary).

[Chn73] J. R. Chen, *On the representation of a larger even integer as the sum of a prime and the product of at most two primes*, Sci. Sinica **16** (1973), 157–176. MR 0434997 (55 #7959)

[Dav00] H. Davenport, *Multiplicative number theory*, third ed., Graduate Texts in Mathematics, vol. 74, Springer-Verlag, New York, 2000, Revised and with a preface by Hugh L. Montgomery. MR 1790423 (2001f:11001)

[dBr81] N. G. de Bruijn, *Asymptotic methods in analysis*, third ed., Dover Publications, Inc., New York, 1981. MR 83m:41028

[DH85] H. G. Diamond and H. Halberstam, *The combinatorial sieve*, Number theory (Ootacamund, 1984), Lecture Notes in Math., vol. 1122, Springer, Berlin, 1985, pp. 63–73. MR 797780 (87b:11089)

[DH97a] _____, *Some applications of sieves of dimension exceeding 1*, Sieve methods, exponential sums, and their applications in number theory (Cardiff, 1995), London Math. Soc. Lecture Note Ser., vol. 237, Cambridge Univ. Press, Cambridge, 1997, pp. 101–107. MR 1635734 (99d:11102)

[DH97b] _____, *On the sieve parameters α_κ and β_κ for large κ*, J. Number Theory **67** (1997), 52–84. MR 1485427 (99b:11106)

[DH99] _____, *Differential inequalities for Iwaniec's q functions*, Number theory in progress, Vol. 2 (Zakopane-Kościelisko, 1997), de Gruyter, Berlin, 1999, pp. 721–735. MR 1689540 (2000f:11123)

[DH01] _____, *A comparison of two sieves. I*, Period. Math. Hungar. **43** (2001), 1–13. MR 1830562 (2002b:11124)

[DH02] _____, *A comparison of two sieves. II*, Number theory for the millennium, I (Urbana, IL, 2000), A K Peters, Natick, MA, 2002, pp. 329–341. MR 1956233 (2004b:11137)

[DHR88] H. G. Diamond, H. Halberstam, and H.-E. Richert, *Combinatorial sieves of dimension exceeding one*, J. Number Theory **28** (1988), 306–346. MR 932379 (89g:11080)

[DHR90a] _____, *A boundary value problem for a pair of differential delay equations related to sieve theory. I*, Analytic number theory (Allerton Park, IL, 1989), Progr. Math., vol. 85, Birkhäuser Boston, Boston, MA, 1990, pp. 133–157. MR 1084179 (92a:11107)

[DHR90b] _____, *Sieve auxiliary functions*, Number theory (Banff, AB, 1988), de Gruyter, Berlin, 1990, pp. 99–113. MR 1106654 (92f:11122)

[DHR93a] _____, *A boundary value problem for a pair of differential delay equations related to sieve theory. II*, J. Number Theory **45** (1993), 129–185. MR 1242713 (94j:11089)

[DHR93b] _____, *Sieve auxiliary functions. II*, A tribute to Emil Grosswald: number theory and related analysis, Contemp. Math., vol. 143, Amer. Math. Soc., Providence, RI, 1993, pp. 247–253. MR 1210519 (94c:11090)

[DHR94a] _____, *A boundary value problem for a pair of differential delay equations related to sieve theory. III*, J. Number Theory **47** (1994), 300–328. MR 1278401 (95e:11100)

[DHR94b] _____, *Estimation of the sieve auxiliary functions q_κ in the range $1 < \kappa < 2$*, Analysis **14** (1994), 75–102. MR 1280531 (95g:11091)

[DHR95] _____, *Monotonicity of the zero of the sieve auxiliary function $\tilde{\chi}_\kappa$ in the range $3/2 < \kappa < 2$*, Analysis **15** (1995), 1–16. MR 1322125 (96c:11107)

[DHR96] _____, *Combinatorial sieves of dimension exceeding one. II*, Analytic number theory, Vol. 1 (Allerton Park, IL, 1995), Progr. Math., vol. 138, Birkhäuser Boston, Boston, MA, 1996, pp. 265–308. MR 1399343 (97e:11112)

[FH00] K. Ford and H. Halberstam, *The Brun–Hooley sieve*, J. Number Theory **81** (2000), 335–350. MR 1752258 (2001d:11095)

[FI78] J. Friedlander and H. Iwaniec, *On Bombieri's asymptotic sieve*, Ann. Scuola Norm. Sup. Pisa Cl. Sci. (4) **5** (1978), 719–756. MR 519891 (80j:10049)

[GPY] D. A. Goldston, J. Pintz, and C. Y. Yıldırım, *Primes in tuples I*, to appear in Ann. of Math.

[GPY06] _____, *Primes in tuples. III. On the difference $p_{n+\nu} - p_n$*, Funct. Approx. Comment. Math. **35** (2006), 79–89. MR 2271608

[GGPY] D. A. Goldston, S. W. Graham, J. Pintz, and C. Y. Yıldırım, *Small gaps between products of two primes*, to appear.

[Grv01] G. Greaves, *Sieves in number theory*, Ergebnisse der Mathematik und ihrer Grenzgebiete (3) [Results in Mathematics and Related Areas (3)], vol. 43, Springer-Verlag, Berlin, 2001. MR 1836967 (2002i:11092)

[Gru88] F. Grupp, *On zeros of functions satisfying certain differential-difference equations*, Acta Arith. **51** (1988), 247–268. MR 971078 (90e:11133)

[GrR88] F. Grupp and H.-E. Richert, *Notes on functions connected with the sieve*, Analysis **8** (1988), 1–23. MR 954455 (89i:11101)

[Hal00] H. Halberstam, *A sieve application*, Asian J. Math. **4** (2000), 831–837, Loo-Keng Hua: a great mathematician of the twentieth century. MR 1870661 (2002i:11093)

[HR74] H. Halberstam and H.-E. Richert, *Sieve methods*, Academic Press [A subsidiary of Harcourt Brace Jovanovich, Publishers], London-New York, 1974, London Mathematical Society Monographs, No. 4. MR 0424730 (54 #12689)

[HR85] _____, *A weighted sieve of Greaves type. II*, Elementary and analytic theory of numbers (Warsaw, 1982), Banach Center Publ., vol. 17, PWN, Warsaw, 1985, pp. 183–215. MR 840478 (87m:11088)

[HW79] G. H. Hardy and E. M. Wright, *An introduction to the theory of numbers*, fifth ed., The Clarendon Press, Oxford University Press, New York, 1979. MR 568909 (81i:10002)

[Hrm96] G. Harman, *On the distribution of αp modulo one. II*, Proc. London Math. Soc. (3) **72** (1996), 241–260. MR 1367078 (96k:11089)

[Hrm07] G. Harman, *Prime-detecting sieves*, Princeton University Press, Princeton, NJ [London Mathematical Society Monographs Series, **33**], 2007. MR 2331072

[H-B97] D. R. Heath-Brown, *Almost-prime k-tuples*, Mathematika **44** (1997), 245–266. MR 1600529 (99a:11106)

[Hen91] P. Henrici, *Applied and computational complex analysis. Vol. 2. Special functions—integral transforms—asymptotics—continued fractions*, John Wiley & Sons Inc., New York, 1991, Reprint of the 1977 original. MR 1164865 (93b:30001)

[Hil86] A. Hildebrand, *On the number of positive integers $\leq x$ and free of prime factors $> y$*, J. Number Theory **22** (1986), 289–307. MR 831874 (87d:11066)

[HlTn93] A. Hildebrand and G. Tenenbaum, *On a class of differential-difference equations arising in number theory*, J. Anal. Math. **61** (1993), 145–179. MR 1253441 (94i:11069)

[HoTs06] K.-H. Ho and K.-M. Tsang, *On almost prime k-tuples*, J. Number Theory **120** (2006), 33–46. MR 2256795 (2007h:11113)

[Hoo94] C. Hooley, *On an almost pure sieve*, Acta Arith. **66** (1994), 359–368. MR 1288352 (95g:11092)

[Iwa76] H. Iwaniec, *The half dimensional sieve*, Acta Arith. **29** (1976), 69–95. MR 0412134 (54 #261)

[Iwa78] _____, *Almost-primes represented by quadratic polynomials*, Invent. Math. **47** (1978), 171–188. MR 0485740 (58 #5553)

[Iwa80] _____, *Rosser's sieve*, Acta Arith. **36** (1980), 171–202. MR 581917 (81m:10086)

[ILR80] H. Iwaniec, J. van de Lune, and H. J. J. te Riele, *The limits of Buchstab's iteration sieve*, Nederl. Akad. Wetensch. Indag. Math. **42** (1980), 409–417. MR 597998 (82a:10054)

[JR65] W. B. Jurkat and H.-E. Richert, *An improvement of Selberg's sieve method. I*, Acta Arith. **11** (1965), 217–240. MR 0202680 (34 #2540)

[Kub64] J. Kubilius, *Probabilistic methods in the theory of numbers*, Translations of Mathematical Monographs, Vol. 11, American Mathematical Society, Providence, R.I., 1964. MR 0160745 (28 #3956)

[Kuh54] P. Kuhn, *Neue Abschätzungen auf Grund der Viggo Brunschen Siebmethode*, Tolfte Skandinaviska Matematikerkongressen, Lund, 1953, Lunds Universitets Matematiska Institution, Lund, 1954, pp. 160–168. MR 0067147 (16,676e)

[Lan03] E. Landau, *Ueber die zu einem algebraischen zahlkörper gehörige zetafunction und die ausdehnung der tschebyschefschen primzahlentheorie auf das problem der vertheilung der primideale*, J. reine angew. Math. **125** (1903), 64–188.

[Lev49] W. J. LeVeque, *On the size of certain number-theoretic functions*, Trans. Amer. Math. Soc. **66** (1949), 440–463. MR 0030993 (11, 83i)

[Lin63] J. V. Linnik, *The dispersion method in binary additive problems*, Translated by S. Schuur, American Mathematical Society, Providence, R.I., 1963. MR 0168543 (29 #5804)

[LS] J. Liu and P. Sarnak, *Integral points on quadrics in three variables whose coordinates have few prime factors*, to appear.

[Mrs06] G. Marasingha, *On the representation of almost primes by pairs of quadratic forms*, Acta Arith. **124** (2006), 327–355. MR 2271248 (2007k:11154)

[Mrc77] D. A. Marcus, *Number fields*, Springer-Verlag, New York, 1977, Universitext. MR 0457396 (56 #15601)

[MV06] H. L. Montgomery and R. C. Vaughan, *Multiplicative number theory I: Classical theory*, Studies in Advanced Mathematics, no. 97, Cambridge University Press, Cambridge, November 2006.

[Olv97] F. W. J. Olver, *Asymptotics and special functions*, AKP Classics, A K Peters Ltd., Wellesley, MA, 1997, Reprint of the 1974 original [Academic Press, New York; MR0435697 (55 #8655)]. MR 1429619 (97i:41001)

[Raw80] D. A. Rawsthorne, *Improvements in the small sieve estimate of Selberg by iteration*, Ph.D. thesis, University of Illinois at Urbana-Champaign, 1980, p. 100.

[Raw82] _____, *Selberg's sieve estimate with a one sided hypothesis*, Acta Arith. **41** (1982), 281–289. MR 668914 (83m:10083)

[Ric69] H.-E. Richert, *Selberg's sieve with weights*, 1969 Number Theory Institute (Proc. Sympos. Pure Math., Vol. XX, State Univ. New York, Stony Brook, N.Y., 1969), Amer. Math. Soc., Providence, R.I., 1971, pp. 287–310. MR 0318083 (47 #6632)

Bibliography

[Ros75] P. M. Ross, *On Chen's theorem that each large even number has the form $p_1 + p_2$ or $p_1 + p_2 p_3$*, J. London Math. Soc. (2) **10** (1975), 500–506. MR 0389816 (52 #10646)

[Sel47] A. Selberg, *On an elementary method in the theory of primes*, Norske Vid. Selsk. Forh., Trondhjem **19** (1947), 64–67. MR 0022871 (9, 271h)

[Sel52] _____, *On elementary methods in prime number-theory and their limitations*, Den 11te Skandinaviske Matematikerkongress, Trondheim, 1949, Johan Grundt Tanums Forlag, Oslo, 1952, pp. 13–22. MR 0053147 (14, 726k)

[Sel91] _____, *Collected papers. Vol. II*, Springer-Verlag, Berlin, 1991, With a foreword by K. Chandrasekharan. MR 1295844 (95g:01032)

[Sng01] J. M. Song, *Sums of nonnegative multiplicative functions over integers without large prime factors. I*, Acta Arith. **97** (2001), 329–351. MR 1823551 (2002f:11130)

[Sng02] _____, *Sums of nonnegative multiplicative functions over integers without large prime factors. II*, Acta Arith. **102** (2002), 105–129. MR 1889623 (2003a:11123)

[Snd07] K. Soundararajan, *Small gaps between prime numbers: the work of Goldston-Pintz-Yıldırım*, Bull. Amer. Math. Soc. (**N.S.**) **44** (2007), no. 1, 1–18. MR 2265008 (2007k:11150

[teR80] H. J. J. te Riele, *Numerical solution of two coupled nonlinear equations related to the limits of Buchstab's iteration sieve*, Afdeling Numerieke Wiskunde [Department of Numerical Mathematics], 86, Mathematisch Centrum, Amsterdam, 1980. MR 585337 (81j:65048)

[Tem77] N. M. Temme, *The numerical computation of special functions by use of quadrature rules for saddle point integrals. I. Trapezoidal integration rules*, Tech. Report TW 164/77, Mathematisch Centrum, Afdeling Toegepaste Wiskunde, Amsterdam, 1977. MR 57 #4483

[Ten01] G. Tenenbaum, *Note on a paper by J. M. Song: "Sums of nonnegative multiplicative functions over integers without large prime factors. I" [Acta Arith. **97** (2001), no. 4, 329–351]*, Acta Arith. **97** (2001), 353–360. MR 1829871 (2002f:11131)

[Tsa89] K. M. Tsang, *Remarks on the sieving limit of the Buchstab–Rosser sieve*, Number theory, trace formulas and discrete groups (Oslo, 1987), Academic Press, Boston, MA, 1989, pp. 485–502. MR 993335 (90f:11082)

[Whe88] F. S. Wheeler, *On two differential-difference equations arising in analytic number theory*, Ph.D. thesis, University of Illinois at Urbana-Champaign, 1988, p. 110.

[Whe90] _____, *Two differential-difference equations arising in number theory*, Trans. Amer. Math. Soc. **318** (1990), 491–523. MR 963247 (90g:11134)

Index

adjoint function, 156, 162, 234
almost-prime numbers, xv, 142, 147, 149, 255
Ankeny–Onishi sieve limit, 144, 146, 228
Ankeny–Onishi sieve method, xv, 20, 71, 74, 79, 191, 227

Bateman–Horn conjecture, 12
block of terms, 46
Bombieri–Vinogradov Theorem, 97, 100, 101, 149, 152
Bradley, D. M., xvi, 152, 233
Brun, V., xv, 6, 13, 20, 27, 71, 122
Brun–Hooley sieve method, 27
Buchstab identity, xv, 20, 28, 33, 62, 71, 95
Buchstab's function, 161

Chen's Theorem, 101, 102
combinatorial function, 27, 70
complementary function, 19

de Bruijn, N. G., 158
DHR sieve method, 79
Dickman's function, 161, 184, 190, 191
dimension of a sieve, xv, 8, 130
divisor closed, 27, 70

$\text{Ein}(\cdot)$ function, 193, 241
Elliott, P. D. T. A., 152
Elliott–Halberstam conjecture, 152
Eratosthenes–Legendre formula, 4, 13

Fourier inversion, 181
Franze, C., xvi
Fundamental Lemma, 29, 35, 36, 42, 43, 62, 70, 74, 78, 81–83, 95, 104, 121
Fundamental Sieve Identity, 19, 21, 26, 27

g-tuple, see prime g-tuples conjecture
Goldston, D. A., 12, 151, 152
Graham, S., xvi
Greaves, G., xvi

Halberstam, H., xvi, 152
Harrington, W. J., 95
Hildebrand, A. J., xvi, 191
Hua, L. K., 158

independent events, 6
iteration methods, 28, 71, 95
Iwaniec inner product, 156, 158, 162, 178, 229, 234
Iwaniec, H., xvi, 12, 27, 95, 151

Jurkat, W. B., 66
Jurkat–Richert sieve method, xv, 28, 71, 95, 122

Landau, E., 40, 126
Laplace transform, 157, 170, 177, 188, 242, 250
LeVeque, W. J., 95
linear sieve, 8
Linnik's dispersion method, 151

Mathematica® mathematical software system, xvi, 233, 239
Mertens' product formula, 6, 41, 44, 57, 63
Mertens' sum formula, 9, 40, 50, 51, 125, 126
method of steepest descent, see saddle point method
modifying factor, 13, 16, 23
Moebius inversion formula, 15

Newton's method, 202, 246
numerical integration, 148, 226, 240, 243

prime g-tuples conjecture, 7, 12, 125, 151
Prime Ideal Theorem, 126
Prime Number Theorem, 6, 126
probabilistic model, 6
Pythagorean triples, 152

quadrature, *see* numerical integration
quasi-prime, 39, 42

Rankin's method, 30, 35
Rawsthorne, D. A., 66
remainder terms, 4, 71, 95
remainder terms, assumptions about, 5, 17, 136
remainder terms, bounds on, 34, 35, 62, 101, 108
Richert's Fundamental Identity, 28
Richert, H.-E., v, xvi, 155
Rosser, J. B., 95
Rosser–Iwaniec inequalities, 163
Rosser–Iwaniec sieve method, xv, 62, 67, 71, 74, 79, 82, 95

saddle point method, 243
Selberg optimal example, 81, 95
Selberg sieve method, xv, 13, 17, 21, 28, 29, 43, 61, 71, 95
Selberg weighted sieve, 150, 151
Selberg, A., 228
sieve, 3
sieving limit, 79, 135, 144, 226, 228
sift, 3
sifting density, 8
Soundararajan, K., 152

Topping-Up Lemma, 44
twin prime conjecture, 6, 97, 99, 101, 150, *see also* prime g tuples conjecture

Vaughan, R. C., 102

Weber, H., 128
Wheeler, F. S., xvi, 152, 233